PRACTICAL CHEMISTRY

Books by A. Holderness and J. Lambert

GRADED PROBLEMS IN CHEMISTRY TO ORDINARY LEVEL

CLASSBOOK OF PROBLEMS IN CHEMISTRY TO ADVANCED LEVEL

PROBLEMS AND WORKED EXAMPLES IN CHEMISTRY TO ADVANCED LEVEL

NEW CERTIFICATE CHEMISTRY

WORKED EXAMPLES AND PROBLEMS IN ORDINARY LEVEL CHEMISTRY

THE ESSENTIALS OF QUALITATIVE ANALYSIS

In conjunction with Dr F. Sherwood Taylor

THE ESSENTIALS OF VOLUMETRIC ANALYSIS

By A. Holderness

INTERMEDIATE ORGANIC CHEMISTRY

ADVANCED LEVEL INORGANIC CHEMISTRY

ADVANCED LEVEL PHYSICAL CHEMISTRY

REVISION NOTES IN ADVANCED LEVEL CHEMISTRY
VOLUME ONE: ORGANIC CHEMISTRY
VOLUME TWO: INORGANIC CHEMISTRY
VOLUME THREE: PHYSICAL CHEMISTRY

ORDINARY LEVEL REVISION NOTES IN CHEMISTRY

All published by Heinemann Educational Books Ltd

PRACTICAL CHEMISTRY

by

JOHN LAMBERT, M.Sc.

Formerly Senior Chemistry Master at King Edward's School, Birmingham

and

T. A. MUIR, B.Sc.

THIRD EDITION

HEINEMANN EDUCATIONAL BOOKS
LONDON

Heinemann Educational Books Ltd
22 Bedford Square, London WCIB 3HH
LONDON EDINBURGH MELBOURNE AUCKLAND
HONG KONG SINGAPORE KUALA LUMPUR NEW DELHI
IBADAN NAIROBI JOHANNESBURG
EXETER (NH) KINGSTON PORT OF SPAIN

ISBN 0 435 65536 1

First published 1947
Reprinted thirteen times
Second Edition 1965
Reprinted four times
Third Edition 1973
Reprinted 1974, 1976, 1977,
1979, 1982

Filmset by Keyspools Ltd, Golborne, Lancs
Printed and bound in Great Britain by
Butler & Tanner Ltd, Frome and London

Preface to the First Edition

The purpose of this work is to provide a course of practical chemistry suitable to the needs of students in the latter part of their Ordinary Level course and throughout their preparation for Advanced Level. By including experimental work in elementary biochemistry, it is hoped that the book will also satisfy the needs of medical students preparing for first M.B. examinations. While it is expected that students will have had three or four years' experience in experimental chemistry, it is realized that some may have had less opportunity than others to cover the preliminary work; the course has therefore been designed to facilitate the work of those who have lacked laboratory experience.

The choice of experiments, and the order in which they are arranged, are based on many years' experience in teaching practical chemistry to senior students in schools; the book has also been divided into parts to correspond with the accepted divisions of the subject in the text-books. As the order chosen by the authors may not conform to the treatment of the theoretical work by other teachers, the division into parts should give a flexibility of presentation to meet individual needs. It is also realized that pressure of work may, on occasions, prevent the teacher from giving individual attention to a student, or to a group of students. The instructions given for each experiment have therefore been fully described, so that a student may work with a minimum of supervision. The amount of practical work covered will depend, very largely, on the limiting factor of the time-table; but with a reasonably generous allowance of time during the two years of preparation for Advanced Level, there should be opportunity to cover, if not the whole, at any rate the majority of the experiments.

Although individual views about the relation between the practical and theoretical aspects of the subject may differ, it is generally agreed that a sound knowledge of practical work, permitting individual observations and experience, is of the utmost importance. Wherever possible in this book, the theoretical considerations have been given as adjuncts to the practical work which is to be performed; but where it is desirable to have a brief introduction of theory prior to the performance of experiments, this latter method has been adopted. It is hoped that this intermingling of theoretical and practical work will prevent the unfortunate tendency to dissociate the practical work of the laboratory from the theoretical knowledge of the lecture-room. It is also hoped that the cross-references from one part of the book to another may help the student to appreciate how closely related are the parts into which the subject is divided for teaching purposes.

Wherever possible, reactions which had been chosen to illustrate properties have been described as test-tube experiments. This has been done for the dual reasons of economy in the use of material and inculcation of the practice of manipulating small quantities. At the same time practice in the use of larger scale apparatus, particularly in organic preparations, has been included in order to familiarize the student with apparatus which has been designed for specific purposes; it will also afford him the satisfaction of proceeding with the preparation of a given compound to a state of reasonable purity. A list of apparatus and material which is required has been given below the title of each experiment, but, to avoid unnecessary repetition, it is assumed that a number of pieces of apparatus and a certain range of material are usually readily accessible in the laboratory; a list of these has been given on page xxvi.

Part I, Physical Chemistry, has been planned to include those topics which may readily be illustrated by the use of simple apparatus. Some work of a revisionary character has also been included. This part is not intended to take the place of a theoretical text-book; but the student who has followed the experimental work with understanding will have become acquainted with much of the physical chemistry which he is required to study. He will also be able to apply his knowledge to reactions which occur in the course of practical work in other parts of the book.

Part II, Inorganic Chemistry, is based on the Periodic Table. The dependence of properties on atomic structure would, in itself, be sufficient justification for arranging the elements, for the purpose of study, in the groups of the Table. There is also the further advantage of maintaining a close relation between the theoretical considerations of chemical properties and the practical observations on which the theory is based. It must, however, be acknowledged that the comparatively small selection of elements which the student is required to study for examination purposes makes it difficult to show, experimentally, the gradation of properties which a more comprehensive study would make evident.

Part III, Organic Chemistry, contains experimental work which exceeds the requirements of many examining boards; but no apology is needed for exceeding minimum demands in favour of a reasonably comprehensive course. If the amount of time does not allow for all the work to be covered, a judicious selection may be made from the preparations which are described.

Part IV, Volumetric Analysis, has stood the test of a very wide experience as a separate publication entitled *The Essentials of Volumetric Analysis*. The text has been revised, and the range of experiments afford a sound ground-work for external examinations.

Part V, Qualitative Analysis, is the scheme of analysis usually followed, but the use of organic reagents in analysis has been included. The value of the present scheme as an illustration of ionic reactions is generally admitted, and the book's intention is to use the suggested organic reagents purely as confirmatory tests.

Part VI, Gravimetric Analysis, is intended to serve as an introduction to the subject. Experience shows that little is gained at this stage by a wide range of gravimetric exercises, and comparatively few illustrations of a fairly simple character have been selected.

Part VII, Biochemistry, has been included primarily for its value to medical students and students of biology; but even those students of chemistry who are not studying biology will benefit from some practical work in this highly important branch of chemistry.

It is hoped that the book will prove useful to teachers and students, and the author will welcome criticisms and suggestions.

Preface to the Third Edition

This new edition has been revised throughout and reset in order to meet the requirements of recent changes in the syllabuses of examination boards, to make alterations in treatment and technique conform with modern practice and to anticipate where possible changes are likely to occur in the near future.

A few experiments have been deleted but many more included in this edition. Ionic equations are given for reactions more often. In Part I a new chapter gives a few electrical experiments and the chapters on solutions and rates of reactions have been expanded; reversible reactions have been separated from reactions that occur principally in one direction. In Part II transition elements have been taken to include all the *d* block elements. In Part III organic preparations can be done as before using apparatus that is corked or else standard-joint glassware. Aliphatic compounds containing a particular functional group are considered alongside their aromatic counterparts. In Part IV, the volumetric analysis, there are new chapters on cerium(IV) salts and complexometric titrations. In Part V the instructions for inorganic qualitative analysis are only given once and it is left to the teacher to decide upon which scale the analysis is to be performed.

SI units have been used throughout. In general the International Union of Pure and Applied Chemistry recommendations have been followed for nomenclature, although not in occasional instances where in my view comprehensibility will suffer.

Alternative names are frequently quoted for inorganic and organic substances because if the nomenclature report of the ASE is accepted by any examining board, the attempt to make naming even more logical may, in the next few years, make the situation more confused and exhausting than in the last few.

I should like to take this opportunity to thank all who have suggested improvements to the two previous editions or who have helped in the preparation of this one. In particular I am indebted to the late Mr A. Holderness and to Mr J. S. Clarke for their valuable suggestions and advice. Finally, I wish to express my sincere thanks to the publisher's managerial staff from whom I have received every assistance and encouragement.

June 1973 J.L.

Contents

PART III ORGANIC CHEMISTRY

PART IV VOLUMETRIC ANALYSIS

PART V QUALITATIVE ANALYSIS

PART VI GRAVIMETRIC ANALYSIS

PART VII BIOCHEMISTRY

APPENDICES

Introduction for the Student

The following general points will assist the student to interpret the directions and arrangement of the book.

1. After the title of each experiment is given a list of the apparatus and materials required. Because ice is not usually readily available, it is given in heavy type thus: **ICE**.

   In giving this list it is assumed that certain simple apparatus and commonly used materials are to hand.

   A list of these is given opposite.

2. A conveniently small quantity of material is termed a spatula-load; this represents approximately 0.25 to 0.5 g and is sufficient to cover the bottom of a test-tube. It will be helpful to note that a test-tube (100×16 mm) holds approximately 12 cm^3.

3. The term 'concentrated', used in connection with solutions, signifies the highest concentration normally available. Dilute solutions of acids and alkalis are those usually available on the bench (usually 1 M, i.e. 1 mol dm^{-3}).

Frequently Required Apparatus and Materials

It is assumed that the following apparatus and materials are easily accessible to the student. Only additional items are specifically mentioned under the title of experiments.

APPARATUS

Beakers, Bunsen burner, balance, burette, carbon block, conical flasks, crucible and lid, evaporating basins, filter funnel and stand, glass rod, glass tubing, ignition tubes (50 × 10 mm), retort stands, pipe-clay triangle, mouth blow-pipe, nichrome wire, pipettes, test-tubes (100 × 16 mm), test-tube rack, holder and brush, tongs, tripod-stand and gauze, wash-bottle, watch-glass.

MATERIALS

Concentrated solutions
Glacial acetic (pure ethanoic, approx. 18 M) acid, ammonia (approx. 20 M, '0.880'), hydrochloric acid (approx. 12 M), nitric acid (approx. 16 M), sodium hydroxide (approx. 5 M), sulphuric acid (approx. 18 M).

Dilute solutions
Hydrochloric, nitric and sulphuric acids, ammonium acetate, ammonium carbonate, ammonium chloride, ammonia, ammonium oxalate, ammonium sulphide, ammonium thiocyanate, barium chloride, bromine water, calcium hydroxide, calcium sulphate, cobalt(II) nitrate, hydrogen peroxide, iron(III) chloride, lead(II) acetate, litmus, methyl orange, mercury(II) chloride, phenolphthalein, potassium hexacyanoferrate(II), potassium hexacyanoferrate(III), potassium chromate, potassium dichromate, potassium iodide, potassium permanganate, silver nitrate, sodium carbonate, sodium hydroxide, disodium hydrogenphosphate, tin(II) chloride.

Solids
Ammonium chloride, ammonium carbonate, copper, copper(II) oxide, copper(II) carbonate, copper(II) sulphate, disodium tetraborate, iron(II) sulphate, iron filings, lead(II) oxide, lead(IV) oxide, trilead tetraoxide, litmus paper, manganese(IV) oxide, sodium carbonate (anhydrous), sodium carbonate (crystalline), sulphur, zinc.

See also pages 371–4.

PART I

Physical Chemistry

1
Relative Atomic and Molecular Masses (including Vapour Density)

Relative atomic masses of elements have been expressed, at any given date in the last 150 years, on some scale adopted for its convenience. By the middle of the nineteenth century, the hydrogen atom was established as the reference unit, mainly because hydrogen is the lightest known element. The basis of the scale of this date was H = 1. On this basis, O = 15.88. This scale continued in use until the early twentieth century, when a change was made to a scale based on O = 16. There were two principal reasons for this change—first, that few elements combine with hydrogen directly in reactions which are experimentally suitable for determination of equivalents, while many elements can be brought into combination conveniently with oxygen; and second, that if an oxygen scale is adopted, any error in the O:H ratio affects only hydrogen, not the other elements, which are referred directly to oxygen.

The adoption of the scale of O = 16 required a slight rise in the values generally, and, in particular, for hydrogen which became 1.008. Correspondingly, the molar volume (V_m) of gases became 22.4 dm^3, instead of the former 22.2 dm^3. At this time, the following relationships were defined.

(*a*) Relative atomic mass of an element

$$= \frac{\text{mass of one atom of the element}}{\frac{1}{16} \text{ of the mass of an oxygen atom}}$$

(*b*) The equivalent of an element is the number of parts by mass of it which combine with, or displace, 8.000 parts by mass of oxygen, 1.008 parts by mass of hydrogen or 35.453 parts by mass of chlorine.

(*c*) The valency of an element is the number of hydrogen atoms which one atom of the element will combine with or displace.

From (*c*), if the valency of an element is v, v atoms of hydrogen combine with 1 atom of the element. If an atom of hydrogen weighs 1.008 units and each unit is u g, v atoms weigh $v \times 1.008 \times u$ g. If an atom of the element weighs A units and each unit is u g, its weight in g is $A \times u$ g.

Therefore, $v \times 1.008 \times u$ g of hydrogen combine with $A \times u$ g of the element.

From (*b*), 1.008 g of hydrogen combine with E g (the equivalent) of the element.

Therefore, $v \times 1.008 \times u$ g of hydrogen combine with $E \times v \times u$ g of the element. Hence

$$A \times u = E \times v \times u$$

or

$$A = E \times v$$

i.e. relative atomic mass = *equivalent* $\times$ *valency*

With the discovery of isotopy (about 1920), a complication arose. Oxygen was found to be an isotopic element, consisting mainly of the isotope $^{16}_{8}O$, but containing small percentages of $^{17}_{8}O$ and $^{18}_{8}O$. Shortly after this discovery, two scales were in operation. The *physical* scale used the relative atomic mass of $^{16}_{8}O$ as the basis, O = 16, of the scale. The *chemical* scale used the weighted average of the relative atomic masses of the three isotopes, as they exist in atmospheric oxygen, as the basis, O = 16, of the scale. This inconvenient situation, with two slightly varying sets of values of relative atomic masses, two slightly different values for V_m of gases and for Avogadro's constant, persisted until 1962, when an internationally agreed scale of relative atomic mass was introduced based upon the value for the isotope of carbon, $^{12}_{6}C$, being taken as precisely 12. By a slight adjustment, the atomic mass values of the physical and chemical scales were brought to this common scale and there is now complete uniformity. Note, however, that the relative atomic mass of the ordinary element, carbon, is not exactly 12; it is slightly higher and slightly variable, because the element contains small, varying percentages of the isotope, $^{13}_{6}C$, as well as the common isotope, $^{12}_{6}C$.

In practice, these recent changes have made very little difference except at the highest levels of accuracy. The change of the chemical scale of O = 16 to the common scale of $^{12}_{6}C = 12$ involved only a few parts per million in the values.

For determination of relative molecular mass by elevation of boiling point, etc., see Chapter 7.

Experiment 1 To find the relative atomic mass of copper

Apparatus: reduction tubes; desiccator.

Material: sample of *dry* copper(II) oxide.

Dulong and Petit's Law states that the atomic heat capacity of a solid element, i.e. the relative atomic mass multiplied by the specific heat capacity, is approximately equal to 27 J K^{-1} mol^{-1}. The approximate relative atomic mass using this law will be used in conjunction with the experimentally determined equivalent to find the valency (which must be a whole number).

Heat some copper(II) oxide in a dish for five minutes, stirring all the while, and allow it to cool in a desiccator. Weigh a clean dry test-tube (a light copper boat may be used to contain the oxide) in which a hole has been blown at one end. Introduce about 2 g of copper(II) oxide and weigh again. Fit up the apparatus as shown in Figure 1, and warm the oxide gently in a stream of town-gas, taking care to have the black powder near the centre of the tube and a gentle stream of town-gas (the flame should not be green—this indicates that particles are being carried over mechanically). When reduction is complete, allow the copper to cool in a current of town-gas and

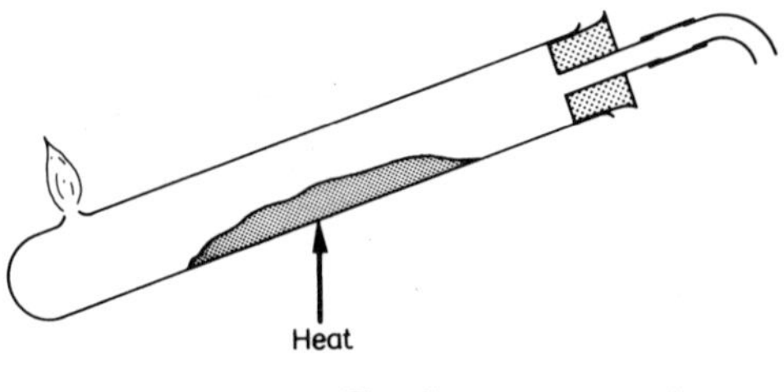

FIG. 1

weigh the test-tube and copper when cold. The reduction, cooling and weighing must be repeated until the mass is constant. Traditional town-gas (coal-gas) contains 50 per cent hydrogen; reduction with natural gas (methane) is slower.

From the weighings find the mass of copper which had combined with 8 g of oxygen, and, assuming the specific heat capacity 0.4 J g^{-1} K^{-1} for copper, calculate a value for the accurate relative atomic mass from your experiment.

Experiment 2 To find the relative atomic mass of mercury

Material: mercury(II) chloride; hypophosphorous (phosphinic) acid or sodium hypophosphite (phosphinate); ethanol; diethyl ether (ethoxyethane).

Determine the specific heat capacity of mercury by the usual physical calorimetric method and hence find the approximate relative atomic mass.

(It is assumed that Dulong and Petit's Law applies to the non-solid element mercury.)

Weigh a clean dry boiling-tube and weigh again with approximately 3 g of mercury(II) chloride crystals. Half fill the boiling-tube with water and immerse in a beaker of water, and warm the latter on a tripod and gauze. Add one-third of the total volume of hypophosphorous acid, or four spatula-loads of sodium hypophosphite. A grey precipitate of mercury is thrown down which soon settles out as globules of mercury. When the mercury is all in the form of globules in the boiling-tube, wash by decantation with water several times.

Dry with ethanol followed by ether (*care!*). Weigh the boiling-tube and contents and from your result calculate the mass of mercury which would have combined with 35.5 g of chlorine. Using this value as the accurate value for the equivalent find the accurate relative atomic mass.
N.B. Valency must be a whole number.

$$2HgCl_2 + H_3PO_2 + 2H_2O \rightarrow 2Hg + 4HCl + H_3PO_4$$

i.e. $$2Hg^{2+} + H_2PO_2^- + 2H_2O \rightarrow 2Hg + PO_4^{3-} + 6H^+$$

Hypophosphorous acid and its salts are very powerful reducing agents.

Experiment 3 To find the relative atomic masses of potassium, silver and chlorine (Stas)

Material: 0.04 M silver nitrate containing free nitric acid; potassium chlorate (potassium chlorate(V)).

These determinations are really equivalent determinations but since the valency of each of the above elements is unity, the relative atomic masses are obtained.

Stas knew that potassium perchlorate (chlorate(VII)), potassium hypochlorite (chlorate(I)) and potassium chlorate (chlorate(V)) had masses of oxygen in the proportion of 4:1:3 combined with equal masses of potassium chloride, and he assumed that potassium chlorate contained six equivalents of oxygen.

(a) DETERMINATION OF THE RELATIVE MOLECULAR MASS OF POTASSIUM CHLORIDE

Weigh out accurately in a clean dry crucible about 1.2 g of potassium chlorate. Replace the lid and heat gently at first and finally strongly for five minutes. Allow to cool and reweigh. Repeat to constant mass. Assuming six equivalents of oxygen have been evolved calculate the equivalent (which is also the relative molecular mass) of potassium chloride.

Specimen results:

Mass of crucible and lid = 10.200 g
Mass of crucible and lid + potassium chlorate = 11.425 g
Mass of crucible and lid + potassium chloride = 10.945 g

$\therefore$ 0.48 g of oxygen was combined with 0.745 g potassium chloride
$\therefore$ 48 g of oxygen was combined with 74.5 g potassium chloride
i.e. the relative molecular mass of potassium chloride.

(b) DETERMINATION OF THE RELATIVE ATOMIC MASS OF SILVER

Transfer the residue from (*a*) completely to a 250 cm^3 flask by putting a short-necked funnel in the flask and emptying the solid into the funnel, and washing out the crucible with distilled water, allowing all the washings to flow into the 250 cm^3 flask. Shake at intervals and make up to the mark. Take 25 cm^3 of this solution (it should be accurately 0.04 M if in part (*a*) 1.225 g were taken) and run in the silver nitrate solution from a burette. The silver nitrate solution contains 4.32 g of silver dissolved in nitric acid and made up to 1 dm^3. The latter is exactly 0.04 M and therefore equivalent (approximately) to the chloride solution volume for volume. Run in 22 cm^3 of the 0.04 M silver nitrate and then boil over the Bunsen burner to coagulate the precipitate of silver chloride. Allow more silver nitrate solution to run in a little at a time, repeating the boiling, and the titration is complete when one drop of silver nitrate no longer produces a cloudy precipitate of silver chloride. Repeat the titration in the same flask (without removing the precipitated silver chloride.

Specimen results:
25 cm^3 potassium chloride (containing 0.0745 g) required 25.0 cm^3 silver nitrate containing 4.32 g dm^{-3} silver.

$$\therefore\ 0.0745 \text{ g potassium chloride} \equiv \frac{4.32 \times 25.0}{1000} \text{ g silver}$$

$$74.5 \text{ g potassium chloride} \equiv 4.32 \times 25.0 \text{ g}$$

$$= 108 \text{ g silver}$$

i.e. equivalent (relative atomic mass) of silver is 108.

(c) DETERMINATION OF THE RELATIVE ATOMIC MASS OF CHLORINE AND POTASSIUM

Filter the precipitate of silver chloride obtained in the last experiment and wash three or four times with a little hot water. Dry the precipitate by allowing the air from a small flame to rise about 60 cm on to the filter paper suspended in a ring.

Open the dried filter paper on a sheet of glazed paper (alternatively, the paper may be ignited in the crucible, and a little concentrated nitric acid added, followed by a little concentrated hydrochloric acid and ignition) and remove all the silver chloride by means of a feather. Transfer the silver chloride to a weighed crucible and warm the crucible gently, then allow it to cool and reweigh it.

Specimen results:

Mass of silver chloride = 0.287 g.

The silver was precipitated from 50.0 cm^3 of the silver nitrate solution by 50.0 cm^3 of the potassium chloride solution. Hence

$$\text{mass of silver} = \frac{50}{1000} \times 4.32 = 0.216 \text{ g}$$

$\therefore$ 0.216 g silver has combined with 0.071 g chlorine

$$1 \text{ g silver has combined with } \frac{0.071}{0.216} \text{ g chlorine}$$

$$108 \text{ g silver has combined with } \frac{0.071}{0.216} \times 108 \text{ g} = 35.5 \text{ g chlorine}$$

i.e. the equivalent (relative atomic mass) of chlorine is 35.5.

But mass of potassium chloride in 50 cm^3 was 0.149 g.
$\therefore$ mass of potassium united with 0.071 g chlorine was

$$(0.149 - 0.071) \text{ g} = 0.078 \text{ g}$$

0.071 g chlorine had combined with 0.078 g potassium.
35.5 g chlorine had combined with 39 g potassium.
i.e. equivalent (relative atomic mass) of potassium is 39.

Experiment 4 To find the relative molecular mass of carbon dioxide[1]

Apparatus: 500 cm^3 flask fitted as shown in Figure 2; Kipp's apparatus for carbon dioxide.

By definition,

$$\text{relative (vapour) density} = \frac{\text{mass of 1 volume of carbon dioxide}^2}{\text{mass of 1 volume of hydrogen}}$$

$$= \frac{\text{mass of 1 molecule of carbon dioxide}}{\text{mass of 1 molecule of hydrogen}}$$

(by Avogadro's law). But, for practical purposes:

relative molecular mass of carbon dioxide

$$= \frac{\text{mass of 1 molecule of carbon dioxide}}{\text{mass of 1 atom of hydrogen}}$$

Since the molecule of hydrogen contains two atoms,

$$\text{relative density} = \frac{\text{mass of 1 molecule of carbon dioxide}}{\text{mass of 2 atoms of hydrogen}}$$

[1] Alternatively, sulphur dioxide may be used direct from a siphon.
[2] Measured under the same conditions of temperature and pressure.

Hence the relative molecular mass is twice the relative density.

Weigh the flask (full of air) and cork and clips, and then pass a stream of dry carbon dioxide through the apparatus for 5–10 minutes.

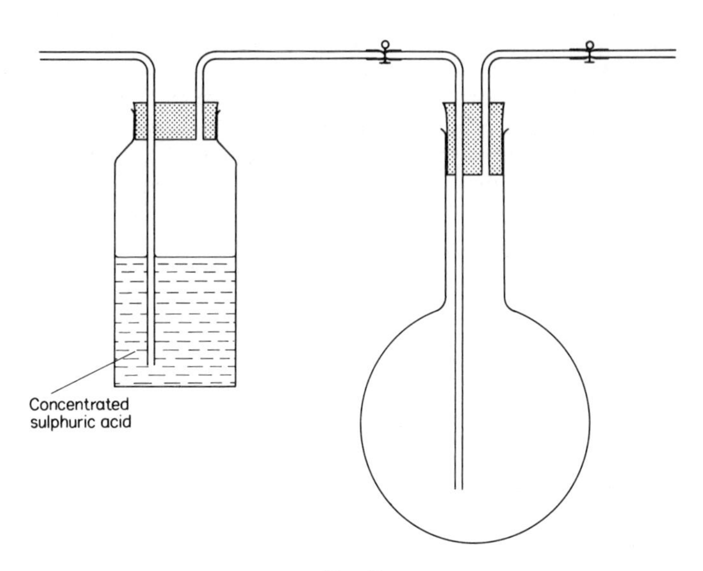

FIG. 2

Close the clips—the one nearer the apparatus first or an increase in pressure might be obtained—and weigh the flask again. Note the temperature and pressure. Now fill the flask with water by connecting the apparatus to a water supply (gently controlled), and empty the water into a measuring cylinder and so find the volume of the flask. Correct this volume to s.t.p. (V cm^3) and hence calculate the mass of:

(*a*) air in the flask,
(*b*) the flask,
(*c*) carbon dioxide,
(*d*) an equal volume of hydrogen.

By definition, the relative density of carbon dioxide is $(c)/(d)$. Hence the relative molecular mass is $2(c)/(d)$.

Note: 1 dm^3 of dry air at s.t.p. weighs 1.293 g: 1 dm^3 of dry hydrogen at s.t.p. weighs 0.090 g.

Experiment 5 To find the relative molecular mass of acetone (Dumas)

Apparatus: Dumas bulb; water-bath to fit; thermometer (100 °C).
Material: acetone (propanone).

This method will give a good degree of accuracy but requires a comparatively large volume of the liquid, the relative molecular mass of which is to be determined.

A glass bulb of about 150 cm^3 capacity is weighed and about 10 cm^3 of acetone is introduced into the bulb by warming the bulb slightly and allowing it to cool with the neck under the surface of acetone. Alternatively, a teat pipette may be used to insert the acetone. The bulb is fixed in a bath of water (Figure 3) which is

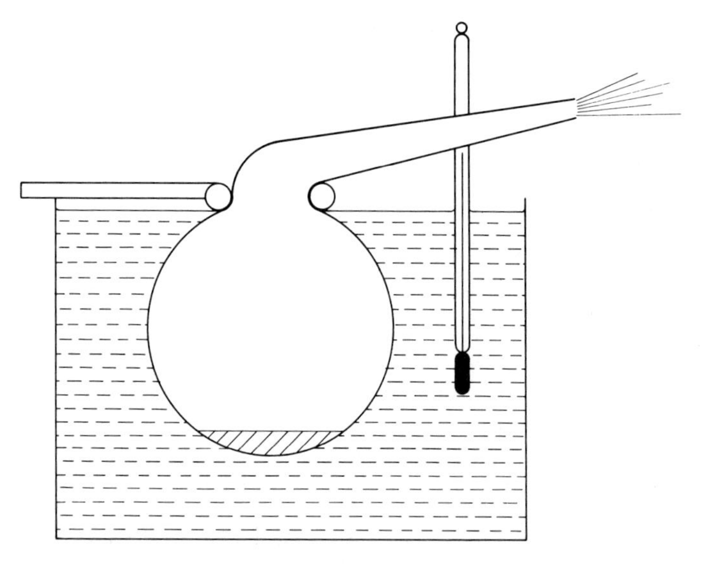

FIG. 3

heated until the temperature is about 25 °C above the boiling point of the acetone (b.p. 61 °C). The acetone vaporizes and expels the air as it fills the bulb. Allow the vapour to impinge on a small flame so that the rate of issue of vapour can be estimated. (The flame can subsequently be used for sealing off.) Just as the vapour ceases to issue, seal off the bulb and record the temperature of the water bath and the atmospheric pressure. Weigh the bulb again. Break off the neck under water and allow the bulb to fill completely with water. Weigh the bulb (also the piece broken off) full of water.

Specimen results:

Mass of bulb and air	27.56 g
Mass of bulb and vapour after sealing	28.08 g
Mass of bulb full of water (and piece broken off)	205.6 g
Temperature of laboratory	17 °C
Temperature of bath	75 °C
Pressure of atmosphere	750 mmHg
1 dm^3 of dry air at s.t.p. weighs	1.293 g
1 dm^3 of hydrogen at s.t.p. weighs	0.09 g

$$\text{Approximate volume of bulb} = (205.6-27.56)\ \text{cm}^3 = 178\ \text{cm}^3$$

178 cm^3 of air at 17 °C and 750 mmHg weigh at s.t.p.

$$\frac{178}{1000}\times\frac{273}{290}\times\frac{750}{760}\times 1.293\ \text{g} = 0.213\ \text{g}$$

$$\therefore \text{true mass of bulb} = 27.347\ \text{g}$$

$$\therefore \text{mass of vapour} = 0.733\ \text{g}$$

But 178 cm^3 of vapour at 75 °C and 750 mmHg would occupy at 0 °C and 760 mmHg a volume of

$$178\times\frac{273}{348}\times\frac{750}{760}\ \text{cm}^3 = 137.5\ \text{cm}^3$$

$\therefore$ 137.5 cm^3 of vapour weigh 0.733 g

$$1\ \text{dm}^3 \text{ of vapour weighs } 0.733\times\frac{1000}{137.5}\ \text{g}$$

$$\therefore \text{relative density} = \frac{0.733\times 1000}{137.5\times 0.09} = 59.2$$

This, when doubled, gives the relative molecular mass.

Experiment 6 To find the relative molecular mass of diethyl ether (Victor Meyer)

Apparatus: Victor Meyer's apparatus.
Material: diethyl ether (ethoxyethane).

This method is much less accurate than Dumas' method, but is very rapid and is sufficiently accurate when it is necessary merely to decide what relation exists between the empirical and molecular formulas.

Note that the volume measured is that of an equal volume of air expelled as the liquid vaporizes. If the vapour diffuses rapidly and enters the measuring tube, it will condense to a liquid and so cause an error.

The apparatus consists (Figure 4) of a long glass tube with a bulb at one end, into which can be dropped a small cylindrical bottle (Hofmann bottle). The long glass tube fits into an outer casing,

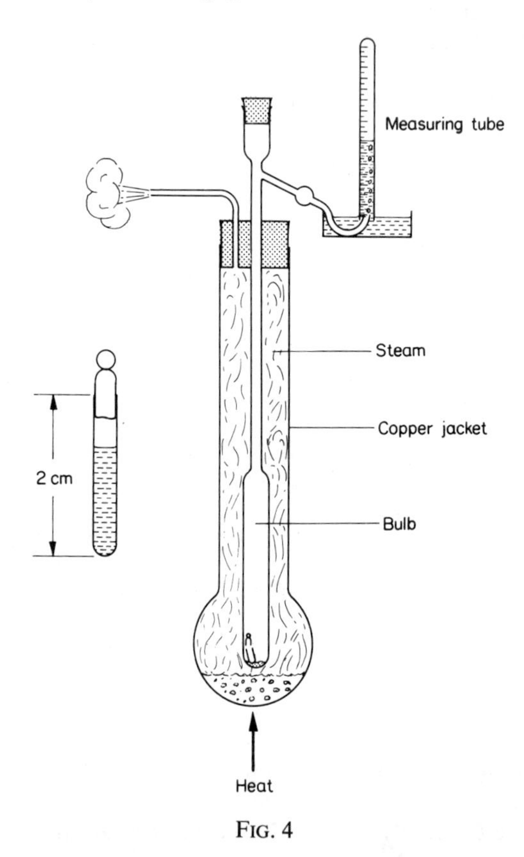

FIG. 4

usually of copper, which contains water (or a liquid boiling above the boiling point of the liquid whose relative molecular mass is being determined) which is raised to boiling point. When the temperature is steady no more bubbles of air will emerge from the side tube (rubber stopper inserted in top). The bottle is meanwhile weighed empty, then filled with ether and weighed again, and the stopper is loosely inserted. The measuring tube is filled with water and placed in position, then the stopper is removed, the bottle is dropped in and the stopper is rapidly and firmly replaced. The pressure of the vapour formed in the bottle blows off its stopper and the vapour displaces its own volume of air. When no more bubbles of air are seen to come off, the volume of air is measured[1] and the temperature and pressure of the air in the room are measured.

Calculation:

Mass of bottle	$= W_1$ g
Mass of bottle + liquid	$= W_2$ g
Volume of displaced air	$= V_1$ cm^3
Temperature of laboratory	$= T$ °C
Atmospheric pressure	$= P$ mmHg
Water vapour pressure at T °C	$= p$ mmHg

Then V_1 cm^3 of air at T °C and $(P-p)$ mmHg pressure would occupy, at 0 °C and 760 mmHg, a volume

$$V_1 \times \frac{273}{273+T} \times \frac{(P-p)}{760} \text{ cm}^3 = V_2 \text{ cm}^3 \text{ (say)}$$

$$1 \text{ dm}^3 \text{ of vapour would weigh at s.t.p. } \frac{1000}{V_2} \times (W_2 - W_1) \text{ g}$$

$$\therefore \text{ relative density } = \frac{1000\,(W_2 - W_1)}{V_2 \times 0.09}$$

This, when doubled, gives the relative molecular mass.

[1] The volume of air may be measured at atmospheric pressure by transferring the tube to a gas-jar containing water and moving the tube so that the water levels inside and outside the tube are the same.

2
Solutions

A solution is a homogeneous mixture of two (or more) substances the proportions of which may vary between certain limits.

The following types of mixtures will be mentioned:

1. gas in gas,
2. gas in liquid,
3. liquid in liquid,
4. solid in liquid.

Experiment 7 To obtain a solubility curve for potassium chlorate

Apparatus: thermometer (100 °C); stirrers; pestle and mortar.

Material: potassium chlorate (potassium chlorate(V)).

This experiment may be performed as a class experiment by different students finding the solubility at different temperatures.

Grind some potassium chlorate to a fine powder in a mortar and add it to water in a boiling-tube fitted up as shown in Figure 5. Heat the water to about 5 °C above the temperature at which the solubility is to be found (say 40 °C) and allow to cool. Stir continuously and maintain the presence of solid in the bottom of the boiling-tube. At the required temperature, quickly decant a portion (about 5 cm^3) of the solution into a weighed dish. Weigh the dish again and evaporate the solution to dryness on the water-bath. When it is apparently dry, transfer the dish to a tripod and gauze and gently warm it with a flame held 7–10 cm away from the dish for a minute. Allow the dish to cool and then weigh it. Repeat the warming process until the mass is constant. Calculate the mass of potassium chlorate necessary to saturate 1 kg of water at the temperature. Collect values from other students and construct a graph.

Whilst waiting for the potassium chlorate solution to evaporate on the water bath, show that calcium hydroxide is *less* soluble in hot water than in cold by warming a test-tube containing some of the filtered saturated solution, when a slight precipitate of the hydroxide will be obtained.

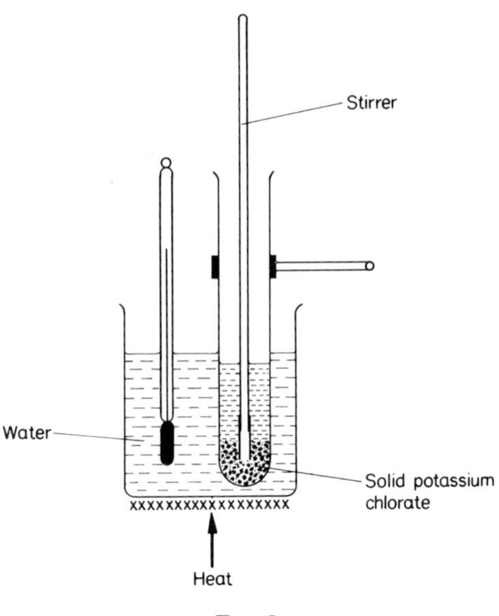

FIG. 5

Experiment 8 Solubility of potassium chlorate (alternative method)

Apparatus: thermometers (100 °C); 10 cm^3 pipette; stirrers.
Material: potassium chlorate (potassium chlorate(V)).

Note: This method will not give accurate results with solids which exhibit supersaturation.

Weigh out into a series of test-tubes the following masses of potassium chlorate:

0.7 g; 1.0 g; 1.5 g; 2.0 g; 3.0 g.

Run in exactly 10 cm^3 (g) of water from a pipette into each. Use the apparatus of Figure 5 and heat, with constant stirring, until the solid has just dissolved. Allow it to cool and note the temperature at which crystals can be seen forming in the test-tube. This temperature can be taken as the temperature at which the solution was just saturated. Proceed in a similar manner with each of the other test-tubes and plot the solubility curve.

Note: The above experiment will usually be carried out as a class experiment, each student making one observation. If a student must obtain all the necessary data for the graph he should proceed in a slightly different manner. Weigh out 3.0 g of the salt into a boiling-tube and add 10 cm^3 of water from a burette. Proceed as above to find one point on the curve. Now add 5 cm^3 of water and continue to find points on the curve until a total of 40 or 45 cm^3 of water has been added.

Experiment 9 Preparation of potassium perchlorate as an example of fractional crystallization

Material: potassium chlorate (potassium chlorate(V)).

Half fill a crucible with potassium chlorate and arrange it firmly in a pipe-clay triangle. Heat gently until it melts and then stir the liquid (just molten) continuously until it becomes pasty (15 minutes). During the period supply sufficient heat to keep the mass just molten.

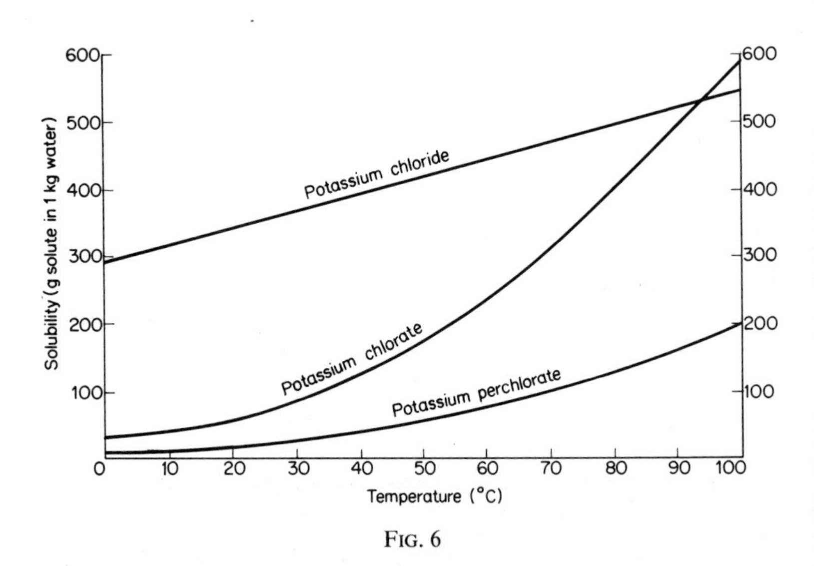

FIG. 6

Allow the crucible to cool, add an equal bulk of water and heat gently until all has dissolved. Pour the solution on to a watch-glass and allow it to cool. The crystals which come out first are almost pure potassium perchlorate (potassium chlorate(VII)) and can be purified further by dissolving in the minimum of hot water and recrystallizing.

$$4KClO_3 \rightarrow 3KClO_4 + KCl$$

The solubility curve, Figure 6, explains the separation.

Experiment 10 Preparation of a supersaturated solution

Material: sodium thiosulphate ('hypo') crystals.

Two-thirds fill a clean boiling-tube with crystals of sodium thiosulphate ($Na_2S_2O_3 \cdot 5H_2O$) and warm carefully until dissolved. Cool the tube under the tap, avoiding undue agitation, for three to five minutes. The tube may be plugged with cotton wool while it is cooling. When cold, hold it in the hand whilst dropping in one small crystal of sodium thiosulphate. As the crystal grows, the tube gets warm and finally the contents are almost completely solid.

Experiment 11 To show the solution of bromine in air

Apparatus: gas-jar.

Material: bromine.

Gases are miscible in all proportions. Thus air contains numerous gases: one gas is always 'soluble' in another.

Take a gas-jar to a fume chamber, allow one drop of liquid bromine to fall to the bottom of it (*care!*) and replace the lid. The bromine can be seen to diffuse rapidly until the mixture is homogeneous. This is in spite of a very large difference in density since bromine vapour is nearly six times as dense as air.

Experiment 12 To determine the composition of the gas boiled out of water

Apparatus: apparatus shown in Figure 7; absorption cup.

Material: pyrogallol (1,2,3-benzenetriol).

Set up the apparatus shown in Figure 7, using a burette to collect the gas, and making sure that all air is excluded from the apparatus. Boil the water in the flask until all air has been expelled, allow it to cool and note the volume of gas. The alkaline pyrogallol is best introduced by means of an absorption cup, which is inserted in a rubber bung which will fit the burette. Fill the cup with crystals of pyrogallol, add a few drops of water to fill the air spaces and close the end with the finger. Have ready a pellet of sodium hydroxide and insert this into the burette (which is *in situ*) followed quickly by the absorption cup. Invert the tube several times and release the cup under water. Note the volume of gas absorbed. This will be found to be approximately one third of the total volume of air boiled out of the water. Now the volume of oxygen dissolved (and therefore boiled out) will vary directly with the partial pressure of oxygen in the air (0.2 atm approximately) and directly with the coefficient of

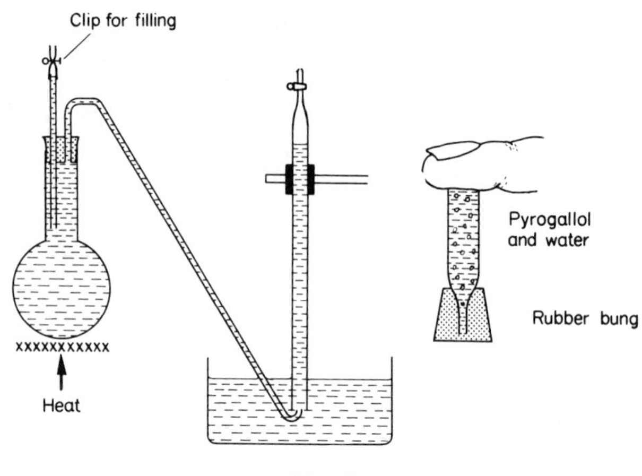

FIG. 7

absorption of oxygen by water; similarly with nitrogen, which occupies 80% (approximately) by volume of the air.

$$\therefore \frac{\text{coefficient of absorption of nitrogen} \times 80}{\text{coefficient of absorption of oxygen} \times 20}$$

$$= \frac{\text{volume of gas remaining in tube}}{\text{volume of gas absorbed by pyrogallol (approx.).}}$$

If the volume of the flask is known, the coefficients can be estimated in terms of volume of gas (oxygen or nitrogen) absorbed by one volume of water at the temperature of the laboratory.

Experiment 13 To determine the solubility of a very soluble gas

Apparatus: ammonia apparatus; test-tube fitted as in Figure 8; gas-jar; string.

Material: 2 M hydrochloric acid; 2 M sodium hydroxide.

Fit a test-tube with a rubber bung through which is a heavy glass rod. Tie a piece of string round the mouth of the test-tube to facilitate weighing. See Figure 8. Weigh the test-tube with bung and glass rod. Into the test-tube put about 15 cm^3 of 0.880 ammonia and pass ammonia gas into it for a few minutes to ensure maximum solubility at the temperature of the laboratory. Cork the tube firmly with the rubber bung (with the glass rod in position touching the bottom of the test-tube) and weigh it again. Put 250 cm^3 of 2 M hydrochloric acid into a gas-jar and in it place the test-tube containing the ammonia. Strike the rod on the bottom of the test-tube so as to break it, and then use the rod and broken tube as a stirrer throughout the rest of the operation. Add methyl orange to the solution and then withdraw about 10 cm^3 of the solution by means of a pipette. Run in the 2 M sodium hydroxide fairly rapidly with stirring until the solution turns yellow, add the 10 cm^3 (with washings) which had been withdrawn and proceed to obtain an accurate end-point.

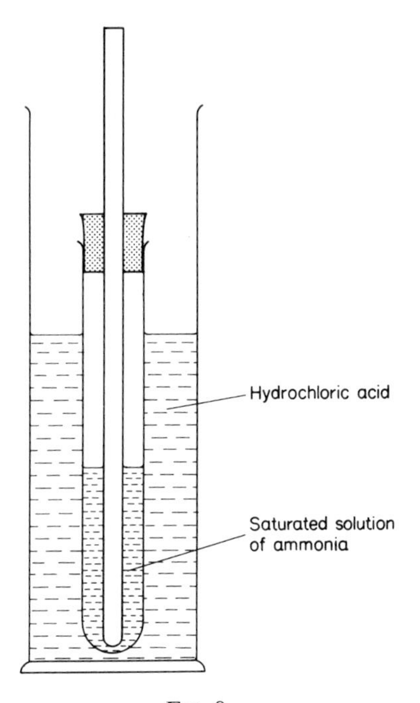

FIG. 8

Specimen results:

Temperature of laboratory	= 15 °C
Mass of test-tube, bung and glass rod	= 21.75 g
Mass of test-tube, bung, rod and ammonia solution	= 37.00 g
Volume of 2 M hydrochloric acid taken	= 250 cm^3
Volume of 2 M sodium hydroxide used to neutralize excess acid	= 100.3 cm^3

149.7 cm^3 2 M acid ≡ ammonia in 15.25 g of solution.

$$\begin{array}{lll} NH_3 + HCl & & \rightarrow NH_4Cl \\ 17\text{ g} & 1\text{ dm}^3\ 1\text{M} & \end{array}$$

$$\therefore\ 149.7\text{ cm}^3\ 2\text{ M ammonia contain } \frac{17 \times 149.7 \times 2}{1000}\text{ g} = 5.09\text{ g ammonia}$$

∴ 10.16 g water dissolved 5.09 g ammonia

∴ 1000 g water would dissolve $\frac{5.09 \times 1000}{10.16}$

$= 50.2$ g at 15 °C

Experiment 14 To obtain a solubility curve for phenol in water

Apparatus: corked test-tubes; thermometer (100 °C).

Material: phenol. *Care must be taken throughout the experiment to avoid contact between phenol and the skin.*

About 7 g of phenol is weighed accurately into a corked test-tube and 3 cm^3 of water are added from a burette. The test-tube is gradually warmed in a beaker of water with continual shaking until the two layers disappear. It is allowed to cool slowly with shaking until the contents suddenly go cloudy. The temperature of the solution is then read off. The contents of the test-tube then represent a saturated solution of water in phenol at this temperature; 1 cm^3 of water is added from the burette and the same procedure is repeated. By continuing in this manner the solubility of water in phenol at various temperatures can be determined. When a temperature of

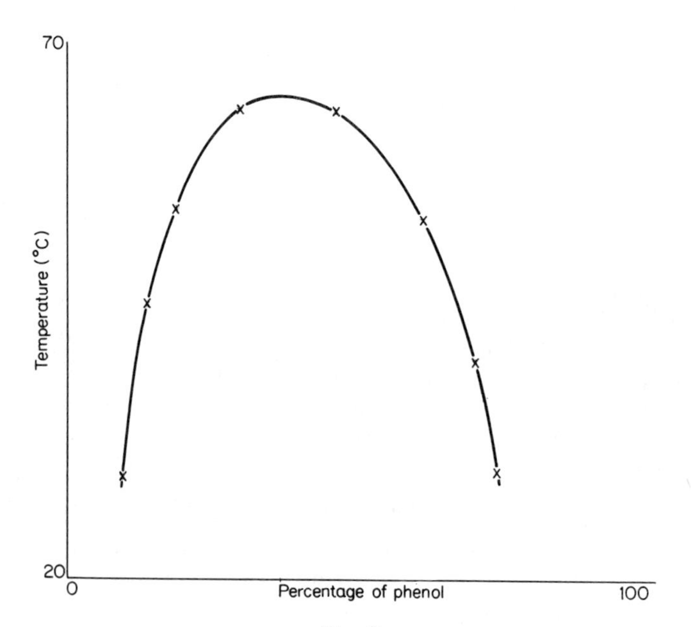

FIG. 9

about 68 °C is reached, a second test-tube is taken and about 10 cm^3 of water measured out into it. A weighed quantity of phenol is then added (about 1 g) and the temperature at which this mixture just becomes a saturated solution is found as above. A further weighed amount of phenol is added and the experiment is repeated until a temperature of about 68 °C is again reached. Plot a graph of the temperature against the percentage of phenol in the mixture (see Figure 9).

Experiment 15 To effect a partial[1] separation of water and ethanol by distillation

Apparatus: distillation apparatus fitted with thermometer (100 °C).
Material: ethanol.

Mix 25 cm^3 of water and 25 cm^3 of ethanol and put them into a distilling flask in which there are some pieces of porous pot. Allow the flask to rest on gauze whilst heated—this minimizes the risk of fracture at the end of the distillation. Fit the flask with a thermometer dipping into the liquid and distil, separating the fractions which pass over through successive ranges of about 4 °C. With care, almost the whole of the liquid can be distilled and the fractions are arranged in order in a test-tube rack. Warm the first fraction and apply a light to the mouth of the test-tube. The alcohol vapour burns and this fraction will burn away almost completely. The final fraction shows no tendency to produce an inflammable gas on heating. Now add a spatula-load of anhydrous copper(II) sulphate to each fraction and shake. There is a gradation of colour from very pale blue to a deep blue solution.

Experiment 16 Determination of the partition coefficient of succinic acid between water and ether

Apparatus: measuring cylinder; pestle and mortar; separating funnel.
Material: succinic acid (butanedioic acid); diethyl ether (ethoxyethane).

If a solid (or liquid) A dissolves in each of two immiscible liquids B and C then, providing A is in the same molecular state in each, and there is insufficient to saturate B or C, A will divide itself between B and C in a fixed proportion. This proportion is expressed as a ratio of the molecular concentrations of A in the two layers and is termed the partition coefficient. The value of the partition coefficient, although

[1] Ethanol and water form a constant boiling point mixture (96% ethanol). The final traces of water, therefore, cannot be removed by distillation.

constant at any one temperature, has a different value at different temperatures.

Grind a little succinic acid in a mortar and weigh out quantities of 1.5 g, 1 g and 0.5 g (these weighings need only be roughly done). Pour into a separating funnel about 50 cm^3 of water and add one quantity of succinic acid. Add about 50 cm^3 of ether, shake until the acid has dissolved and allow to stand for a few minutes. Run off nearly all the water layer and titrate 25 cm^3 of this against 0.5 M sodium hydroxide using phenolphthalein as an indicator (about 29 cm^3 will be required for the initial addition of 1 g of succinic acid). Discard the boundary portion. Take 25 cm^3 of the ether layer by means of a measuring cylinder and titrate this against the 0.1 M alkali. Shake well after each addition of alkali. (The volume of 0.1 M alkali required will be approximately the same as the amount of 0.5 M alkali required for the water layer.) Obtain the distribution ratio

$$K = \frac{5 \times \text{volume of 0.5 M alkali for 25 cm}^3 \text{ water layer}}{\text{volume of 0.1 M alkali for 25 cm}^3 \text{ ether layer}}$$

Repeat the process with the other quantities of the succinic acid.

NOTES ON EXTRACTION BY ETHER (AND OTHER LIQUIDS)

Extraction by ether is an application of the Distribution Law (or Partition Coefficient).

If a substance A is soluble in two liquids which are themselves immiscible, the substance distributes itself in a ratio of concentrations which is constant for the system.

Suppose that 11 g of A are dissolved in 1 dm^3 of water, and that 1 dm^3 of ether is available. The question to be discussed is whether it is more profitable to use all the ether in one extraction, or to take 500 cm^3 for an extraction and then, after separating off the used ether, to add the remaining 500 cm^3 for a further extraction. Let us assume that the solubility of A in ether is ten times as great as its solubility in water; then

$$\frac{\text{Solubility of A in ether}}{\text{Solubility of A in water}} = \frac{10}{1}$$

(*a*) On shaking 11 g of A in 1 dm^3 of ether and 1 dm^3 of water, 10 g of A will dissolve in ether and 1 g in water; 10 *g of A will be extracted* and 1 g left in the water.

(*b*) On shaking 11 g of A in 500 cm^3 of ether and 1 dm^3 of water, let x g dissolve in the ether.

$$\frac{\text{Concentration of A in ether (g dm}^{-3})}{\text{Concentration of A in water (g dm}^{-3})} = \frac{2x}{11-x} = \frac{10}{1}$$

From this, $x = 9.16$ g, i.e. 9.16 g of A extracted by the first 500 cm^3 of ether.

1.84 g of A remain in the water. Shake with the second 500 cm^3 of ether, and let y g of A dissolve in the ether.

$$\frac{\text{Concentration of A in ether (g dm}^{-3})}{\text{Concentration of A in water (g dm}^{-3})} = \frac{2y}{1.84-y} = \frac{10}{1}$$

From this, $y = 1.53$ g, i.e. 1.53 g of A extracted by the second 500 cm^3 of ether.

Total of A extracted $= 9.16+1.53 = 10.69$ g.

It therefore follows that if the ether were used in three or more portions, the total extraction would approximate closely to the mass of A originally present.

Experiment 17 To determine the partition coefficient of iodine between carbon tetrachloride and water

Apparatus: separating funnel; weighing bottle; measuring cylinders.

Material: carbon tetrachloride (tetrachloromethane); iodine; 0.01 M sodium thiosulphate.

Weigh accurately (in a stoppered weighing bottle) 1 g of iodine, and add to 50 cm^3 of carbon tetrachloride in a separating funnel. Add 50 cm^3 of water and shake well. Leave to stand for some time and then run off the lower (carbon tetrachloride) layer. Take 25 cm^3 of the water layer and titrate against the sodium thiosulphate (see p. 339). Repeat the experiment with different masses of iodine.

Specimen results:

Let the volume of sodium thiosulphate used for 25 cm^3 of solution be 4.5 cm^3.

$\therefore$ Iodine solution is $\dfrac{4.5}{50\times 100}$M and contains $\dfrac{4.5\times 254}{50\times 100}$ g dm^{-3} of iodine

$$= 0.0114 \text{ g iodine in 50 cm}^3 \text{ of solution}$$

$\therefore$ mass of iodine in the 50 cm^3 carbon tetrachloride

$$= 1-0.0114 \text{ g} = 0.9886 \text{ g}$$

$$\therefore \frac{\text{concentration of iodine in carbon tetrachloride}}{\text{concentration of iodine in water}} = \frac{0.9886}{0.0114} = \frac{87}{1}$$

The partition coefficient will be found to be approximately constant over the range of experiments, showing that iodine is in the same molecular state in the two liquids.

Experiment 18 To find the partition coefficient of glacial acetic acid between water and carbon tetrachloride

Apparatus: separating funnels; 1 cm^3 pipette and filler; measuring cylinder.

Material: carbon tetrachloride (tetrachloromethane); 1 M and 0.1 M sodium hydroxide.

Put 1 cm^3 of glacial acetic (pure ethanoic) acid into 20 cm^3 of water and 20 cm^3 of carbon tetrachloride in a separating funnel and shake for three minutes. Allow to settle and run off the lower layer of carbon tetrachloride, discarding the boundary layer. Titrate 10 cm^3 of the aqueous layer with 1 M alkali using phenolphthalein as an indicator. Titrate 10 cm^3 of the carbon tetrachloride layer using 0.1 M alkali. Repeat the experiment using 2 cm^3 and then 3 cm^3 of acetic acid. Allow plenty of time when titrating against the carbon tetrachloride solution and shake between each addition. You will observe from the figures for the titration that

$$\frac{\text{concentration of acetic acid in water}}{\text{concentration of acetic acid in carbon tetrachloride}}$$

is not constant.

Specimen results (at 15 °C)

Vol. of acetic acid added cm^3	*Vol. of 0.1 M NaOH for 10 cm^3 CCl_4 layer cm^3*	*Vol. of 1 M NaOH for 10 cm^3 H_2O layer cm^3*	*$\frac{\text{Concentration in } H_2O}{\sqrt{(\text{concentration in } CCl_4)}}$ $mol^{\frac{1}{2}}\ dm^{-\frac{3}{2}}$*
1	0.8	9.05	14.3
2	2.9	17.45	14.1
3	6.1	25.70	14.7

The volumes of alkali for the organic layer must be divided by 10 when calculating the partition coefficient.

If the acetic acid exists in the carbon tetrachloride layer as double molecules then it can be deduced theoretically that

$$\frac{\text{concentration in water}}{\sqrt{(\text{concentration in carbon tetrachloride})}} = \text{a constant}$$

The observed results from the above experiment do give a constant in accordance with this expression, proving the double-molecular condition of acetic acid in carbon tetrachloride.

Experiment 19 To find the formula of the complex ion formed by copper(II) with ammonia

Apparatus: separating funnel; pipette filler.

Material: 0.1 M copper(II) sulphate; 0.5 and 0.05 M hydrochloric acid; chloroform (trichloromethane); 1 M ammonia.

Put about 75 cm^3 each of ammonia solution and of chloroform into the separating funnel and shake it for several minutes. Obtain the separate layers, discarding the boundary layer. Titrate 25 cm^3 of each layer, using methyl orange as the indicator: the lower (chloroform) layer with the 0.05 M acid and the aqueous layer with the 0.5 M acid. Record the temperature. Calculate the partition coefficient of ammonia between water and chloroform.

Then put 25 cm^3 of ammonia solution and 25 cm^3 of copper(II) sulphate with 75 cm^3 of chloroform in the separating funnel and shake it for several minutes. This time only the chloroform layer can be titrated but, using the partition coefficient, the concentration of ammonia in the water layer can be calculated. Thus the quantity of ammonia that has combined with a known quantity of copper ions can be found.

$$Cu^{2+} + xNH_3 \rightleftharpoons Cu(NH_3)_x^{2+}$$

The calculation is best done in terms of the number of millimoles of a material, i.e. the volume in cm^3 multiplied by the concentration in mol dm^{-3}.

Experiment 20 The distillation of immiscible liquids (steam distillation)

Apparatus: see Figure 10.

Material: chlorobenzene.

Put about 100 cm^3 water and 40 cm^3 chlorobenzene in flask B and heat flasks A and B. When the temperature of the contents of flask B reach about 80 °C, turn off the Bunsen burner beneath flask B. Record the temperature as distillation proceeds and continue passing steam from flask A into flask B until about three-quarters of the chlorobenzene has distilled, then stop. Record the volumes of the separate layers of the distillate and the atmospheric pressure.

Then, if the two immiscible liquids are X and Y, by Naumann's equation

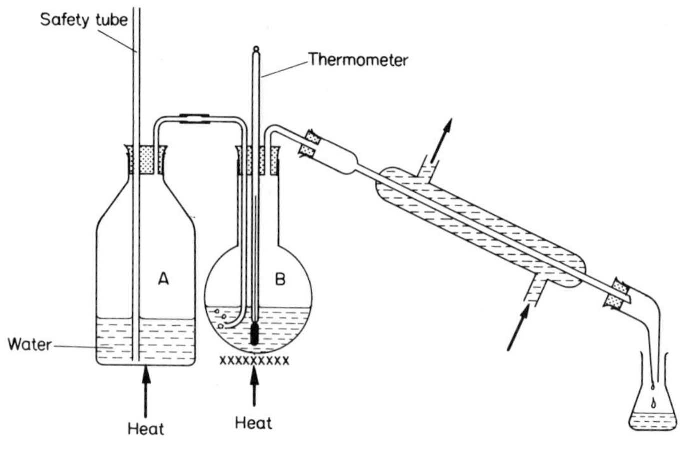

FIG. 10

$$\frac{\text{vapour pressure of X}}{\text{vapour pressure of Y}}$$

$$= \frac{\text{mass of X}}{\text{relative molecular mass of X}} \times \frac{\text{relative molecular mass of Y}}{\text{mass of Y}}$$

The density of chlorobenzene is 1.1 g cm^{-3}. This experiment can be used to determine its relative molecular mass. See also Experiment 267, p. 263.

Experiment 21 The distillation of completely miscible liquids

Apparatus: see Figure 39, but thermometer must dip into acid.
Material: 1 M hydrochloric acid; 1 M sodium hydroxide.

Put exactly 150 cm^3 of the hydrochloric acid into the flask and arrange to collect 10 cm^3 samples of the distillate in marked test-tubes. Heat the acid until it boils and then continue to heat moderately until the flask is just empty. Record the average temperature during the collection of each sample of distillate. Titrate each sample of the distillate with the alkali using methyl orange as the indicator; the titres are very low until towards the end of the experiment.

Calculate the number of millimoles of hydrochloric acid in each sample: there are 150 in the flask at the start of the experiment. Hence deduce the concentrations of the liquid in the flask and in the distillate at each stage. Plot a graph of these concentrations against the temperature and deduce the composition and temperature of the constant boiling point mixture.

Experiment 22 A eutectic mixture

Apparatus: a test-tube suspended in a beaker of water.

Material: 2-nitrophenol; 4-nitrotoluene (methyl-4-nitrobenzene).

Find the melting point of each of the organic substances. Mix 0.2 g of one substance with 1 g of the other and find the melting point of the mixture; then mix 0.4 g with 1 g and again find the melting point. Next reverse the quantities and find the melting points of the mixtures obtained. Plot a graph of melting point against percentage composition of the mixture by mass and deduce the eutectic temperature and the composition of the eutectic mixture. Further mixtures can be made up and studied to clarify any difficulties in deciding the shape of the graph around the central portion.

Experiment 23 Paper chromatography—radial method

Apparatus: two glass plates about 15 cm square, one having a hole at its centre.

Cut a wick by making two parallel incisions on a piece of filter paper from the circumference to the centre. Put a drop of screened methyl orange or black ink (e.g. Parker's Super Quink) at the centre of the paper and put a solvent such as a mixture of butanol, ethanol and concentrated ammonia and water (percentages by volume 60, 20, 3 and 17 respectively) in a beaker (see Figure 11).

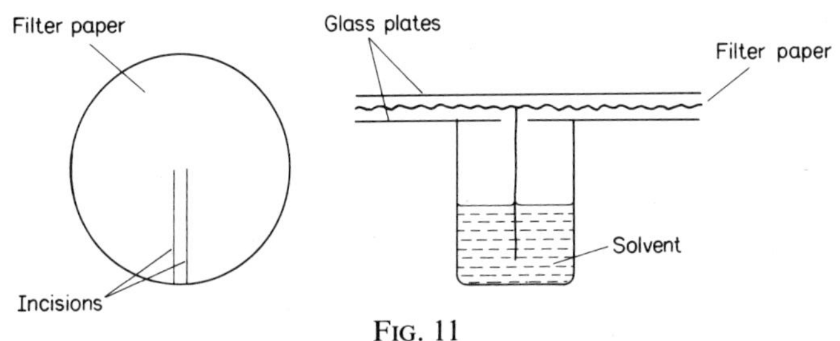

FIG. 11

After a while the substance used to spot the paper will spread out into a series of concentric circles. Allow this to continue until the solvent front reaches nearly to the circumference of the paper, then dry the paper.

The R_f value of a substance is given by the ratio

$$\frac{\text{distance moved by substance}}{\text{distance moved by solvent}}$$

Calculate the R_f values of the components of the substance studied.

Experiment 24 Paper chromatography—ascending method

Apparatus: 1 or 2 dm^3 beaker and a cover.

The previous experiment can be modified so that the substances in a mixture are not spread out in a circle, but are separate spots, by using the method of ascending chromatography. The solvent is put to a depth of 1 cm in a tall vessel, e.g. 1 or 2 dm^3 beaker, and the substance(s) for analysis spotted on the chromatography paper (e.g. 20 × 20 cm) at 2 cm intervals at a distance of 2 cm from one end. The paper is then coiled into a smooth cylinder and clipped: it should be stood in the solvent with the spots just above the surface, and a cover for the beaker reduces losses of solvent by evaporation.

Proceeding in this manner, the substances involved in one particular analytical group may be separated. After drying the paper to remove the solvent, it may be necessary to spray it with some reagent with which the colourless cations involved form coloured derivatives. For example, in analytical groups II and IV hydrogen sulphide solution should be sprayed on to the paper. If indicators are being studied, the paper, after drying, should be sprayed or fumed with ammonia and then with hydrochloric acid.

Experiment 25 Thin layer chromatography

Apparatus: Glass plates about 15 cm square.

Material: powdered silica gel containing calcium sulphate as a binding agent.

Make the silica gel into a thick cream with water and spread it out in a thin layer on a sheet of glass, using a glass rod as a rolling pin. After a brief pause to allow some solidification, bake the plates in an oven at 90 °C for half an hour. Spots of the substance(s) to be analysed can then be put 2 cm from one edge and the plate stood vertically in a vessel containing solvent to a depth of 1 cm. Comparison of this method with that of paper chromatography is of interest.

Experiment 26 Ion exchange

Apparatus: glass columns with taps at the lower ends.

Material: cation and anion exchange resins. A soap solution can be made by putting 10 g of sodium stearate (octadecanoate) in 1 dm^3 of water and ethanol (1:1 by volume).

Permanent hardness in water is caused by substances such as magnesium and calcium sulphates, and they can be replaced in solution by running tap-water through a cation exchange resin (in its sodium form). Temporary hardness in water is caused by substances such as magnesium and calcium hydrogencarbonates, and the water can be softened by running it through an anion exchange resin (in its chloride form). The emergent samples of water, like the original water, can be titrated with soap solution to determine the degree of hardness.

By using two columns, in the hydrogen and hydroxide forms respectively, demineralized water may be obtained: this is comparable with distilled water except that it may contain organic impurities unless steps are taken to filter them out.

A cation exchange resin of the domestic type is usually a sodium compound, and regeneration is achieved by washing it with sodium chloride solution and discarding the resultant calcium chloride solution. The hydrogen and hydroxide resins are regenerated by hydrochloric acid and sodium hydroxide respectively. A column—which can be a burette packed with the resin—should never be allowed to run dry when in use.

Experiment 27 Polarimetry

Apparatus: polarimeter.
Material: maltose.

For monochromatic light, e.g. the yellow light of sodium (589 nm), at a constant temperature

$$[\alpha] = \frac{\theta}{lc}$$

where $\theta°$ is the observed rotation of the plane of polarization of plane polarized light,
l is the path length of the light through the solution (in decimetres),
c is the concentration of the solution (in g cm^{-3}),
$[\alpha]$ is the optical or specific rotation.

A solution, e.g. 10% m/V maltose in water, is put in the flat bottomed polarimeter tube. Light passes from one polaroid through the solution and then to a second polaroid. The second polaroid is rotated until the light is cut off and the angle measured.

Solutions that are more dilute can then be studied and the optical rotation of the solute found by plotting a graph of θ against c, l being a known and constant distance.

(See also Experiment 72, p. 66.)

Experiment 28 Colorimetry

Apparatus: colorimeter.
Material: 1 M copper(II) sulphate, 1 M ammonia.

Choose a coloured filter for a given solution such that very little light is transmitted, i.e. a filter of a complementary hue to the colour of the solution.

violet is complementary to yellow-green
blue is complementary to yellow
green-blue is complementary to orange
blue-green is complementary to red
green is complementary to purple

In a colorimeter, light from a lamp is channelled through the solution in a test-tube and the light that is transmitted falls on a photo-conductive device (photo-electric cell) permitting a current which is roughly proportional to the intensity of the light. Lambert (1760) discovered that the intensity of light decreased exponentially as the thickness of a medium increased arithmetically. Beer (1852) extended these observations to solutions and their laws can be summarized as

$$\log_{10}\left(\frac{I_0}{I_t}\right) = kc$$

where I_0 and I_t are the intensities of the incident and transmitted light—these are assessed by the current passing,
c is the concentration of the solution (in mol dm^{-3}),
k is a constant for a particular substance.

Beer's Law may be verified by studying a series of solutions of copper(II) sulphate of differing concentrations: plot a graph of $\log_{10}(I_0/I_t)$ against concentration. A straight line may be obtained but a smooth curve is a satisfactory calibration chart enabling solutions of unknown concentration to be studied.

Secondly, the formula of a complex ion may be determined by mixing increasing quantities (4, 5, 6 ... cm^3) of 1 M ammonia solution with a fixed quantity of 1 M copper(II) sulphate (2 cm^3) solution and sufficient water to make the total volume constant each time (say 20 cm^3). Study each mixture in the colorimeter. From a graph of $\log_{10}(I_0/I_t)$ against the volume of ammonia solution the formula of the complex ion can be deduced.

3
The Colloidal State

True solutions contain particles of solute of approximately molecular size. The solution is transparent to the ultramicroscope and the particles diffuse rapidly. There is a gradation in size from the molecule through colloidal sizes to such large aggregates of particles that they settle out under the action of gravity (suspensions).

By careful control of the size of the particles of a substance, most materials can be prepared in the colloidal state.

Experiment 29 The ultramicroscope—Brownian movement

Apparatus: microscope; square cover slips; lantern.

Construct a small cubical box made of cover slips and on it mount a microscope as shown in Figure 12. Arrange for a narrow beam of light to strike the box at right angles to the field of view. Puff a little

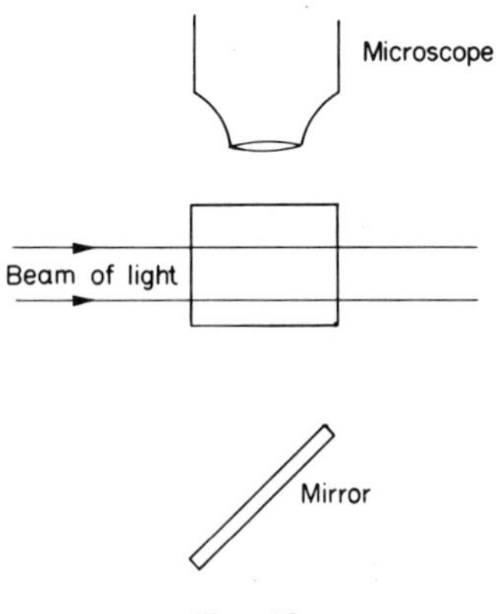

FIG. 12

smoke (e.g. from burning toluene) into the box and view the field by means of the high-power objective. The particles are seen to be in constant irregular motion, and even if a convection current is present the relative motion of the particles (seen as small circles of light) shows the Brownian movement clearly.

Experiment 30 To show dialysis

Apparatus: glass tube; string.
Material: cellophane.

Fit up a cellophane membrane round the end of the glass tube using a tight binding of string, and arrange them in a boiling-tube containing water as shown in Figure 13. Add potassium hexacyanoferrate(II) solution in excess to 2–3 cm^3 of iron(III) chloride solution, and pour the mixture into the glass tube. After five minutes, test for the presence of hexacyanoferrate(II) ions in the boiling-tube by pouring off a little of the solution and adding iron(III) chloride.

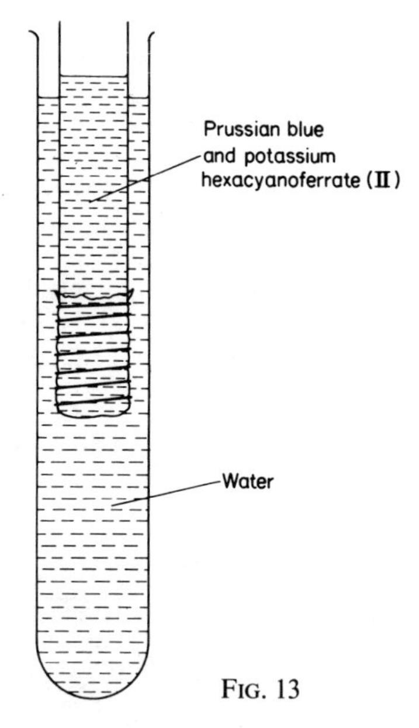

FIG. 13

It will be found that the hexacyanoferrate(II) ions have passed easily through the membrane, whereas the Prussian blue (which is colloidal) obviously remains in the inner tube.

Experiment 31 Preparation of a colloidal solution of antimony(III) sulphide

Material: antimony potassium oxide tartrate (tartar emetic); starch solution.

Pour about 20 drops of yellow ammonium sulphide into a boiling-tube nearly full of water. Put enough of the tartrate to cover the bottom of a test-tube in another boiling-tube and fill with water. The colloidal solution can be made by mixing equal volumes of the two solutions. Carry out the following experiments with portions of the solution.

(*a*) Show that the solution filters unchanged.

(*b*) Boil some of the solution in a test-tube. After a while coagulation takes place.

(*c*) Add a small quantity of sodium chloride. Precipitation takes place.

(*d*) Add some iron(III) hydroxide solution prepared as in Experiment 32.
Coagulation takes place because the particles in the two solutions are oppositely charged, iron(III) hydroxide solution being positively charged and antimony(III) sulphide negatively charged.

(*e*) Add 2 cm^3 of starch solution to 2 cm^3 of the antimony sulphide solution. Dilute 2 cm^3 of the antimony sulphide solution with 2 cm^3 of water to act as 'control'. Add a solution of sodium chloride to both. The chloride solution coagulates the 'control' but no effect is observed in the solution where it is protected by the starch.

Experiment 32 Preparation and electrophoresis of colloidal iron(III) hydroxide

Apparatus: source of 24 V d.c.; U-tube; copper electrodes.

Material: solid iron(III) chloride.

Prepare a fresh solution of iron(III) chloride by dissolving a spatula-load of the solid in a test-tube half full of water. Divide into two parts. Add two or three drops of ammonia solution to one part and notice the reddish brown liquid which contains colloidal iron(III) hydroxide. The iron(III) chloride serves to show the contrast. Show that the liquid will pass through a filter paper. Add more ammonia to the test-tube containing the iron(III) hydroxide and the hydroxide will be precipitated.

Put the iron(III) hydroxide solution in the curved portion of a U-tube and then simultaneously pour distilled water into each of the two upright portions. Insert copper electrodes into the water and apply 24 V: movement of the solution towards one electrode or the other indicates that the particles are charged. Decide what is the charge on the colloid.

The solution will also show dialysis, as in Experiment 30.

Experiment 33 Preparation of colloidal sulphur

Apparatus: siphon of sulphur dioxide; Woulfe bottle.

Sulphur dioxide is much more soluble in water than is hydrogen sulphide. To obtain suitable solutions for producing a sulphur solution proceed as follows. Saturate some distilled water with hydrogen sulphide which has been washed by passing it through water in a Woulfe bottle. Saturate a smaller quantity of distilled water with sulphur dioxide from a siphon. Dilute this solution approximately twenty-five times with distilled water and add an equal volume of the hydrogen sulphide solution. A clear yellow colloidal solution of sulphur is formed. This will easily pass through the pores of a filter paper and is coagulated by solutions of electrolytes, e.g. sodium chloride, magnesium chloride or aluminium chloride. The Hardy–Schulze rule is illustrated.

Experiment 34 Preparation of rosin sol by dispersion

Material: powdered rosin; ethanol.

Weigh about 2.5 g of rosin and dissolve in about 50 cm^3 of alcohol on a water-bath. Filter into a 100 cm^3 flask and fill to the mark with ethanol. This is a true solution of rosin in ethanol.

Mix 25 cm^3 of solution with about 500 cm^3 of water to obtain an opalescent colloidal solution. This is a very stable colloidal solution and may be used to show the simpler properties of colloids.

(*a*) A portion of the solution should be treated with an electrolyte, e.g. sodium chloride, magnesium chloride or aluminium chloride, the last giving an immediate precipitate of the coagulated colloid. (The Hardy–Schulze rule is illustrated.)

(*b*) The solution shows dispersion of light when a beam is focused into it (Tyndall effect).

(*c*) The solution can be used to demonstrate the Brownian movement.

(*d*) Colloids pass readily through filter paper.

Experiment 35 Preparation of an olive oil emulsion

Material: olive oil; soap solution; gelatine solution.

In three separate test-tubes shake 5–10 drops of olive oil with the following substances:

(*a*) water,

(*b*) soap solution,

(*c*) soap solution and gelatine solution.

The final volume of liquid in each of the three cases should be approximately the same. Leave them to settle and note that the oil separates out from the water very quickly, from the soap solution slowly, and from the soap solution with gelatine very slowly or not at all.

The soap solution acts as an emulsifier, whilst the gelatine protects the colloid from coagulation.

Experiment 36 The preparation and properties of silica gel

Material: 1 M hydrochloric acid; sodium silicate solution (6% *m/V*).

Add the acid from a burette to the silicate solution together with two drops of phenolphthalein in a beaker until the pink colour has just been discharged. This gives silica gel: a liquid is dispersed in a solid.

Examine portions of the gel by the following tests:

(*a*) heating and then cooling,
(*b*) diluting with water,
(*c*) for the Tyndall effect as in Experiment 34,
(*d*) for Brownian motion as in Experiment 29,
(*e*) for elasticity, i.e. if it is slightly deformed does it regain its shape?
(*f*) for fluidity, i.e. if an irregular lump is put in a test-tube does it soon conform to the shape of the bottom of the tube?

Experiment 37 The thixotropy of gels

Material: bentonite.

Bentonite (and other clays), many paints, salad cream etc., form gels which break down on mechanical agitation. Shake a boiling-tube containing water and bentonite: the liquid can be heard moving when it is shaken, but even when the tube is turned upside down none may fall out.

4
Oxidation and Reduction

Oxidation, in its simplest sense, may be defined as the chemical addition of oxygen to an element or compound, e.g.

$$2Mg + O_2 \rightarrow 2MgO$$
$$2CO + O_2 \rightarrow 2CO_2$$

The term *oxidation* may be extended to include the removal of hydrogen from a compound by a reaction with oxygen or any other reagent.

$$2H_2S + O_2 \rightarrow 2S + 2H_2O$$
$$H_2S + Cl_2 \rightarrow S + 2HCl$$

In both cases, hydrogen sulphide is oxidized.

The term may be further extended to include reactions in which an increase of valency occurs.

$$2FeCl_2 + Cl_2 \rightarrow 2FeCl_3$$

or

$$2Fe^{2+} + Cl_2 \rightarrow 2Fe^{3+} + 2Cl^-$$

Divalent iron is oxidized to trivalent iron.

These aspects of oxidation are unified in the electronic theory. Consider the oxidation of iron(II) chloride to iron(III) chloride. An iron(II) ion is an atom of iron which has lost two valency electrons.

$$\text{Fe (atom)} \rightarrow Fe^{2+} + 2e^-$$

Similarly, an iron(III) ion results from the atom losing three electrons.

$$\text{Fe (atom)} \rightarrow Fe^{3+} + 3e^-$$

Thus, oxidation of an iron(II) salt, no matter how it is brought about, is the process of removing an electron from the iron(II) ion.

$$Fe^{2+} \rightarrow Fe^{3+} + e^-$$

Consider, also, the oxidation of hydrochloric acid to chlorine. The acid is fully ionized and the process of oxidation is essentially the change from chloride ion to chlorine atom. The chloride ion occurs when a chlorine atom has gained an electron in the change of electron pattern from 2,8,7 to 2,8,8. To change the chloride ion of hydrochloric acid to a chlorine atom, an electron must be removed.

$$Cl^- \rightarrow \text{Cl (atom)} + e^-$$

Oxidation is, therefore, the process of removing one or more electrons. The substance which causes oxidation is the *oxidizing agent* or *oxidizer*. During oxidation, the oxidizing agent gains the electrons which the substance which is being oxidized loses; the oxidizer is reduced. Thus when a solution of iron(III) chloride is added to a solution of tin(II) chloride, the iron(III) salt is reduced and the tin(II) salt oxidized.

$$Sn^{2+} + 2Fe^{3+} \rightarrow Sn^{4+} + 2Fe^{2+}$$

i.e.

$$Sn^{2+} \rightarrow Sn^{4+} + 2e^- \quad \text{(oxidation)}$$

$$2Fe^{3+} + 2e^- \rightarrow 2Fe^{2+} \quad \text{(reduction)}$$

Tin(II) is *oxidized* to tin(IV) by electron loss; iron(III) is *reduced* to iron(II) by electron gain.

In the following examples, it will be noted that:

(*a*) the substance which is being oxidized loses electrons;
(*b*) the substance which is being reduced gains electrons.

Experiment 38 To show that bromine, concentrated nitric acid, potassium permanganate and hydrogen peroxide are oxidizing agents

Material: aqueous solution of hydrogen sulphide.

Fill each of four test-tubes to a depth of 2 cm with:
(*a*) acidified potassium iodide solution,
(*b*) acidified iron(II) sulphate solution,
(*c*) hydrogen sulphide solution,
(*d*) concentrated hydrochloric acid.

Add two or three drops of bromine water to each of the test-tubes in turn, and warm if necessary to complete the reaction. Repeat the experiments using concentrated nitric acid, hydrogen peroxide or potassium permanganate (manganate(VII)) solution instead of bromine water.

An oxidizing agent will react to give the following results:

(i) It will turn the iodide solution brown, iodine being formed; black crystals of iodine may be precipitated, e.g. using hydrogen peroxide:

$$2H^+ + 2I^- + H_2O_2 \rightarrow 2H_2O + I_2$$

$$2I^- \rightarrow I_2 + 2e^- \quad \text{(iodide ion oxidized)}$$

$$2H^+ + H_2O_2 + 2e^- \rightarrow 2H_2O \quad (H_2O_2 \text{ reduced})$$

(ii) It will turn the green iron(II) salt yellow, forming iron(III) salt, e.g. using bromine:

$$2Fe^{2+} + Br_2 \rightarrow 2Fe^{3+} + 2Br^-$$

$$Fe^{2+} \rightarrow Fe^{3+} + e^- \quad \text{(iron(II) ion oxidized)}$$

$$Br_2 + 2e^- \rightarrow 2Br^- \quad \text{(bromine reduced)}$$

The presence of iron(III) salt may be confirmed by adding a slight excess of sodium hydroxide solution and obtaining a brown precipitate of iron(III) hydroxide.

(iii) It will precipitate sulphur from hydrogen sulphide as a white or pale yellow suspension, e.g. using concentrated nitric acid:

$$\underbrace{2H^+ + S^{2-}}_{\substack{\uparrow\downarrow \\ H_2S}} + 2H^+ + 2NO_3^- \rightarrow S + 2H_2O + 2NO_2$$

$$S^{2-} \rightarrow S + 2e^- \quad \text{(sulphide ion oxidized)}$$

$$4H^+ + 2NO_3^- + 2e^- \rightarrow 2H_2O + 2NO_2 \quad \text{(nitric acid reduced)}$$

(iv) It will liberate chlorine from hydrochloric acid, damp litmus paper being bleached, e.g. using potassium permanganate:

$$MnO_4^- + 8H^+ + 5Cl^- \rightarrow Mn^{2+} + 4H_2O + 2\tfrac{1}{2}Cl_2$$

$$5Cl^- \rightarrow 2\tfrac{1}{2}Cl_2 + 5e^- \quad \text{(chloride ion oxidized)}$$

$$MnO_4^- + 8H^+ + 5e^- \rightarrow Mn^{2+} + 4H_2O \quad \text{(permanganate ion reduced)}$$

Experiment 39 To compare potassium chlorate and potassium persulphate as oxidizing agents

Material: potassium chlorate (chlorate(V)); potassium persulphate (peroxodisulphate(VI))

Arrange in pairs, in a test-tube rack, test-tubes containing approximately equal volumes of solutions (*a*) to (*d*) in Experiment 38. Perform the experiment with two of the test-tubes containing the same reagent, e.g. acidified potassium iodide, by adding approximately equal amounts (half a spatula-load) of potassium chlorate and potassium persulphate respectively. Note the action, warm if necessary, and observe in which case the reaction appears to take place the more readily. Repeat with pairs of test-tubes containing reagents (*b*) to (*d*) above. Both solids are powerful oxidizing agents.

Note: The persulphate ion oxidizes by accepting electrons to become sulphate ions, e.g. using potassium iodide,

$$S_2O_8^{2-} + 2I^- \rightarrow 2SO_4^{2-} + I_2$$

$$S_2O_8^{2-} + 2e^- \rightarrow 2SO_4^{2-} \quad \text{(persulphate ion reduced)}$$

Experiment 40 To show that nitrous acid may act as either an oxidizing or a reducing agent

Material: sodium nitrite (nitrate(III)).

(*a*) Make a solution of sodium nitrite in water and add it gradually to a solution of potassium iodide acidified with dilute sulphuric acid. Iodine will be liberated showing that the nitrous acid, produced by the action of the dilute acid on the sodium nitrite, has oxidized the potassium iodide. The nitrous acid has itself been reduced to nitrogen oxide (which on coming into contact with the oxygen of the air forms brown fumes of nitrogen oxide).

$$2NO_2^- + 2I^- + 4H^+ \rightarrow I_2 + 2NO + 2H_2O$$

That is, in acting as an oxidizing agent, nitrous acid gains electrons and is reduced to nitrogen monoxide:

$$2NO_2^- + 4H^+ + 2e^- \rightarrow 2H_2O + 2NO$$

(*b*) Acidify a solution of potassium permanganate with dilute sulphuric acid in a test-tube and pour in a solution of sodium nitrite until the colour of the permanganate is just discharged. Note the absence of brown fumes; the solution, which contains nitric acid, can be tested by the nitrate test (see p. 125). The potassium permanganate has oxidized the nitrous acid to nitric acid, itself being reduced to manganese(II) salts. See Experiment 138(g), p. 124, for an explanation.

$$2MnO_4^- + 6H^+ + 5NO_2^- \rightarrow 2Mn^{2+} + 3H_2O + 5NO_3^-$$

Nitrous acid here acts as a reducing agent; it loses electrons and is oxidized to nitric acid.

$$NO_2^- + H_2O \rightarrow NO_3^- + 2H^+ + 2e^-$$

Experiment 41 To show that hydrogen sulphide and sulphurous acid are reducing agents

Apparatus: sulphur dioxide siphon.

Material: potassium iodate.

(*a*) Bubble hydrogen sulphide from a Kipp's apparatus into an acidified potassium permanganate solution (very dilute) in a boiling-tube. The colour of the permanganate is discharged but a milky precipitate of sulphur remains:

$$2MnO_4^- + 6H^+ + 5H_2S \rightarrow 2Mn^{2+} + 8H_2O + 5S\downarrow$$

(*b*) Repeat (*a*) using sulphur dioxide from a siphon or sulphurous (sulphuric(IV)) acid in place of hydrogen sulphide. The colour of the permanganate is discharged but no precipitate of sulphur is formed. The sulphurous acid has been oxidized to sulphuric (sulphuric(VI)) acid:

$$2MnO_4^- + 6H^+ + 5SO_3^{2-} \rightarrow 2Mn^{2+} + 3H_2O + 5SO_4^{2-}$$

(*c*) Bubble hydrogen sulphide for ten minutes through a dilute solution of iron(III) chloride acidified with a few drops of hydrochloric acid. The colour will change from yellow to green. Boil the solution in a dish for two minutes to expel hydrogen sulphide, filter through a double filter paper to remove sulphur, and add sodium hydroxide solution in excess to the filtrate. A dirty green precipitate of iron(II) hydroxide will be obtained, showing that the iron(III) ion has been reduced to iron(II) ion:

$$2Fe^{3+} + H_2S \rightarrow 2Fe^{2+} + 2H^+ + S\downarrow$$

(*d*) Bubble sulphur dioxide for some time through a dilute solution of iron(III) chloride. The liquid turns red (due to a complex sulphite). Transfer the solution to a dish and boil for a few minutes on a tripod and gauze. The resulting solution will be pale green or colourless. Add sodium hydroxide solution in excess to a portion, when a dirty green precipitate of iron(II) hydroxide will show that reduction is complete.

$$2Fe^{3+} + SO_3^{2-} + H_2O \rightarrow 2Fe^{2+} + SO_4^{2-} + 2H^+$$

(*e*) Dissolve a spatula-load of potassium iodate (iodate(V)) in water in a boiling-tube and pass a brisk stream of sulphur dioxide through it. Iodine will be deposited as black crystals.

$$IO_3^- + 3SO_3^{2-} \rightarrow I^- + 3SO_4^{2-}$$

followed by

$$5I^- + IO_3^- + 6H^+ \rightarrow 3I_2\downarrow + 3H_2O$$

If the stream of sulphur dioxide continues for a minute or two the solution goes clear due to the formation of hydrogen iodide.

$$I_2 + SO_3^{2-} + H_2O \rightarrow 2I^- + SO_4^{2-} + 2H^+$$

Experiment 42 To measure the electrode potentials of some metals

Apparatus: as in Figure 14; selection of metals.

Material: 1 M solutions of metals as their sulphates or chlorides.

Hydrogen, at one atmosphere pressure, is bubbled over a piece of platinum to give a hydrogen electrode. The platinum should be coated with finely divided platinum, making it black rather than silver-coloured, by electrolysing platinum(IV) chloride solution between platinum electrodes first in one direction then in the opposite direction.

A clean strip of the metal is dipped into the solution of the metal sulphate (or chloride). Suitable metals include iron, nickel,

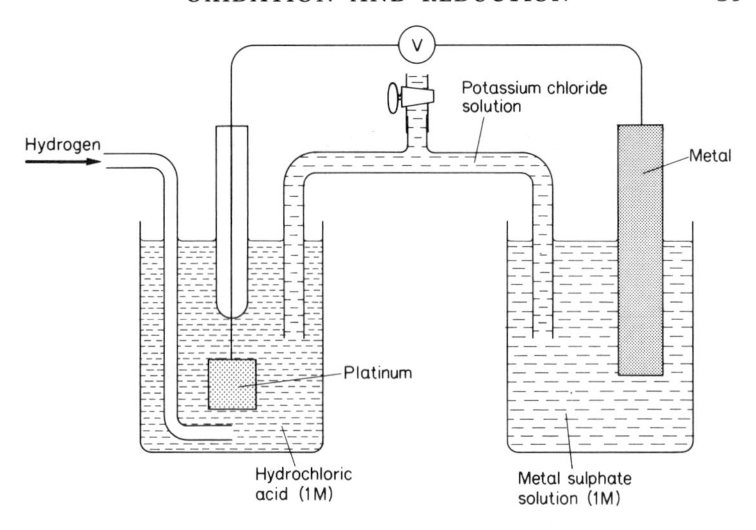

FIG. 14

aluminium, copper and zinc.

A simpler version of this experiment, which only gives the relative values and not the absolute values but which has the advantage that it can be extended to the metals in groups IA and IIA, is as follows. Attach a wire from a voltmeter to a sheet of copper and place a piece of filter paper soaked in copper(II) sulphate solution on the copper. Put the metal to be studied on to the filter paper and touch the metal with the other wire from the voltmeter.

Experiment 43 To assess the relative redox powers of a series of reagents

Take a selection of reagents in approximately 0.1 M solution, e.g. potassium permanganate (manganate(VII)), potassium dichromate (dichromate(VI)), hydrogen peroxide, iron(III) chloride, tin(II) chloride and iron(II) sulphate. Add each to the other and if observation does not indicate whether a reaction has taken place or not, try further tests, e.g. the addition of sodium hydroxide in the case of iron salts. Tabulate your results and try to place the reagents in an order such that the most powerful oxidizing agent is at the top, the substances which show either type of behaviour are in the centre and the most powerful reducing agent is at the bottom.

5
Electrical Conductance

Experiment 44 To study the variation of conductance with dilution

Conductance bridges are now available commercially which obviate the student's need to set up a Wheatstone bridge circuit. A dipping electrode made of two pieces of platinized platinum, and held a fixed distance apart in a glass cylinder, is attached to the bridge.

By starting with a concentrated (saturated) solution of hydrochloric acid and progressively diluting with water until very dilute, it is possible to obtain a set of readings of conductance (measured in ohm^{-1} or siemens, abbreviation S).

Multiplying the conductance by the cell constant (distance apart divided by area of cross-section of the electrodes) gives the electrolytic conductivity of a substance in solution.

Graphs of molar conductivity (the electrolytic conductivity of a solution multiplied by the volume in cm^3 containing a mole of electrolyte), first against the concentration and second against the square root of the concentration, should be plotted.

An alternative substance to study is glacial acetic acid (pure ethanoic acid).

Experiment 45 To perform a conductance titration

The measurement of the conductance of a solution can be used to find the end-point of a titration. A complication is that the addition of the solution from the burette dilutes the solution in the beaker; this is usually countered by adding a concentrated solution from the burette to a dilute (or diluted) solution in the beaker so that the total volume does not alter appreciably.

Put 25 cm^3 of saturated barium hydroxide solution (approximately 0.1 M) into a beaker and add about 100 cm^3 of distilled water. From a burette add 1 cm^3 portions of 0.5 M sulphuric acid, measuring the conductance after each addition. Plot a graph of conductance against volume of acid added and hence deduce the accurate concentration of the barium hydroxide solution.

Alternative experiments are to titrate silver nitrate solution with sodium chloride solution or iron(II) sulphate solution with barium chloride solution.

6
Acids, Alkalis and Salts

Note: The theoretical considerations of pH value are set out in Chapter 32.

Experiment 46 To make solutions of pH values 3 to 11 using buffer solutions

Material: universal indicator; 0.1 M acetic (ethanoic) acid; 0.1 M sodium acetate (ethanoate); 0.1 M hydrochloric acid; 0.1 M sodium hydroxide; 0.1 M disodium hydrogenphosphate(V).

A buffer solution is one whose pH value does not materially alter for small additions of acid or alkali. A typical buffer is one made up of a mixture of sodium acetate and acetic acid, the former being highly ionized, the latter only partially. If hydrogen ions are added to the solution, the following reaction takes place:

$$H^+ + Ac^- \leftrightarrow HAc$$

thus removing hydrogen ions from the solution. If an alkali is added, more acetic acid dissociates to form hydrogen ions which combine with the hydroxyl ions to form water.

$$HAc \leftrightarrow H^+ + Ac^-$$
$$H^+ + OH^- \leftrightarrow H_2O$$

and again the hydroxide ions are removed from the solution. [Ac represents the radical CH_3 COO]

For preparing solutions of hydrogen ion concentrations of 10^{-3} to 10^{-6} mol dm^{-3}, i.e. pH 3 to 6, the following are used.

(*a*) 0.1 M acetic acid.

(*b*) 0.1 M sodium acetate solution (13.6 g dm^{-3} of crystalline sodium acetate, $CH_3COONa \cdot 3H_2O$). The solutions are mixed as follows:

pH	*Volume (cm³)* *0.1 M CH_3COOH*	*Volume (cm³)* *0.1 M CH_3COONa*
3	982.3	17.7
4	847.0	153.0
5	357.0	643.0
6	52.2	947.8

For solutions with hydrogen ion concentrations between 10^{-7} and 10^{-11} mol dm^{-3}, i.e. pH 7 to 11, the following are required.

(*a*) A 0.1 M solution of disodium hydrogenphosphate, made by dissolving 35.8 g dm^{-3} of the crystalline salt $Na_2HPO_4 \cdot 12H_2O$.
(*b*) 0.1 M hydrochloric acid.
(*c*) 0.1 M sodium hydroxide.

The mixtures are made as follows:

pH	*Volume (cm³)* *0.1 M Na_2HPO_4*	*Volume (cm³)* *0.1 M HCl*	*Volume (cm³)* *0.1 M NaOH*
7	756.0	244.0	—
8	955.1	44.9	—
9	955.0	5.0	—
10	996.4	—	3.6
11	965.3	—	34.7

A convenient method of exhibiting the changes in colour shown by the indicator in solutions of various pH's is to arrange three rows of nine test-tubes or boiling-tubes in a wooden stand, covered with white paper, so that the three rows can be seen simultaneously one above the other, and to place in each tube the same volume of solution according to the following scheme:

pH	*3*	*4*	*5*	*6*	*7*	*8*	*9*	*10*	*11*
Top row	MO	MO*	MO	MO	—	PP	PP*	PP	PP
Second row	MR	MR	MR	MR*	MR	—	TP	TP	TP*
Third row	—	L	L	L	L*	L	L	—	—

Methyl orange, MO. Phenolphthalein, PP. Litmus, L. Methyl red, MR. Thymolphthalein, TP.

A few drops of the indicators are then added as shown above, and the colour changes observed. The asterisks in the table above show the pH's at which the transition colours are most marked.

To show that the solutions are comparatively stable, make a solution of 0.001 M hydrochloric acid (pH = 3) and 0.001 M sodium hydroxide (pH = 11), and add a few drops of universal indicator to each. Pour out approximately the same volume of buffer solutions of pH = 3 and pH = 11 made as indicated above and add a few drops of universal indicator to these two solutions. Now add a drop of acid or alkali in turn to each of the four solutions and observe the change and estimate the alteration in pH value. It will be seen that there is a rapid change in the 0.001 M acid and alkali but little change in the buffer solutions.

Experiment 47 To show the effect of a buffer salt

Material: a universal indicator; 0.1 M sodium hydroxide solution; 0.1 M hydrochloric acid; sodium acetate (ethanoate).

Take a measured volume (10 or 25 cm^3) of the sodium hydroxide solution by means of a pipette, add two drops of universal indicator and titrate the mixture with the acid. Note the colour changes indicating the rapid change of pH about the equivalence point.

Take the same measured volume of alkali and add a spatula-load of sodium acetate, then two drops of indicator, and titrate as before. Notice that, although hydrogen ions are being added, the green colour of the indicator persists; the pH at that point remains constant over a long period of addition of hydrogen ion.

The buffer salt, sodium acetate, is highly ionized and furnishes acetate ions. The hydrogen ions provided by the hydrochloric acid, instead of increasing the pH, are used to form molecular acetic acid.

$$\text{NaAc is } Na^+ + Ac^-$$
$$H^+ + Ac^- \leftrightarrow HAc$$

Only when a large excess of hydrogen ions has been added does the pH of the solution decrease. In this example, a buffer salt is essentially a highly ionized salt of a weak acid.

Similarly, when a highly ionized salt of a weak base is present, the addition of strong alkali does not at once increase the pH of the mixture. Thus, ammonium chloride furnishes ammonium ions with which the added hydroxide ions of the strong alkali form ammonia and water and only after an excess of alkali has been added does the pH of the solution rise. (See Experiment 46 and Chapter 32.)

Experiment 48 The change in pH near the equivalence point

Material: a universal indicator solution; 0.1 M solutions of hydrochloric acid, sodium hydroxide and acetic (ethanoic) acid.

(*a*) Take 10 cm^3 of sodium hydroxide solution by means of a pipette, add two drops of universal indicator and titrate the mixture with hydrochloric acid. Note the rapid change of colour from green-blue (pH about 8.5) to orange-red (pH about 4), showing that, in the titration of a strong alkali with a strong acid, the equivalence point is indicated, with negligible error, by any indicator.

Repeat using the individual low pH indicator, methyl orange, and the high pH indicator, phenolphthalein, and note the slight difference in titres.

(*b*) Take 10 cm^3 of sodium hydroxide solution, add two drops of universal indicator, and titrate with acetic acid. Note that when the equivalence point is reached, the pH is approximately 8.5. Note also the considerable excess of acid necessary to approach the orange colour of pH 4, showing that only a high pH indicator is efficient in the titration of a strong alkali with a weak acid. Repeat using the individual indicators phenolphthalein and methyl orange to confirm this conclusion. (See also Chapter 32.)

Experiment 49 To find the pH value of solutions of certain salts

Material: a universal indicator; aluminium chloride; sodium sulphite (sulphate(IV)).

A normal salt is one in which the replaceable hydrogen atoms of an acid have been completely replaced by a metal. A normal salt is not necessarily a neutral salt since hydrolysis may occur. Thus, sodium carbonate, a normal salt, is alkaline in solution, whereas ammonium chloride is acidic. The section on *hydrolysis*, Chapter 32, explains this.

Arrange several test-tubes in a rack, half fill each with water, and add a few drops of universal indicator. To the test-tubes add a spatula-load of one of the following salts: sodium carbonate, sodium sulphite, sodium chloride, ammonium chloride, aluminium chloride, disodium tetraborate, iron(II) sulphate. (Use solutions of these reagents where available.) Note the pH value according to the colour produced and explain the reactions.

Warm the solutions. This increases the hydrolysis in some cases producing an even greater divergency from neutrality.

7
The Colligative Properties of Dilute Solutions

When a substance dissolves in a liquid, the substance is called a solute and the liquid a solvent. If a dilute solution of a given solute in a given solvent is separated from a concentrated solution of the same solute and solvent by a semipermeable membrane (i.e. a membrane permeable to the solvent but not permeable to the solute), solvent passes through the membrane from the dilute solution to the concentrated one until the concentrations are equalized. This movement of solvent generates the *osmotic pressure*. When a pure solvent is separated by a semipermeable membrane from a solution of a substance in that solvent, the pressure exerted by the solvent passing into the solution can be measured. The presence of a solute in a solvent also has the effect of lowering the freezing point of the solvent and of raising the boiling point. The actual value of the osmotic pressure, or the number of degrees by which the freezing point is lowered or the boiling point raised, depends upon the concentration of the solution. Provided the solutions considered are sufficiently dilute, the laws governing these phenomena are accurate, the effects being directly proportional to the number of particles present in a fixed amount of solvent. 'Particles' includes molecules of the solute and ions into which molecules have dissociated. Thus 2 moles of the non-ionizable solute urea will have twice the effect of 1 mole of urea in a given amount of a solvent, and 1 mole of urea (60 g) will have the same effect as 1 mole of cane sugar (342 g). An ionizable substance, if fully ionized, will furnish as many particles (ions) as its constitution allows; thus 1 mole of sodium chloride will give two moles of ions:

$$NaCl \text{ is } Na^{+} + Cl^{-}$$

1 mole of barium chloride, when fully ionized, gives three moles of ions:

$$BaCl_2 \text{ is } Ba^{2+} + 2Cl^{-}$$

A mole of barium chloride will have three times the effect of 1 mole of urea.

OSMOTIC PRESSURE

Osmotic pressure effects can be shown qualitatively in the laboratory by means of simple apparatus, but quantitative measurements are not usually possible because of difficulties of technique.

One mole of a non-electrolyte dissolved in water and made up to 22.4 dm^3 of solution causes an osmotic pressure, at 0 °C, of 760 mmHg (1 atmosphere or 101 325 Pa).

Experiment 50 Copper(II) hexacyanoferrate(II) as a semipermeable membrane

Apparatus: glass tubing drawn out to form a capillary.

Material: potassium hexacyanoferrate(II); 2 M copper(II) sulphate.

Make a concentrated solution of potassium hexacyanoferrate(II) and allow it to drop from a fine capillary tube into the dilute solution of copper sulphate. The drop formed will be surrounded by a layer of copper hexacyanoferrate(II):

$$2CuSO_4 + K_4Fe(CN)_6 \rightarrow 2K_2SO_4 + Cu_2Fe(CN)_6\downarrow$$

$$2Cu^{2+} + Fe(CN)_6^{4-} \rightarrow Cu_2Fe(CN)_6\downarrow$$

The drop, being denser than the copper sulphate solution, will sink, but after a short time will rise by virtue of alteration in density caused by osmosis.

Experiment 51 Prussian blue as a semipermeable membrane

Material: solid iron(III) chloride.

Half fill an evaporating basin with a solution of potassium hexacyanoferrate(II) (about 5% m/V) and drop in a small lump of iron(III) chloride. At the surface of the iron(III) chloride a layer of Prussian blue will form and will act as a semipermeable membrane. Inside the membrane there will be a highly concentrated solution of iron(III) chloride, and water will pass through from the dilute potassium hexacyanoferrate(II) solution. The layer of Prussian blue will swell, due to the dilution of the iron(III) chloride.

Experiment 52 Vegetable cells as a semipermeable membrane

Material: large potato.

Cut a large peeled potato into the shape shown in Figure 15, and fill the cavity about half full of sodium chloride solution. Stand it in a vessel and add enough water to bring the level equal to that of the sodium chloride solution. Note the level and leave until the following

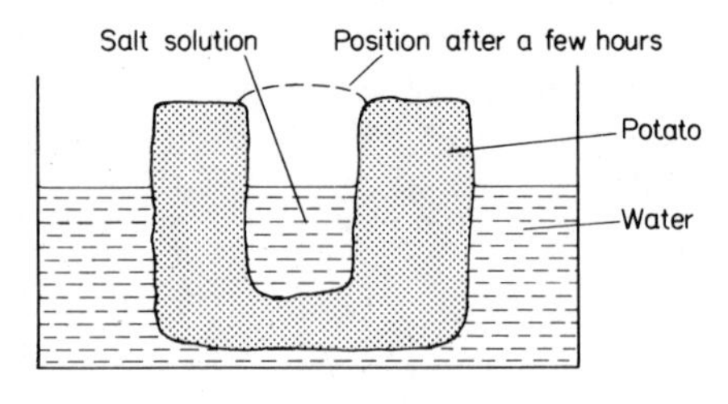

FIG. 15

day. The sodium chloride solution will then have grown in bulk by passage of water molecules through the membrane.

Experiment 53 Cellophane as a semipermeable membrane

Apparatus: short length of glass tube (about 18 mm diameter); fine string; cellophane.

Take a piece of cellophane about 8 cm square, and fit it round one end of the tube, tying it firmly in position with string. Fill the tube with sugar (sucrose) solution coloured with red ink, and close the other end of the tube with a well-fitting rubber bung fitted with a long piece of glass tubing (Figure 16). Set up the tube in water and mark the level. The level in the tubing rises appreciably after a few minutes and shows a considerable rise overnight.

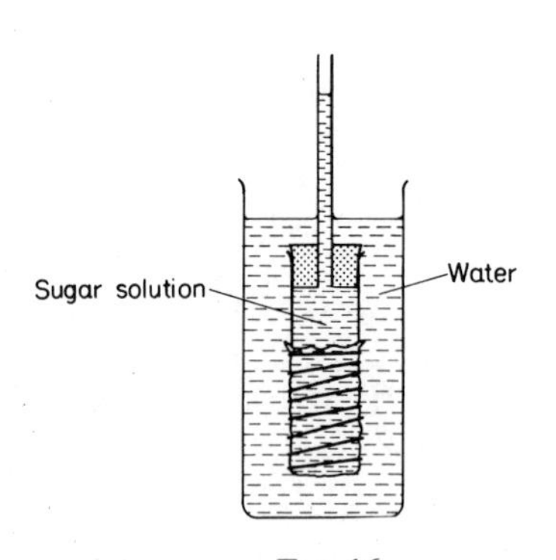

FIG. 16

ELEVATION OF THE BOILING POINT

A solution has a lower vapour pressure than its pure solvent and will therefore boil at a higher temperature. The lowering of the vapour pressure can be shown to be proportional to the osmotic pressure, and, if the solution is dilute, the lowering of the vapour pressure is accompanied by a proportional elevation of the boiling point (and also a proportional lowering of the freezing point). One mole of any non-ionizable solute when dissolved in 1 kg of a solvent elevates the boiling point by a constant number of degrees, the number being specific for each solvent; thus, 1 mole of any non-electrolyte when dissolved in 1 kg of water elevates the boiling point by 0.52°C, i.e. the solution would boil at 100.52°C. Calculations of relative molecular masses of solutes are based on this statement, but it must be remembered that the experimental data must be obtained by using dilute solutions. It would be more accurate to state that if n is a small fraction, then n moles of a non-electrolyte in 1 kg of water would elevate the boiling point by $0.52n$°C at 1 atmosphere pressure.

Experiment 54 To determine the molal elevation of the boiling point of water

Apparatus: Landsberger (or modified Landsberger) apparatus; steam generator; weighing bottle.

Material: sugar (sucrose) or urea.

It is assumed that sugar (sucrose) is a non-electrolyte and has a relative molecular mass of 342.

The modified Landsberger apparatus consists of a boiling-tube with a small hole blown about 1 cm from the top and fitted loosely

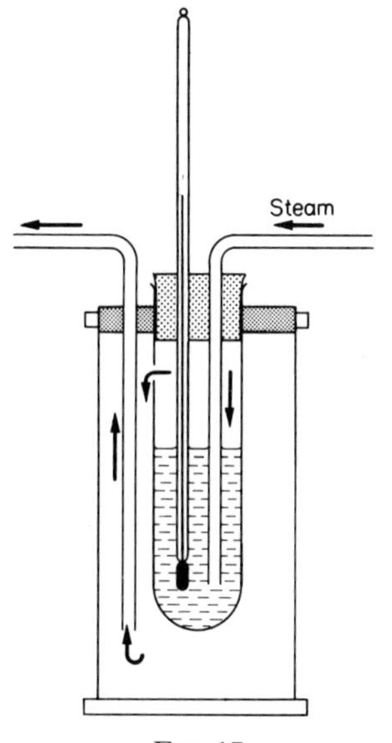

FIG. 17

with a bung which carries a glass tube and thermometer capable of being read accurately to 0.1 °C. The whole fits into a gas-jar, a cork holding the boiling-tube in place (see Figure 17). With careful handling the apparatus will last for several determinations. Weigh accurately about 7 g of sugar in a weighing bottle. Half fill the boiling-tube with water and pass in steam. Note the temperature when the mercury has become steady; this is the boiling point of water at the pressure of the atmosphere. The bung in the boiling-tube must be loosely fitting throughout, or the tube will crack. Remove this bung, slide in the sugar, and determine the new boiling point. Weigh the solution immediately, and subtract the mass of solute to find the mass of water.

Specimen results:

Mass of sugar	=	6.9 g
Temperature of boiling solvent	=	100.1 °C
Temperature of boiling solution	=	100.4 °C
Mass of solution	=	41.9 g
Mass of water	=	35.0 g

6.9 g of sugar in 35 g of water caused an elevation of 0.3 °C.
6.9 g of sugar in 1 kg of water caused an elevation of

$$\frac{0.3 \times 35}{1000}\ ^{\circ}\mathrm{C}$$

342 g (1 mole) of sugar in 1 kg of water caused an elevation of

$$\frac{342 \times 0.3 \times 35}{1000 \times 6.9}\ ^{\circ}\mathrm{C} = 0.52\ ^{\circ}\mathrm{C}$$

Note: The error in the thermometer reading being large, the volume in cm^3 of the final solution may be taken as the mass (in grammes) of the solvent, without serious additional error.

Experiment 55 To find the apparent degree of ionization of potassium chloride at a given dilution

Apparatus: as for Experiment 54.
Material: potassium chloride.
It is assumed that the elevation of boiling point constant for water is 0.52°C.

If there were no ionization, 1 mole of potassium chloride would have an effect equal to that of 1 mole of sugar, i.e. 74.5 g of potassium chloride in 1 kg of water would cause an elevation of the boiling point of 0.52 °C. But ionization does, in fact, occur and the observed elevation is therefore higher than this value.

$$\mathrm{KCl\ is\ K^{+} + Cl^{-}}$$

$$\frac{\text{observed elevation}}{\text{calculated elevation}} = \frac{\text{number of particles of solute effectively present}}{\text{number of particles present if there had been no ionization}}$$

Consider 1 mole of a binary electrolyte (a uni-univalent electrolyte, potassium chloride). Let the degree of ionization (i.e. the fraction of a mole present as ions under the conditions of the experiment) be x. Then in solution there would be $1-x$ moles of undissociated potassium chloride, x moles of potassium ions, and x moles of chloride ions, giving a total number of moles of $1-x+2x$, or $1+x$. If ionization had not occurred there would have been 1 mole only. Hence

$$\frac{\text{observed elevation}}{\text{calculated elevation}} = \frac{1+x}{1}$$

In this expression the observed value is obtained experimentally, the calculated elevation is based on the constant obtained in Experiment 54 and x can therefore be determined.

Specimen results:

Mass of potassium chloride	=	1.02 g
Boiling point of solvent	=	100.1 °C
Boiling point of solution	=	100.4 °C
Mass of water	=	42.8 g

Calculation of elevation if ionization had not occurred:

74.5 g of potassium chloride in 1 kg water cause an elevation of 0.52 °C.

1.02 g of potassium chloride in 42.8 g of water cause an elevation of

$$\frac{0.52 \times 1.02 \times 1000}{74.5 \times 42.8} = 0.167\ ^\circ\text{C}$$

The observed elevation was 0.30 °C.

$$\frac{0.30}{0.167} = \frac{1+x}{1}$$

$$\therefore x = 0.8 \text{ (approx.)}$$

$$\text{or degree of ionization} = 80\%$$

DEPRESSION OF THE FREEZING POINT

Experiment 56 To show the effect of concentration on the freezing point of a solution

Apparatus: thermometer (preferably reading −10 to 50 °C); jam-jar or a plastic bottle with the top cut off.

Material: **ICE**; sodium chloride; urea.

Take 10 cm^3 (10 g) of distilled water by pipette. Transfer to a test-tube. Weigh accurately three or four portions of 0.32 g each of urea. Fill the jar with crushed ice, add a handful of salt and make a hole large enough to insert the test-tube. Place the test-tube and contents into the ice-salt mixture, insert the thermometer into the water and allow it to cool. Carefully scrape the side of the test-tube with the thermometer (to prevent supercooling) and note the temperature when the water begins to freeze.

Remove the test-tube and warm it with the hand to melt traces of ice. Add the first portion of urea, allow time for it to dissolve, insert the test-tube into the ice-salt mixture and proceed as before to find when ice begins to form. Note the temperature and, from this, the depression of the freezing point of water.

Remove the test-tube, warm it, add the second portion of urea and repeat the experiment, noting the new depression of the freezing point. Repeat after adding the third and fourth portions. Results should show that (i) each portion causes a depression of approximately 1 °C; (ii) the depression of freezing point is, consequently, proportional to the concentration of urea in the solution. Although errors occur when using this simple apparatus, reasonably satisfactory results may be expected.

The results may be used to find the relative molecular mass of urea on the basis that 1 mole of a non-ionizing solute in 1 kg of water causes a depression of the freezing point of 1.86 °C.

Let 0.32 g of urea in 10 g of water cause a freezing-point depression of 1 °C. Let the freezing-point depression given by 1 mole of non-ionized solute in 1 kg of water be 1.86 °C. 0.32 g of urea in 10 g of water = 32 g in 1 kg of water.

32 g of urea in 1 kg of water cause a depression of 1 °C. Therefore 32×1.86 g of urea in 1 kg of water cause a depression of 1.86 °C.

Hence the relative molecular mass of urea is 60.

Experiment 57 To find the molal depression of the freezing point of water

Apparatus: accurate thermometers (−10 to 50°C); stirrers.
Material: **ICE**; sodium chloride; urea.

Measure 30 cm^3 (g) of water by means of a burette into a clean boiling-tube. Place this in a beaker and pack round it alternate layers of ice and salt, making five or six layers of each (Figure 18). Insert a thermometer (capable of being read accurately to 0.1 °C) and a stirrer. Note the temperature when pure water begins to freeze. If the indicated temperature is much below 0 °C before freezing occurs there is probably supercooling and the true freezing point is the value obtained when ice has separated out and the thermo-

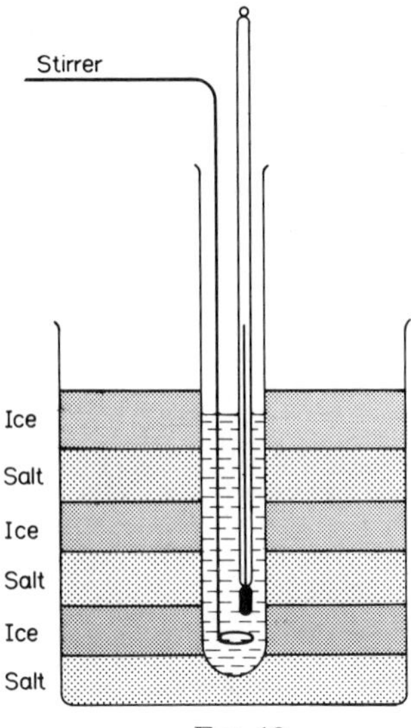

FIG. 18

meter reading is steady. Have ready an accurately weighed quantity of urea (about 1 g) in a weighing bottle. Warm the boiling-tube in the hand to melt the ice and then slide the urea into the water and stir to dissolve. Find the freezing point of the solution. Given that the relative molecular mass of urea is 60, use the results to calculate the lowering of the freezing point when 1 mole is dissolved in 1 kg of water.

Note: (i) Thorough stirring is necessary throughout the experiment to avoid supercooling.

(ii) This experiment is capable of reasonably good results because of the much greater value of this constant than the corresponding boiling point constant, and because the freezing point is much steadier during the period of observation than is the boiling point. The degree of ionization of an electrolyte is found from freezing-point data by a similar method to that given in Experiment 55 and a suitable variation would be to use the ternary electrolyte barium chloride (about 1 g in 30 g of water).

Experiment 58 To determine the relative molecular mass of naphthalene (Rast)

Apparatus: ignition tubes; thermometer (360 °C); olive oil.
Material: camphor, naphthalene.

Note: The freezing point constant of camphor has a high value, i.e. 40 °C mol^{-1} kg^{-1}, and this method is suitable for a small quantity of solute.

Weigh an ignition tube empty and then with about 0.5 to 1.0 g of camphor. Add about one-tenth of the camphor's mass of naphthalene, weigh, and insert a thermometer in the tube with a piece of rubber tubing to act as a cork (Figure 19). Place the ignition tube in

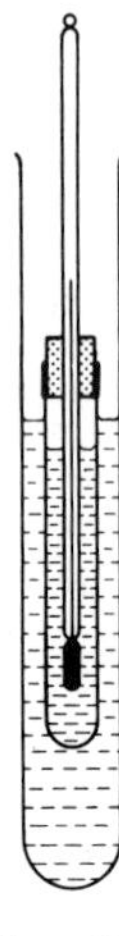

FIG. 19

a boiling-tube containing olive oil and warm gently until the white solids melt. Remove the ignition tube and adjust the thermometer so that it dips into the molten contents. Replace the tube and allow it to cool while in the oil. Note the temperature when the camphor becomes cloudy and take this as the freezing point. Use another ignition tube to find the freezing point of pure camphor.

Specimen results:

Freezing point of pure camphor	=	175 °C
Freezing point of solution	=	124 °C
Mass of ignition tube	=	3.02 g
Mass of ignition tube + camphor	=	3.83 g
Mass of ignition tube + camphor + naphthalene	=	3.96 g

51 °C is the depression caused by 0.13 g naphthalene in 0.81 g camphor

51 °C is the depression caused by $\frac{0.13 \times 1000}{0.81}$ g naphthalene in 1000 g camphor

40 °C is the depression caused by $\frac{0.13 \times 1000 \times 40}{0.81 \times 51}$ g naphthalene in 1000 g camphor

Hence the relative molecular mass of naphthalene is 126 (correct value 128).

8
Thermochemistry

A chemical change is usually accompanied by an evolution or an absorption of energy (e.g. heat), and the following definitions indicate how these energy changes are compared by specifying the quantities involved; certain physical changes also produce an energy change.

The heat of solution of a substance is the energy evolved or absorbed when 1 mole of the substance is dissolved in such a volume of water that any increase in dilution brings about no further heat change.

The heat of neutralization of an acid is the energy evolved when *one mole* of the acid is neutralized by *one mole* of a base; the reactants must be stated.

Experiment 59 To find the heat of solution of sodium thiosulphate crystals

Apparatus: calorimeter and stirrer; pestle and mortar; thermometer (100 °C).

Material: sodium thiosulphate crystals.

Into a calorimeter of known mass put a quantity of warm water (at a temperature 5–10 °C above that of the laboratory) and weigh again. Add finely powdered crystals of sodium thiosulphate, stirring vigorously until the temperature falls approximately as far below the room temperature as the warm water was originally above it. When a batch of added crystals has just dissolved, take the temperature and weigh the calorimeter and contents.

Calculation:

mass of copper calorimeter $= a$ g
mass of copper calorimeter + water $= b$ g
mass of copper calorimeter + water + crystals $= c$ g

Hence

mass of water $= (b-a)$ g
mass of crystals $= (c-b)$ g

Suppose:

temperature of warm water $= t_1$ °C
final temperature of solution $= t_2$ °C

Assuming the specific heat capacity of sodium thiosulphate solution to be the same as that of water, then the heat absorbed (A joules) is

$$[(t_1 - t_2) \times a \times 0.42] + [(t_1 - t_2) \times (b - a) \times 4.2] \text{ joules}$$

heat lost by calorimeter   heat lost by water.

Then $(c - b)$ g of sodium thiosulphate absorbed A joules of heat when dissolved.

Hence 248 g (1 mole) absorbed $\frac{A}{c-b} \times 248$ joules when dissolved.

This value is the heat of solution (assuming further dilution causes no heat change).

Experiment 60 To determine the heat of neutralization of sodium hydroxide by hydrochloric acid

Apparatus: glass calorimeter; thermometer (110 °C) and stirrer; measuring cylinder; large dish.

Material: solutions of sodium hydroxide, hydrochloric acid and acetic acid (1 M).

Into a glass calorimeter (beaker surrounded with non-conducting material) put 100 cm^3 of sodium hydroxide solution, insert a thermometer (reading to 0.1 °C) and allow the beaker to stand in a large dish of cold water until the temperature is steady. Measure out 100 cm^3 of hydrochloric acid into a second beaker and allow it to stand under the same conditions as the alkali, when the steady temperature should be the same. Put the beaker containing the alkali into the container (see Figure 20) and pour the acid quickly but

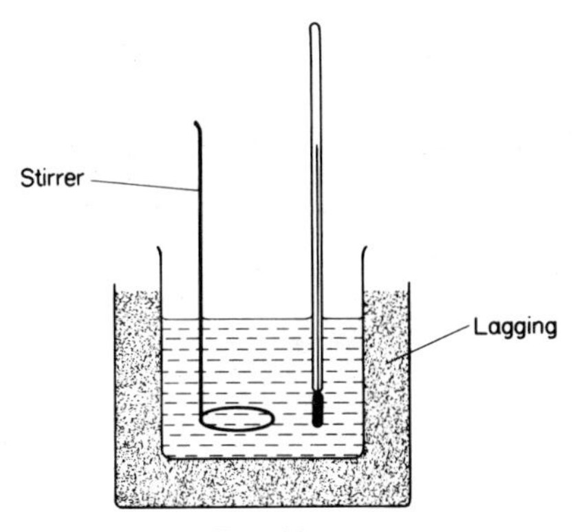

FIG. 20

without splashing into the alkali solution, stir and record the highest temperature reached. (A correction can be made for cooling during this time by recording subsequent cooling.) Assuming the specific heat capacity of the solutions to be the same as that of water, calculate the heat which would be evolved if 1 dm^3 of sodium hydroxide (1 M) were neutralized by 1 dm^3 of hydrochloric acid. This value should be 57 kJ.

The experiment may be repeated using solutions of any strong acids and alkalis, when the value should be the same since the action

$$Na^+ + OH^- + H^+ + Cl^- \rightarrow Na^+ + Cl^- + H_2O$$

is merely the ionic action in each case:

$$OH^- + H^+ \rightarrow H_2O$$

Repeat the experiment using acetic acid in place of hydrochloric acid. Note the lower result for the heat of neutralization. Acetic acid is a weak acid and mainly in the molecular form; part of the energy which should be evolved as heat of neutralization is used to complete the ionization of the molecules; the heat actually evolved is, therefore, lower.

Experiment 61 To determine the heat of combustion of a liquid

A simple device which is useful for *comparing* the heats of combustion of two inflammable liquids—but does not give highly accurate answers—is to use two tins as shown in Figure 21.

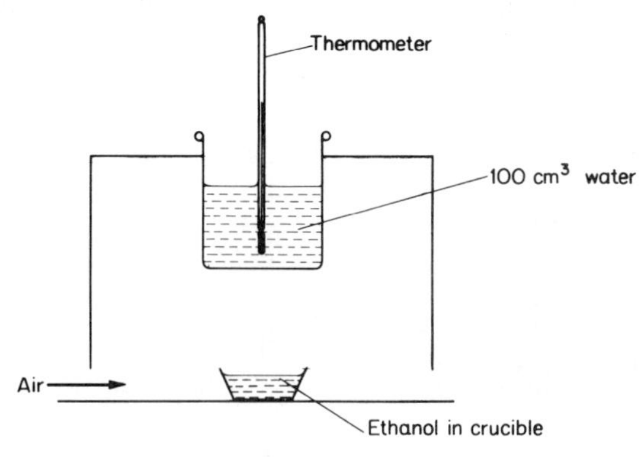

FIG. 21

Into the crucible put 2 cm^3 of ethanol (or acetone (propanone), diethyl ether (ethoxyethane) etc.) and clamp the tins so that there is an air gap of a few centimetres. After ignition of the fuel, stir the water with the thermometer and measure the rise in temperature. Calculate the energy evolved if one mole of the fuel is burnt; the specific heat capacity of water is 4.2 J g^{-1} K^{-1} and of copper 0.42 J g^{-1} k^{-1}.

9
The Rates of Chemical Reactions

Experiment 62 To show that manganese(IV) oxide is a catalyst in the thermal decomposition of potassium chlorate

Apparatus: deep sand tray.

Material: potassium chlorate (chlorate(V)); manganese(IV) oxide.

Mix a little manganese(IV) oxide with about four times its bulk of potassium chlorate and place it in an ignition tube. Into each of two other tubes put an approximately equal bulk of manganese(IV) oxide and potassium chlorate respectively. Surround each with sand on the sand tray so that they are close together and vertical. Commence to heat. After about one minute, test for oxygen by means of a glowing splint; it will be found to be coming off steadily from the mixture. The compounds in the other two tubes do not decompose for some considerable time.

To show that the mass of manganese(IV) oxide remains unchanged, the original mixture may be made by weighing 0.5 g of manganese(IV) oxide and 2 g of potassium chlorate. When the reaction is complete, wash the contents into a beaker, filter, wash, dry and weigh the residue and show there is no loss in mass. Both potassium chloride and potassium chlorate are soluble in water. A qualitative test for the presence, at the end of the experiment, of manganese(IV) oxide, is to warm some of the residue in a test-tube in a fume chamber with concentrated hydrochloric acid and to identify the chlorine evolved.

Experiment 63 Catalytic decomposition of hydrogen peroxide

Material: 20-volume hydrogen peroxide; manganese(IV) oxide.

(*a*) Fill a test-tube with hydrogen peroxide to a depth of about 2 cm and add a little manganese(IV) oxide. Test for oxygen with a glowing splint. There is a rapid evolution of oxygen and the manganese(IV) oxide suffers no discernible change.

$$2H_2O_2 \rightarrow 2H_2O + O_2$$

(*b*) Take two test-tubes and fill each approximately 2 cm deep with hydrogen peroxide. Make one alkaline with a little sodium hydroxide solution and the other acid with approximately an equal volume of dilute sulphuric acid. Immerse both in a beaker half full of hot water. The alkaline solution rapidly begins to decompose and the oxygen can be tested for by means of a glowing splint. The acidic solution remains unaffected.

(*c*) *Quantitative.* Set up an apparatus to collect a gas at atmospheric pressure: there are at least two arrangements possible, one involving a syringe and manometer, another involving a burette or gas-tube in a tall vessel. Into a conical flask put 100 cm^3 of water by means of a measuring cylinder and then sprinkle in 0.2 g of manganese(IV) oxide. Suspend a small test-tube in the flask by means of a thread, the tube containing 5 cm^3 of 20-volume hydrogen peroxide. At a convenient time release the thread and gently shake the flask for the remainder of the experimental time to facilitate the escape of oxygen from the solution. Record the volume of oxygen at times such as 1, 2, 4, 6, 8, 10, 15, 20 ... minutes or record the times at which 10, 20, 30, V_t, ... cm^3 of oxygen have been released. Record the temperature at which the experiment is conducted.

Plot a graph of the volume of the gas against the time and comment upon its shape. Then plot a graph of $\log(V_\infty - V_t)$ against the time t: if this graph is a straight line the reaction is of the first order and the velocity constant k can be calculated from the slope of the graph (which is $-0.434k$).

Experiment 64 Catalytic oxidation of methanol and ammonia

Apparatus: platinum spiral, oxygen cylinder.
Material: methanol.

Fill a beaker to a depth of about 2 cm with methanol and warm it gently by means of a small flame. Prepare a platinum spiral by winding it round a glass rod, leaving a length above the spiral so that it will be just above the alcohol when lowered into it.

Heat the spiral strongly in the Bunsen burner and transfer the glowing spiral to the beaker (Figure 22). It will continue to glow and

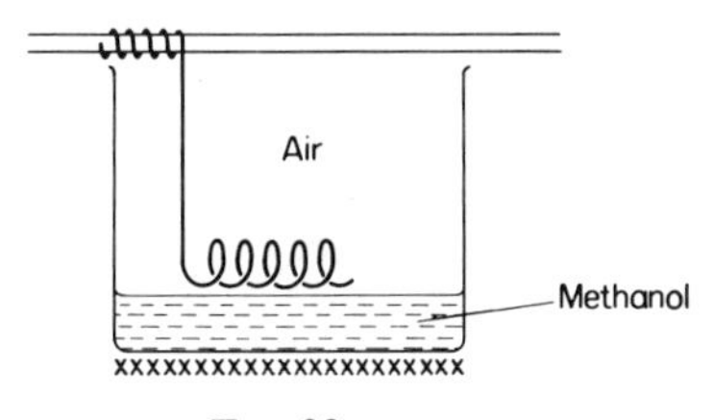

FIG. 22

the smell of formaldehyde (methanal) is quickly apparent. Have ready a piece of asbestos to place over the beaker in the event of the alcohol catching fire.

$$2CH_3OH + O_2 \rightarrow 2HCHO + 2H_2O$$

The experiment may be repeated with concentrated ammonia solution ('0.880') in place of methanol and a current of oxygen from a cylinder sent through the mixture, when brown fumes of nitrogen dioxide or white fumes of ammonium nitrate and nitrite will be seen:

$$4NH_3 + 7O_2 \rightarrow 4NO_2 + 6H_2O$$

See also Experiment 223.

Experiment 65 To show that carbon monoxide will not burn in dry air

Apparatus: large flask; carbon monoxide apparatus fitted with calcium chloride tube.

Take a flask with a wide neck and pour 20–30 cm^3 of concentrated sulphuric acid into it, close the mouth with a cork and swill the acid round. Allow to stand for ten minutes. Meanwhile set up a carbon monoxide apparatus in a fume chamber as shown in Figure 23, with

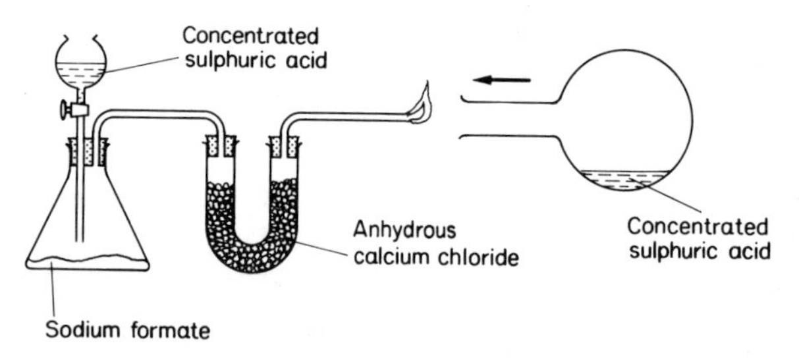

FIG. 23

a long jet so that the jet will pass into the interior of the flask. *After all the air has been expelled from the apparatus,* light the jet and slide the flask towards the carbon monoxide apparatus, allowing the burning jet to pass into the flask. The flame is extinguished. A similar experiment using a flask containing undried air may be set up. In this latter case the jet of carbon monoxide will continue to burn. In this reaction moisture is acting as a catalyst.

Experiment 66 To show that dry hydrogen sulphide and dry sulphur dioxide will not react

Apparatus: sulphur dioxide siphon; gas-jars; Woulfe bottle; calcium chloride tube.

Collect sulphur dioxide in a dry gas-jar after passing the gas slowly through a wash-bottle containing concentrated sulphuric acid to dry it. Collect a gas-jar of hydrogen sulphide, after drying with a calcium chloride tube (phosphorus pentaoxide is a better, but rather inconvenient, drying agent). Invert the jar containing sulphur dioxide over the jar containing the hydrogen sulphide and remove the covers. No reaction takes place. Leave for a minute or two and then pour a few drops of water into the lower gas-jar and quickly replace the upper one. Sulphur is precipitated at once:

$$2H_2S + SO_2 \rightarrow 2H_2O + 3S\downarrow$$

Experiment 67 To show that bromine catalyses the oxidation of sulphur to sulphuric acid

Material: bromine; flowers of sulphur.

Put a spatula-load of flowers of sulphur into each of two dishes and add approximately 5 cm^3 of concentrated nitric acid to each, but to one also add one drop of bromine (*care!*). The whole reaction is best performed in the fume chamber. Warm each for two to three minutes, decant the solution from each into a test-tube, and add hydrochloric acid followed by barium chloride. In the case of the solution to which bromine has been added there is a copious deposit of barium sulphate.

One explanation of catalysis is that intermediate compounds are formed which are more readily decomposed. In this example an explanation is given by the equations

$$2S + Br_2 \rightarrow \underset{\text{sulphur monobromide}}{S_2Br_2}$$

$$2S_2Br_2 + 2H_2O \rightarrow SO_2 + 4HBr + 3S$$

$$SO_2 + 2HNO_3 \rightarrow H_2SO_4 + 2NO_2$$

$$2HBr + 2HNO_3 \rightarrow 2H_2O + 2NO_2 + Br_2$$

Experiment 68 Catalytic decomposition of hypochlorites

Materials: sodium hypochlorite (chlorate(I)); a cobalt(II) salt.

Warm a test-tube half full of sodium hypochlorite solution and notice that there is no decomposition. Add a few drops of cobalt(II) sulphate solution, and there is an immediate evolution of oxygen which will rekindle a glowing splint.

$$2NaOCl \rightarrow 2NaCl + O_2$$

Experiment 69 Dependence of rate of reaction on concentration

Apparatus: measuring cylinders; 1 dm^3 graduated flasks; burette; 10 cm^3 pipette; stop-watch.

Material: magnesium ribbon; potassium iodate (iodate(V)); hydrated sodium sulphite (sulphate(IV)); starch solution; 2 M sulphuric acid.

(*a*) *Qualitative, using magnesium-acid reaction.* Measure into each of four beakers 10 cm^3 of concentrated hydrochloric acid. Leave the first beaker unaltered, but to the second, third and fourth add respectively 10 cm^3, 30 cm^3, and 70 cm^3 of water, thus giving solutions of concentrations in the ratios 1:0.5:0.25:0.125. Measure four equal lengths (about 10 cm each) of magnesium ribbon, and drop one piece into each beaker simultaneously. Observe that the rate of reaction (assessed by $time^{-1}$) is approximately proportional to the concentrations of the acid.

(*b*) *Quantitative, using iodide-iodate reaction.* Make up solution A containing 5 g dm^{-3} of potassium iodate, and solution B containing 10 g dm^{-3} of hydrated sodium sulphite. Place 30 cm^3 of A in a measuring cylinder and dilute to 200 cm^3 with water. Transfer to a 600 cm^3 beaker and add a little starch solution. In another measuring cylinder, place 30 cm^3 of B, 2.5 cm^3 of sulphuric acid and dilute to 200 cm^3 with water. Pour this solution into the beaker containing the diluted solution A, simultaneously starting the stop-watch. The time is recorded when a dark blue coloration appears. The experiment is repeated using 5 cm^3, 7.5 cm^3, 10 cm^3, 12.5 cm^3 and 15 cm^3 of 2 M acid in place of the 2.5 cm^3 employed above, and a table is drawn up of time (seconds) to concentration of acid (cm^3). On plotting a graph of reciprocal of time against volume of acid, a straight line results.

$$IO_3^- + 3HSO_3^- \rightarrow I^- + 3HSO_4^- \quad (1)$$
$$5I^- + IO_3^- + 6H^+ \rightarrow 3I_2\downarrow + 3H_2O \quad (2)$$

Reaction (1) is slow and (2) is rapid but does not take place until (1) is complete. Hence presence of iodine indicates the end of reaction (1).

Experiment 70 The catalytic effect of the hydrogen ion

Apparatus: thermostat (or trough of warm water); 5 cm^3 pipette; thermometer; 2 cm^3 pipette.

Material: ethyl acetate (ethanoate); phenolphthalein indicator; methyl formate (methanoate); 0.1 M sodium hydroxide.

Place two flasks containing 100 cm^3 of 0.5 M acids—hydrochloric acid and acetic acid respectively—in the thermostat (or trough of water) at 25 °C. Place also a flask containing a supply of ethyl acetate in the thermostat. Leave the flasks to attain the temperature of the water.

Titrate 2 cm^3 of each acid separately against 0.1 M sodium hydroxide. Add 5 cm^3 of ethyl acetate to the hydrochloric acid and leave standing in the thermostat for 15 minutes, then titrate 2 cm^3 of the solution against the sodium hydroxide solution, using the indicator phenolphthalein. In the meantime, and at a five minute interval, treat the acetic acid with 5 cm^3 of ethyl acetate and titrate after 15 minute intervals likewise.

Tabulate the resulting titrations every fifteen minutes for each solution over a period of about two hours, and finally make titrations of each after at least three hours.

The increased titration with time is due to the formation of acetic acid by hydrolysis of ethyl acetate.

$$CH_3COOC_2H_5 + H_2O \rightarrow CH_3COOH + C_2H_5OH$$

A comparison of the titrations after a given time is a measure of the rate of hydrolysis in each case; the rate in the presence of a mineral acid (where the hydrogen ion concentration is appreciable) is greater than in the case of acetic acid, which has a low hydrogen ion concentration. The catalytic action of the hydrogen ions in the original acid is augmented by those furnished by the acetic acid produced by the hydrolysis, but the concentration from this source is small. It is, however, an example of *autocatalysis*.

The hydrolysis of methyl formate follows a similar path and can be studied quantitatively by plotting a graph of $\log(V_\infty - V_t)$ against time t. A straight line verifies that the hydrolysis is of the first order, and from the slope of the graph ($-0.434k$) the velocity constant k can be calculated.

Experiment 71 The rate of reaction of sodium thiosulphate with nitric acid

Material: 1 M sodium thiosulphate; 0.1 M nitric acid.

$$S_2O_3^{2-} + 2H^+ \rightarrow S\downarrow + H_2O + SO_2\uparrow$$

The precipitation of the sulphur causes the solution to become opaque, and an assessment of the rate can be made by measuring the time for some writing on a piece of paper under the reaction vessel to become obscured. Set up three burettes for the solutions involved.

Volumes (cm^3)	*Nitric acid* (0.1 M)	*Sodium thiosulphate* (1 M)	*Water*
a	5	5	0
b	5	4	1
c	5	3	2
d	5	2	3
e	5	1	4

Put the acid into a small beaker and the sodium thiosulphate and water into a second small beaker. At a convenient time pour the acid into the sodium thiosulphate and stir well: note the time taken to precipitate sufficient sulphur to obscure the writing. Record the temperature.

Plot a graph of the rate of reaction (assessed by time^{-1}) against the concentration of the sodium thiosulphate (assessed by the volume taken). A straight line verifies that the rate of the reaction is of the first order with respect to the concentration of the sodium thiosulphate.

Experiment 72 To study the rate of hydrolysis of a sugar

Apparatus: polarimeter.

If a polarimeter is available (see also Experiment 27) the rate of hydrolysis of a sugar such as sucrose or maltose when catalysed by a dilute solution (2 M) of a strong acid may be studied. Weigh out approximately 10 g of the sugar, shake it rapidly with 25 cm^3 of water and then, in quick succession, add 15 cm^3 of 2 M hydrochloric acid and then sufficient water to make the total volume 50 cm^3. Put it in the polarimeter tube and obtain a reading as soon as possible. Take further readings on the polarimeter at times such as 5, 10, 15, 20, 25, 30, 45, 60 ... minutes. Record the temperature.

Plot a graph of $\log(\theta_\infty - \theta_t)$, where θ is the angle of rotation, against time t. If this is a straight line the hydrolysis is of the first order, and from the slope of the graph ($-0.434k$) the velocity constant k can be calculated.

Experiment 73 To determine the energy of activation of a reaction

Apparatus: thermometer.

Material: 0.02 M potassium permanganate; 0.05 M oxalic acid made up in 0.5 M sulphuric acid.

Reactions between oppositely charged ions are often very fast but the reaction between similarly charged ions, e.g. permanganate (manganate(V)) and oxalate (ethanedioate), may proceed at a rate which is measurable.

$$2MnO_4^- + 16H^+ + 5C_2O_4^{2-} \rightarrow 2Mn^{2+} + 8H_2O + 10CO_2\uparrow$$

The time is taken for the disappearance of the purple colour of the potassium permanganate when it reacts with the oxalate solution.

Using burettes, put 10 cm^3 portions of the permanganate and of the oxalate in separate boiling-tubes and put the tubes in a beaker of warm water at 50 °C. When the solutions have attained this temperature, pour one into the other and measure the reaction time, t. Repeat the experiment at temperatures of 60, 70, 80 and 90 °C. Plot a graph of $\log t$ against the reciprocal of the temperature (in K). The slope of the graph, according to the Arrhenius equation, is $0.052E$, where E J mol^{-1} is the activation energy.

Experiment 74 To study the iodination of acetone

Material: Iodine in aqueous potassium iodide solution (0.02 and 0.2 M respectively), solution of acetone (propanone) in water (1 M) and 1 M sulphuric acid, 0.5 M sodium hydrogencarbonate, 0.01 M sodium thiosulphate.

The dependence of the rate upon the concentration of iodine is determined in experiments in which the acetone concentration is so large that it is almost unchanged in the course of the reaction.

Put 50 cm^3 of the iodine solution in one conical flask and 25 cm^3 each of the acetone solution and the acid in a second flask. Into other flasks put 10 cm^3 portions of the sodium hydrogencarbonate solution, which serves to neutralize the hydriodic acid produced, thus permitting the titration of excess iodine with sodium thiosulphate.

$$I_2 + CH_3COCH_3 \rightarrow CH_3COCH_2I + HI$$

At a convenient time pour the iodine solution into the acidified acetone and stir the mixture well. At times such as 5, 10, 15, 20 ... minutes transfer a 10 cm^3 portion of the reaction mixture into a portion of the alkali, and, after any effervescence has ceased, titrate with the thiosulphate using starch as an indicator. Record the temperature.

Plot a graph of the titres against the time and deduce the probable dependence of the rate of reaction upon the iodine concentration.

Experiment 75 The oxidation of an iodide by a persulphate

Material: 0.5 M potassium iodide; 0.02 M potassium persulphate (peroxodisulphate(VI)); 0.025 M sodium thiosulphate; starch solution.

$$2I^- + S_2O_8^{2-} \rightarrow I_2 + 2SO_4^{2-}$$

This reaction can be proved as follows to be first order with respect to the persulphate ions. The reaction rate is followed by adding portions of sodium thiosulphate and measuring the time taken for the iodine released to react with these and then to give the familiar blue coloration with starch.

Into one flask put 50 cm^3 of the iodide solution, 25 cm^3 of water and a small quantity of the starch. Into a second flask put 25 cm^3 of the persulphate and 50 cm^3 of water. Have the supply of sodium thiosulphate in a burette and add 10 cm^3 of it to the second flask. At a convenient time pour the contents of one flask into the other and stir the mixture thoroughly. When a blue colour appears, record the time and add 3 cm^3 of the thiosulphate: proceed in this manner until 28 cm^3 have been added (i.e. $10 + 6 \times 3$). Record the temperature of the solution.

Finally the equivalence of the iodide and persulphate solutions, i.e. the end-point if a titration could be done 24 hours later, must be found: maintain a mixture of 50 cm^3 of the iodide solution and 25 cm^3 of the persulphate solution at about 60 °C for five minutes before titrating them, in the presence of starch, with the sodium thiosulphate (V_∞). Plot a graph of log ($V_\infty - V_t$) against the time t: a straight line is obtained.

Experiment 76 To show that reactions may proceed through transition states

(*a*) Acidify some potassium dichromate solution in a boiling tube with dilute sulphuric acid and then add some hydrogen peroxide solution. Observe the sequence of colours. Repeat the experiment, away from any flames, adding in addition some diethyl ether (ethoxyethane). The unstable intermediate compound is thought to be a chromium peroxide and it is more soluble in, and is stabilized by, the ether.

(*b*) Hydrogen peroxide is not decomposed even when boiled with potassium sodium tartrate (2,3-dihydroxybutanedioate) solution. Observe the absence of effervescence when the solutions are heated and then add cobalt(II) chloride solution until the mixture is pink. Record what happens thereafter.

Experiment 77 The rate of hydrolysis of 2-bromo-2-methylpropane

Apparatus: conductance bridge and dipping electrode.

The conductance bridge was employed for experiments in Chapter 5; it can also be used to follow the hydrolysis of a covalent material which yields ions.

Material: 2-bromo-2-methylpropane.

In a small beaker put about 80 cm^3 of ethanol and 20 cm^3 of water and check that its conductance is very low. At a convenient time add a few drops of 2-bromo-2-methylpropane and measure the conductance (G_t) at one-minute intervals for a while and then at longer intervals until it reaches a constant value (G_∞). Record the temperature.

Plot a graph of $\log(G_\infty - G_t)$ against the time t: the reaction is of the first order and so this should be a straight line; from the slope of the graph ($-0.434k$) the velocity constant k can be calculated.

10
Reversible Reactions

A reversible reaction is one which can proceed in either direction by altering the conditions of the reaction. One of these conditions is the relative concentrations of the substances involved. According to the Law of Mass Action the velocity of a chemical change is directly proportional to the 'active mass' of the reactants, and active mass in a homogeneous system is usually expressed in mol dm^{-3} of a substance. Other conditions which can influence the direction are temperature and pressure.

In many chemical reactions a point is reached where action apparently ceases in the forward direction although some of the reacting substances remain unchanged. If A and B represent the reacting substances, and C and D represent the resulting substances, an equilibrium is reached with some A and B unchanged and a definite quantity of C and D formed. The explanation, based on the Kinetic Theory, is that initially A and B react at a rate which depends on their concentrations (in mol dm^{-3}) and since a change in the concentration of *either* A or B will produce a corresponding change in the rate of the reaction, the rate of the forward action may be proportional to the *product* of the concentrations of A and B. The dependence must be found *experimentally*; it cannot be *proved* theoretically.

$$\text{Rate of reaction of A and B} \propto [A]\,[B] = k_1[A]\,[B]$$

The square brackets symbolize concentration of substance in mol dm^{-3}.

As soon as A and B react, their concentrations will decrease, and the rate of reaction will progressively decrease. At the same time, the reaction between A and B will have formed some of C and D, and the concentrations of these will progressively increase, and will in turn react to form A and B. The rate of reaction may be proportional to the product of their concentrations. Again the dependence must be found experimentally.

$$\text{Rate of reaction of C and D} \propto [C]\,[D] = k_2[C]\,[D]$$

When there is apparently no further action, an equilibrium has been reached with the rate of reaction of A and B forming C and D equal to the rate of C and D forming A and B.

At equilibrium,

$$k_1[A]\,[B] = k_2[C]\,[D]$$

or

$$\frac{[C]\,[D]}{[A]\,[B]} = \frac{k_1}{k_2} = K \text{ (called the equilibrium constant)}$$

The expression means that, at equilibrium, the product of the concentrations of C and D, divided by the product of the concentrations of A and B, has a definite value. Whatever the experimental results show the rates of forward and backward reactions to depend on, the equilibrium constant always has a value predictable from the equation. Hence, if at equilibrium the concentration of, say, A, is increased by addition of more of it, the concentrations of B, C, and D will assume such new values that the value of the expression (K) will remain unchanged. Clearly this will involve the combination of some A and B to form more C and D, that is, the previous equilibrium concentration of B will be decreased and those of C and D will be increased. Examples of the effect of addition of one of the reacting substances are given in Experiment 78.

Experiment 78 To show the effect of alteration of concentration

Material: bismuth chloride; antimony chloride.

(*a*) *Hydrolysis of bismuth chloride.* Put a little bismuth chloride into a test-tube and add about 1 cm^3 of water. A white substance, bismuth oxychloride (chloride oxide), is formed.

$$BiCl_3 + H_2O \rightleftharpoons BiOCl\downarrow + 2HCl$$

Add drops of concentrated hydrochloric acid until the white precipitate disappears. Add a few more drops of water and notice the reappearance of the white suspension.

On first adding water, equilibrium was reached when definite concentrations of bismuth oxychloride and hydrochloric acid had been formed:

$$\frac{[HCl]^2}{[BiCl_3]\,[H_2O]} = K \text{ (equilibrium constant)}$$

On adding concentrated hydrochloric acid, some bismuth oxychloride and some hydrochloric acid reacted to form more bismuth chloride and water. By these changes the value of the expression assumed the original mathematical value of K.

(*b*) *Hydrolysis of antimony chloride.* Repeat the above experiment using antimony chloride in place of bismuth chloride. The reactions and the explanations are similar to those for bismuth.

$$SbCl_3 + H_2O \rightleftharpoons \underset{\substack{\text{antimony}\\\text{oxychloride}}}{SbOCl\downarrow} + 2HCl$$

(*c*) *The increased concentration of ammonium ion from ammonium chloride reduces the OH^- concentration in ammonia solution.* Add a drop of phenolphthalein to a dilute solution of ammonia. The solution goes pink. Add solid ammonium chloride a little at a time, and the colour will diminish and finally disappear. (See p. 47.)

$$NH_3 + H_2O \rightleftharpoons NH_4^+ + OH^-$$

Experiment 79 To show the effect of temperature on chemical equilibrium

Apparatus: as shown in Figures 24 and 25.

Material: lead nitrate.

(*a*) *Thermal dissociation of ammonium chloride.* Heat about a spatula-load of ammonium chloride in the bottom of a dry test-tube with a piece of damp litmus in the mouth. The solid will sublime and condense part of the way up the tube, leaving a clear space where the tube is hot. The litmus will turn blue. The clear space contains the products of decomposition of ammonium chloride, i.e. ammonia and hydrogen chloride—both colourless gases. The ammonia, being the less dense of the two gases, diffuses more rapidly, reaches the litmus first and turns it blue. In the cooler parts of the tube recombination to form ammonium chloride occurs.

$$NH_4Cl \underset{\text{cooling}}{\overset{\text{heating}}{\rightleftharpoons}} HCl + NH_3$$

An alternative method of showing the thermal decomposition of ammonium chloride is to heat some in a sloping glass tube (Pébal's experiment, Figure 24).

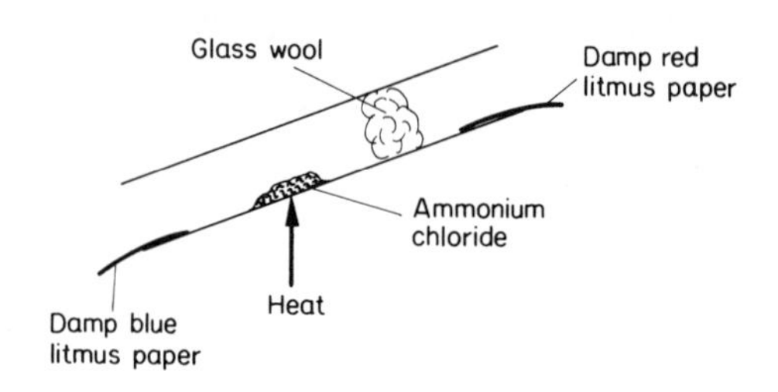

FIG. 24

(*b*) *Action of heat on dinitrogen tetraoxide.* Fit up the apparatus shown in Figure 25, and heat the tube containing lead nitrate, and also the centre of the long horizontal tube. The colour of the gas in the hot part of the tube is a darker brown than in the cooler part because of the presence of more nitrogen dioxide molecules which are brown.

$$N_2O_4 \rightleftharpoons 2NO_2$$

pale yellow   brown

Wet filter paper

Darker brown

Heat

Heat

Sodium hydroxide solution

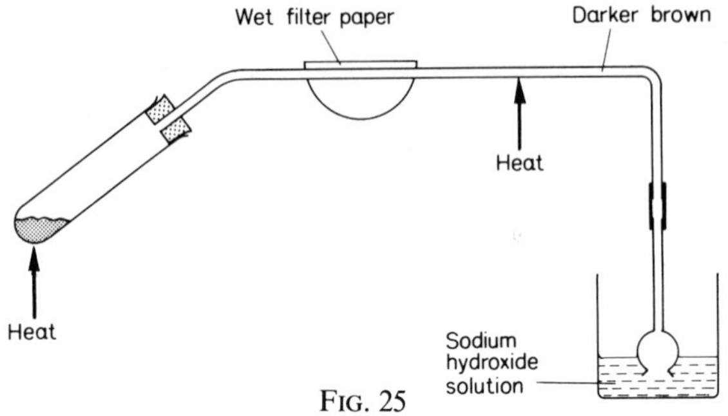

FIG. 25

Experiment 80 To find the equilibrium constant for the formation of ethyl acetate

Material: glacial acetic acid; ethanol; ethyl acetate; 1 M hydrochloric acid.

The reaction of ethanol with acetic (ethanoic) acid to give ethyl acetate (ethanoate) and water is reversible; it is catalysed by strong acids, e.g. hydrochloric acid. Into a set of boiling tubes partially immersed in a thermostat (even a tin full of water will suffice in this particular instance because the heat of reaction is zero) put various mixtures of the reagents and/or products. A control tube containing only the catalyst should be included. A pipette filler should be attached to the pipette for this purpose and the tubes should be corked.

Volume (cm^3)	*Hydrochloric acid* (1 M)	*Water*	*Ethyl acetate*	*Glacial acetic acid*	*Ethanol*
Control	1	9	0	0	0
a	1	3	3	3	0
b	1	3	3	0	3
c	1	3	0	3	3
d	1	0	3	3	3
Density in $g\ cm^{-3}$	1	1	0.92	1.05	0.79

The mixtures should be left at least three days to reach equilibrium. Then 1 cm^3 samples should be taken from each experimental tube (*a–d*) and poured into about 100 cm^3 of water to 'freeze' the equilibrium. However, the entire contents of the control tube should be employed, mixed as before with 100 cm^3 of water.

The samples are titrated against 0.2 M sodium hydroxide solution using phenolphthalein as the indicator. The titration measures the concentration of the hydrochloric and acetic acids in the sample, and one-tenth of the titration of the control tube should be subtracted from this to obtain a measure of the concentration of the acetic acid alone. There may be a greater or lower concentration of acetic acid in a particular tube than at the start. The number of millimoles of water, ester, acetic acid and acohol at the start of the experiment in each tube should be calculated and then, in accordance with the titrations, the numbers of millimoles of these four substances at the end of the experiment.

Finally, the numbers of millimoles (or in a general case the concentrations) are inserted in the equation

$$K_c = \frac{[\text{ester}][\text{water}]}{[\text{acid}][\text{alcohol}]}$$

and the values of K_c so obtained are averaged.

The temperature should be recorded.

PART II

Inorganic Chemistry

With each group of elements in the Periodic Table is given a table comparing the properties of the elements in that group.

The student is advised to use the experimental work as a basis for showing these similarities and gradations in properties as well as for noting the chief differences.

The elements and their more important compounds are considered in the order of the Periodic Table (see p. 76). The properties of hydrogen should be looked for under the various headings, e.g. reduction, diffusion, hydrogen chloride, hydrogen sulphide etc.

THE PERIODIC TABLE

							1 H	2 He									
3 Li	4 Be											5 B	6 C	7 N	8 O	9 F	10 Ne
11 Na	12 Mg											13 Al	14 Si	15 P	16 S	17 Cl	18 Ar
19 K	20 Ca	21 Sc	22 Ti	23 V	24 Cr	25 Mn	26 Fe	27 Co	28 Ni	29 Cu	30 Zn	31 Ga	32 Ge	33 As	34 Se	35 Br	36 Kr
37 Rb	38 Sr	39 Y	40 Zr	41 Nb	42 Mo	43 Tc	44 Ru	45 Rh	46 Pd	47 Ag	48 Cd	49 In	50 Sn	51 Sb	52 Te	53 I	54 Xe
55 Cs	56 Ba	57 La*	72 Hf	73 Ta	74 W	75 Re	76 Os	77 Ir	78 Pt	79 Au	80 Hg	81 Tl	82 Pb	83 Bi	84 Po	85 At	86 Rn
87 Fr	88 Ra	89 Ac†															
			58 *Ce	59 Pr	60 Nd	61 Pm	62 Sm	63 Eu	64 Gd	65 Tb	66 Dy	67 Ho	68 Er	69 Tm	70 Yb	71 Lu	
			90 †Th	91 Pa	92 U	93 Np	94 Pu	95 Am	96 Cm	97 Bk	98 Cf	99 Es	100 Fm	101 Md	102 No	103 Lw	

11
Group IA of the Periodic Table

Li	Na	K	*Rb*	*Cs*
Lithium	Sodium	Potassium	*Rubidium*	*Caesium*

Elements in italics are not usually studied in an elementary course. Sodium and potassium show a very close similarity, and their corresponding compounds are alike both in appearance and in their reactions; these points may have already been noted in elementary work. Their electronic structure and the positions they occupy at the beginning of periods in the Periodic Table account for this close resemblance.

Copper and silver are transition elements and, although a few points of similarity exist to the alkali metals, they are not considered until later (see p.97).

	Lithium	*Sodium*	*Potassium*
Metal	←— All very reactive and strongly metallic —→		
	←— All attack water —→		
	←— All form stable salts —→		
Oxides	←— All oxides strongly basic —→		
Hydroxide	←— All alkaline —→		
Chloride	←— All very soluble in water —→		
Iodide	←— All very soluble in water —→		
Flame coloration of compounds	Red, invisible through blue glass.	Golden yellow, invisible through blue glass.	Lilac, visible through blue glass.

There are no fully reliable, simple precipitation tests for sodium and potassium ions in solution. Lithium, like the first member of each of the other groups in the Periodic Table, is slightly out of step with the other members of its group.

LITHIUM

Experiment 81 Reactions of lithium ions

Material: a lithium salt, e.g. lithium chloride, in solution.

(*a*) Add a solution of sodium carbonate: the white precipitate formed demonstrates that lithium carbonate is much less soluble than sodium carbonate.

$$2Li^+ + CO_3^{2-} \rightarrow Li_2CO_3\downarrow$$

(*b*) Add disodium hydrogenphosphate solution: again a white precipitate is formed, this time of a lithium phosphate.

(*c*) Perform the flame test: a bright red colour is given which is invisible through blue glass.

(*d*) Based on these observations, predict where lithium would be precipitated in the usual qualitative analysis scheme of separation into groups of insoluble compounds. Check whether your prediction is correct.

12
Group IIA of the Periodic Table

Be	Mg	Ca	*Sr*	Ba	*Ra*
Beryllium	Magnesium	Calcium	*Strontium*	Barium	*Radium*

Elements in italics are not usually studied in an elementary course.

Magnesium links the alkaline earths (calcium, strontium and barium) with zinc, which in turn is similar in many respects to cadmium. The elements in the sub-group, zinc, cadmium, and mercury, do not show many points of similarity but calcium, strontium and barium are more closely related.

The properties of zinc, cadmium and mercury will be considered separately as transition elements; see p. 102

	Magnesium	*Calcium*	*Barium*
Metallic character	←——— Activity increases ———→ All strongly metallic, divalent in compounds.		
Metal on water	Burns in steam.	Reacts in cold water.	Reacts in cold water.
Carbonate	Decomposes at 750 °C.	Decomposes at 825 °C.	Decomposes at 1842 °C.
Oxide	Slight action with water.	Slakes on addition of water with evolution of heat.	As for calcium but much greater evolution of heat.
Hydroxide	Slightly alkaline.	Slightly soluble and alkaline.	Even more alkaline.
Halides Nitrates Sulphates Chromates	Solubility in water decreases with increasing relative atomic mass of metal.		
Flame coloration of compounds	—	Brick red.	Green.

MAGNESIUM

Experiment 82 Reactions of magnesium and its compounds

Apparatus: as indicated in (*f*).

Material: magnesium ribbon; magnesium sulphate; sodium nitrite (nitrate(III)).

(*a*) Burn 10 cm of magnesium ribbon in the air, over a piece of paper. The white solid is the oxide. Transfer the oxide to some water in a beaker, put in a piece of red litmus paper and boil. The litmus paper gradually turns blue, showing the oxide to be feebly alkaline.

$$2Mg + O_2 \rightarrow 2MgO$$

(*b*) Add ammonium carbonate solution to a solution of magnesium sulphate in water. A white precipitate is obtained which is a basic carbonate.

(*c*) Add ammonia solution to a solution of magnesium sulphate in water. A white precipitate of magnesium hydroxide is obtained.

$$Mg^{2+} + 2OH^- \rightarrow Mg(OH)_2\downarrow$$

If (*b*) and (*c*) are repeated after the addition of ammonium chloride to the magnesium sulphate solution, there is no precipitate. Hence magnesium does not appear in either Group III or V of the analytical tables. The reason is that the increased concentration of ammonium ion, due to the presence of ammonium chloride, suppresses the ionization of, for example, the ammonia, leaving insufficient hydroxyl ions to attain the solubility product of magnesium hydroxide.

$$NH_3 + H_2O \rightleftharpoons NH_4^+ + OH^-$$

(*d*) Add ammonium chloride and ammonia to a solution of magnesium sulphate in water. Add a solution of disodium hydrogenphosphate. There is a white crystalline precipitate of ammonium magnesium phosphate-6-water.

$$Mg^{2+} + HPO_4^{2-} + NH_3 + 6H_2O \rightarrow MgNH_4PO_4{\cdot}6H_2O$$

(*e*) Heat a few magnesium sulphate crystals on a charcoal block and allow them to cool. Moisten the white mass with cobalt nitrate solution and heat again. A pink mass is obtained on cooling.

(*f*) Take a 250 cm^3 flask fitted with a cork and delivery tube which in turn is connected to a ‘U’ tube and the latter to a piece of combustion tube. Mix about four spatula-loads each of ammonium chloride and sodium nitrite in the flask and add 30 cm^3 of water. Place a length of magnesium ribbon loosely in the combustion tube. Heat the flask cautiously until action begins, then remove the flame from the flask and heat the combustion tube. Nitrogen is formed (see Experiment 138(*h*)) which combines with magnesium to form magnesium nitride, Mg_3N_2. The purpose of the 'U' tube is to condense steam and prevent it passing into the combustion tube. Transfer the white nitride to a test-tube, add water and boil. Test for ammonia with litmus paper.

$$Mg_3N_2 + 6H_2O \rightarrow 3Mg(OH)_2 + 2NH_3$$

CALCIUM

Experiment 83 Reactions of calcium and its compounds

Material: calcium; calcium chloride.

(*a*) Put a flake of calcium on gauze on a tripod and direct a Bunsen flame on to it. It will burn brilliantly with a red flame and leave a white product, calcium oxide. Allow the solid to cool, transfer it to a test-tube and add a few drops of water. There is a vigorous exothermic reaction, and when a piece of red litmus paper is dipped into the solution it turns blue immediately. Note that the solid is not very soluble in water.

$$2Ca + O_2 \rightarrow 2CaO$$
$$CaO + H_2O \rightarrow Ca(OH)_2$$

(*b*) Add ammonium carbonate solution to a solution of calcium chloride in water. The white precipitate is calcium carbonate.

$$Ca^{2+} + CO_3^{2-} \rightarrow CaCO_3\downarrow$$

(*c*) Add ammonium oxalate (ethanedioate) solution to a solution of calcium chloride. A white precipitate of calcium oxalate (soluble in dilute hydrochloric acid but insoluble in acetic acid) is obtained.

$$Ca^{2+} + C_2O_4^{2-} + H_2O \rightarrow CaC_2O_4{\cdot}H_2O\downarrow$$

Note: Solutions of calcium salts give no precipitate with potassium chromate (cf. barium) and no precipitate with calcium sulphate solution (cf. strontium).

(*d*) Add disodium hydrogenphosphate solution to a solution of calcium chloride. White calcium phosphate is precipitated.

$$3Ca^{2+} + 2PO_4^{3-} \rightarrow Ca_3(PO_4)_2\downarrow$$

This precipitate is soluble in dilute hydrochloric, nitric or acetic acid, as also are the corresponding phosphates of strontium and barium.

(*e*) Moisten a little calcium chloride with concentrated hydrochloric acid and perform the flame test. Note the brick-red flame and observe the green colour when seen through blue glass.

STRONTIUM

Experiment 84 Reactions of strontium compounds

Material: strontium nitrate.

(*a*) Perform reactions given in Experiment 83(*b*), (*c*), and (*d*) with strontium nitrate solution instead of calcium chloride. Precipitates of strontium carbonate, strontium oxalate and strontium phosphate are obtained under similar conditions to those in which the corresponding compounds of calcium are obtained.

(*b*) Add calcium sulphate solution to a solution of strontium nitrate in water. Warm and allow to stand. A white precipitate of strontium sulphate (which is much more insoluble than calcium sulphate) is thrown down.

$$Sr^{2+} + SO_4^{2-} \rightarrow SrSO_4\downarrow$$

(*c*) Perform the flame test with strontium nitrate and observe the crimson coloration of the flame. When it is viewed through blue glass, there is no change.

BARIUM

Experiment 85 Reactions of barium compounds

Material: barium chloride.

(*a*) Perform reactions given in Experiment 83(*b*), (*c*), and (*d*) with barium chloride solution instead of calcium chloride. Precipitates of barium carbonate, barium oxalate and barium phosphate are obtained under similar conditions to those in which the corresponding compounds of calcium are obtained. Experiment 84(*b*) performed with barium chloride in place of strontium nitrate also shows barium sulphate to be very insoluble in water.

(*b*) Add potassium chromate solution to a solution of barium chloride. A yellow precipitate of barium chromate is obtained.

$$Ba^{2+} + CrO_4^{2-} \rightarrow BaCrO_4\downarrow$$

(*c*) Perform the flame test and show that barium compounds colour the flame with flashes of green.

13
The Transition Elements

Sc	Ti	V	Cr	Mn
Scandium —	Titanium —	Vanadium —	Chromium —	Manganese —

Fe	Co	Ni	Cu	Zn
Iron —	Cobalt —	Nickel —	Copper —	Zinc

(Elements in italics are not usually studied in an elementary course) Also the corresponding elements in periods five to seven.

VANADIUM

Experiment 86 The oxidation states of vanadium

Material: 0.1 M ammonium or sodium metavanadate (vanadate(V)).

The starting solution, which contains pentavalent vanadium, is almost colourless. To about 25 cm^3 of the vanadate solution add about 5 cm^3 of sulphuric acid (1 M); note the change in colour, and then warm the solution to about 70 °C before adding a spatula-load of sodium metabisulphite (disulphate(IV)). Shake the solution and observe another change in colour. Double the volume of the mixture by adding dilute sulphuric acid and add a spatula-load of zinc dust: a further change in colour is seen. Finally, add more zinc dust and *lightly* cork the flask; agitate it periodically and yet another colour change may be seen.

The following statements should be converted into balanced equations by the student and the oxidation number corresponding to each colour and formula allocated.

$$VO_3^- + H^+ \rightarrow VO_2^+ + H_2O$$

$$VO_2^+ + SO_3^{2-} + H^+ \rightarrow VO^{2+} + SO_4^{2-} + H_2O$$

$$VO_2^{2+} + Zn + H^+ \rightarrow V(H_2O)_6^{3+} + Zn^{2+}$$

$$V(H_2O)_6^{3+} + Zn + H^+ \rightarrow V(H_2O)_6^{2+} + Zn^{2+}$$

CHROMIUM

Chromium is a metal but some of its compounds resemble corresponding sulphur compounds, e.g. the sulphates and chromates of potassium are isomorphous and sulphuryl chloride (sulphur dichloride dioxide) and chromyl chloride (chromium(VI) dichloride dioxide) show similarities. The reactions of chromium and its compounds also show points of similarity with aluminium (Group III) and trivalent iron (Group VIII).

Experiment 87 Reactions of chromium compounds

Material: potassium chromate; potassium dichromate; chromium(III) potassium sulphate; supply of sulphur dioxide; diethyl ether; ethanol; sodium peroxide.

(*a*) *Dry reactions.* Heat a little chromium compound on a carbon block and show that a green residue of chromium(III) oxide (Cr_2O_3) is left.

Heat in a borax bead. Note the emerald-green colour in both oxidizing and reducing flames.

(*b*) *Reactions in solution.* Make a stock solution of chromium(III) potassium sulphate (chrome alum), and use a test-tube about a quarter full for each of the following tests:

(i) Make the solution alkaline with ammonia solution and boil. The green-blue precipitate is chromium(III) hydroxide.

$$Cr^{3+} + 3OH^- \rightarrow Cr(OH)_3\downarrow$$

Show that the precipitate is easily soluble in acids.

(ii) Add sodium hydroxide solution. Recognize the precipitate of chromium(III) hydroxide, and show that it is soluble in excess of the reagent to give a green solution of sodium chromite (chromate(III)).

$$Cr(OH)_3 + OH^- \rightarrow CrO_2^- + 2H_2O$$

(iii) Add a solution of sodium carbonate (or ammonium sulphide). Note again the precipitate of chromium(III) hydroxide (like aluminium, the carbonate and sulphide of chromium are rapidly hydrolysed in solution).

(*c*) *Reactions of dichromates.* Make a stock solution of potassium dichromate. Use a test-tube about a quarter full of the solution for each test.

(i) Add one drop of sodium hydroxide solution. Note the change of colour from orange to yellow, due to the formation of the chromate ion.

$$Cr_2O_7^{2-} + 2OH^- \rightarrow 2CrO_4^{2-} + H_2O$$

To the yellow solution add two or three drops of dilute acid. The orange colour is due to the formation of the dichromate ion.

$$2CrO_4^{2-} + 2H^+ \rightarrow Cr_2O_7^{2-} + H_2O$$

(ii) Add a few drops of dilute sulphuric acid, then pass sulphur dioxide through the solution. The change of colour to green is due to the reduction of potassium dichromate to chromium(III) sulphate, the sulphurous acid being oxidized to sulphuric acid.

$$Cr_2O_7^{2-} + 8H^+ + 3SO_3^{2-} \rightarrow 2Cr^{3+} + 3SO_4^{2-} + 4H_2O$$

Reduce acidified solutions of potassium dichromate by hydrogen sulphide and by ethanol (see Experiments 224 and 240). Show that the relevant equations depend on the ionic equation

$$Cr_2O_7^{2-} + 14H^+ + 6e^- \rightarrow 2Cr^{3+} + 7H_2O$$

(iii) Acidify a solution of potassium dichromate, add enough ether (*care!*) to give a layer about 2 cm deep above the solution. Add a drop of a dilute aqueous solution of hydrogen peroxide. The blue colour is due to chromium peroxide (see Experiments 76 and 153).

Note: For the preparation of chromyl chloride see Experiment 165.

(*d*) *Reactions of chromates.* Make a stock solution of potassium chromate, and use small portions for each of the following experiments.

(i) Add a drop of silver nitrate solution. The brick-red precipitate is silver chromate (see p. 101).

(ii) Add to three portions of the solution, solutions of lead(II) acetate, barium chloride, and mercury(I) nitrate respectively. The precipitates are the chromates of the metals.

(iii) Acidify a portion of the solution. Pass hydrogen sulphide into the solution to reduce the chromate to a chromium salt.

$$2CrO_4^{2-} + 10H^+ + 3H_2S \rightarrow 2Cr^{3+} + 8H_2O + 3S\downarrow$$

(iv) Acidify a portion of the solution and pass in sulphur dioxide. Sulphurous (sulphuric(IV)) acid has reduced the solution of chromate (yellow) to the chromium(III) salt (green).

$$2CrO_4^{2-} + 10H^+ + 3SO_3^{2-} \rightarrow 2Cr^{3+} + 5H_2O + 3SO_4^{2-}$$

(*e*) *Oxidation of chromium compounds to chromates.* Prepare a dilute solution of chromium(III) potassium sulphate. Add about half a spatula-load of sodium peroxide, then boil the solution. The yellow colour of the solution indicates the presence of sodium chromate.

$$2Cr^{3+} + 3H_2O_2 + 10OH^- \rightarrow 2CrO_4^{2-} + 8H_2O$$

Test for the chromate ion by acidifying the solution with acetic (ethanoic) acid and adding a solution of lead(II) acetate (see p. 118).

Experiment 88 Preparation of chromium(VI) oxide, CrO_3

Apparatus: measuring cylinder; glass wool; water-bath.

Material: finely ground potassium dichromate.

Dissolve 25 g of potassium dichromate in 50 cm^3 of boiling water. Cool the solution to room temperature, and *when cold add very gradually* 35 cm^3 of concentrated sulphuric acid. Leave for a few hours, then decant the liquid from the potassium hydrogensulphate crystals. Heat the liquid to 85 °C and add 25 cm^3 of dilute sulphuric acid. Evaporate on a water-bath until crystals form on the surface, then set aside to crystallize. Filter through glass wool, preferably with suction, and evaporate the filtrate for a further crop of crystals. To remove traces of sulphuric acid wash the crystals while still in the filter with concentrated nitric acid, in which the chromium(VI) oxide is not soluble. Transfer the crystals to a dry evaporating basin and heat in an air oven at 130 °C.

$$K_2Cr_2O_7 + 2H_2SO_4 \rightarrow 2KHSO_4 + 2CrO_3 + H_2O$$

Experiment 89 Preparation of potassium chromate

Material: chromium(III) oxide; potassium hydroxide; potassium nitrate.

Into a small crucible put a spatula-load of chromium(III) oxide and add five pellets of potassium hydroxide and half a spatula-load of potassium nitrate. Heat the crucible directly over a Bunsen burner, and after the contents have become molten, stir them with a glass rod and then allow them to cool. When the crucible is cold again, add about 5 cm^3 of water and warm the crucible gently. Decant the yellow solution into a test-tube and centrifuge it to remove any suspended matter. Decant the clear solution into a small beaker and boil off half the solvent, pour the hot solution on to a watch-glass and leave it to crystallize.

$$2Cr_2O_3 + 8KOH + 6KNO_3 \rightarrow 4K_2CrO_4 + 4H_2O + 6KNO_2$$

Experiment 90 Preparation of potassium dichromate

Materials: potassium chromate; 2 M sulphuric acid.

Dissolve 15 g of potassium chromate in 50 cm^3 of the sulphuric acid and evaporate to about half the volume.

$$2K_2CrO_4 + H_2SO_4 \rightarrow K_2SO_4 + K_2Cr_2O_7 + H_2O$$

On cooling, crystals of potassium dichromate separate out. Recrystallization from hot water yields a purer product. Yield: about 10 g.

MANGANESE

It will be seen that manganese appears in the same group of the Periodic Table as the halogens but it is not in the same sub-group. Manganese is typically metallic in character and the similarity between it and the halogens is almost entirely confined to the oxide, Mn_2O_7, which shows similarities to Cl_2O_7 in the corresponding perchlorates (chlorate(VII)) and permanganates (manganate(VII)).

We shall investigate the properties of manganese(II) salts, manganates and permanganates.

Manganese(II) salts (e.g. $MnCl_2$ manganese(II) chloride) correspond to MnO—manganese divalent.
Manganates (e.g. K_2MnO_4 potassium manganate) correspond to MnO_3—manganese hexavalent.
Permanganates (e.g. $KMnO_4$ potassium permanganate) correspond to Mn_2O_7—manganese heptavalent.

Manganese also forms some unstable manganese(III) salts, e.g. manganese(III) sulphate $Mn_2(SO_4)_3$, corresponding to Mn_2O_3 and shows a valency of four in MnO_2.

Experiment 91 Reactions of manganese(II) salts

Material: Use 2 cm^3 of a solution of manganese(II) chloride for each test. Note that hydrated manganese salts are pink when crystalline but the colour is not evident in solution.

(*a*) Add a few drops of yellow ammonium sulphide solution to a solution of manganese(II) chloride. A flesh-coloured precipitate of manganese(II) sulphide is obtained.

$$Mn^{2+} + S^{2-} \rightarrow MnS\downarrow$$

This same precipitate is obtained if hydrogen sulphide is bubbled into an *alkaline* solution of a manganese(II) salt but no precipitate is obtained with an acidic solution.

(*b*) Add sodium hydroxide solution a drop at a time (using a teat pipette) to a solution of manganese(II) chloride. A white precipitate (rapidly turning brown due to atmospheric oxidation) of manganese(II) hydroxide is obtained. Keep on adding the sodium hydroxide solution and note that the precipitate is *not* soluble in excess (cf. zinc hydroxide, Experiment 115, p. 103).

$$Mn^{2+} + 2OH^- \rightarrow Mn(OH)_2\downarrow$$

(*c*) Repeat (*b*) using ammonia solution with same observations.

(*d*) Repeat (*b*) after first adding one spatula-load of solid ammonium chloride to the manganese(II) chloride solution. No precipitate occurs. The ammonium ion introduced depresses the ionization of the ammonia (see p. 44).

(*e*) Add 1 cm^3 of sodium hydroxide solution followed by 2 cm^3 of bromine water (many oxidizing agents will do, e.g. sodium peroxide, hypochlorite or hypobromite) and warm. The divalent oxide or hydroxide is oxidized to the higher tetravalent oxide, manganese(IV) oxide, which comes down as a dark brown precipitate. If the manganese(II) salt is warmed with excess of the oxidizing agent or a powerful oxidizing agent is used, then the permanganate is obtained. Boil a little of the manganese(II) chloride solution with a spatula-load of lead(IV) oxide and 1 cm^3 of concentrated nitric acid. Dilute with water and filter—the solution comes through showing the pink permanganate colour.

Experiment 92 Preparation and reaction of manganates

Material: solid potassium hydroxide; potassium nitrate.

Heat on a crucible lid a piece of potassium hydroxide, a few crystals of potassium nitrate and a little manganese(IV) oxide until the whole mass has fused. Allow to cool, and add a little water and filter. A deep green solution of potassium manganate (manganate(V)) is obtained.

$$4KOH + 2MnO_2 + 2KNO_3 \rightarrow 2K_2MnO_4 + 2H_2O + 2KNO_2$$

The solution is unstable and is readily hydrolysed by dilute acids, and even by considerable dilution, into a permanganate (manganate(VII)).

$$3K_2MnO_4 + 2H_2O \rightarrow 2KMnO_4 + MnO_2 + 4KOH$$

Dilute the green solution about ten times with water and boil—the purple colour of the permanganate is observed on allowing the solution to settle.

Experiment 93 Preparation of potassium permanganate

Apparatus: clean sand-tray; 500 cm^3 flask; glass-wool; mortar and pestle.

Material: potassium hydroxide (powder); potassium chlorate (chlorate(V)); supply of carbon dioxide.

Take 10 g of potassium hydroxide on the tray and warm gently until it melts. Add 1 g of potassium chlorate, mixing with a glass rod. Add 7.5 g of manganese(IV) oxide very gradually while continuing to stir. When all has been added, heat to bright red heat for about fifteen minutes. Allow to cool, break off the melt and grind to a fine powder in a mortar. Transfer the powder to a 500 cm^3 flask, add 200 cm^3 of water and heat on a gauze. During the boiling pass carbon dioxide into the solution. At first the solution is green and contains potassium manganate:

$$6KOH + 3MnO_2 + KClO_3 \rightarrow 3K_2MnO_4 + KCl + 3H_2O$$

Later it becomes purple potassium permanganate:

$$3K_2MnO_4 + 2CO_2 \rightarrow 2KMnO_4 + MnO_2 + 2K_2CO_3$$

After boiling for ten minutes allow the solution to cool and decant it through glass-wool in a filter. Wash the flask and return the filtrate to it. Boil for a further ten minutes while carbon dioxide passes through, then withdraw a drop of solution and place it on a filter paper. If the drop has a green centre, continue the boiling until it is complete. Finally cool and filter. Transfer the solution to a large evaporating basin and evaporate until crystals appear, then set aside to crystallize.

Experiment 94 Reactions of permanganates

Material: materials mentioned in section (*c*) on p. 90.

Permanganates (manganate(VII)) are very powerful oxidizing agents, especially in acidic solution. When the permanganate ion oxidizes in acid solution, the negatively charged permanganate ion becomes the positively charged manganese(II) ion by accepting electrons, as is typical of oxidizing agents.

$$MnO_4^- + 8H^+ + 5e^- \rightarrow Mn^{2+} + 4H_2O$$

In neutral or alkaline solution the reaction is represented by

$$MnO_4^- + 2H_2O + 3e^- \rightarrow MnO_2 + 4OH^-$$

The quantitative oxidizing actions of potassium permanganate are given in Chapter 34.

(*a*) To a solution of potassium permanganate add 1 cm^3 of dilute sulphuric acid followed by 1 or 2 cm^3 of hydrogen peroxide. The oxygen evolved may be recognized by the rekindling of a glowing splint. If the permanganate has not been in excess the solution will be decolorized.

$$2MnO_4^- + 6H^+ + 5H_2O_2 \rightarrow 2Mn^{2+} + 8H_2O + 5O_2$$

In electronic terms:

$$MnO_4^- + 8H^+ + 5e^- \rightarrow Mn^{2+} + 4H_2O$$

(permanganate ion reduced)

$$H_2O_2 \rightarrow 2H^+ + O_2 + 2e^-$$

(hydrogen peroxide oxidized)

(*b*) Add 1 cm^3 of concentrated hydrochloric acid to a solution of potassium permanganate. There is an evolution of chlorine which can be recognized by its smell or by its bleaching action on damp litmus.

$$2MnO_4^- + 16H^+ + 10Cl^- \rightarrow 2Mn^{2+} + 8H_2O + 5Cl_2\uparrow$$

In electronic terms, the chloride ion is oxidized to chlorine.

$$2Cl^- \rightarrow Cl_2 + 2e^-$$

(*c*) Try the effect of potassium permanganate solution on acidic solutions of any of the following materials: iron(II) sulphate, tin(II) chloride, sulphurous acid or sodium sulphite, sodium nitrite, sodium thiosulphate, sodium phosphite (phosphate(III)) or hypophosphite (phosphinate). Oxidation occurs in all these cases as well as in many others not listed.

IRON, COBALT AND NICKEL

The elements in this group (Group VIII) occur in the middle of long periods. Thus iron, cobalt and nickel, the original transition elements, link manganese with copper; ruthenium, rhodium and palladium link technetium with silver; osmium, iridium and platinum link rhenium with gold. In this course only iron, cobalt and nickel are considered. The order of relative atomic masses does not show the gradation of properties of the three metals quite as well as the order in which they are considered below, namely, iron, cobalt and nickel, the order of their atomic numbers.

	Iron	*Cobalt*	*Nickel*
Appearance	←	All tough metals, grey colour	→
Relative atomic mass	55.5	58.93	58.7
Melting point (°C)	1535	1480	1450
Density ($kg\ m^{-3}$)	7900	8800	8900
Magnetic susceptibility	high	less	least
Valency	2 and 3	2 and 3	usually 2 but sometimes 3

table continued

	Iron	*Cobalt*	*Nickel*
Compounds			
Oxides	FeO. Basic. Easily oxidized to Fe_2O_3. Fe_2O_3. Basic.	CoO. Basic. Stable but oxidized when heated to Co_3O_4. Co_2O_3. Basic.	NiO. Basic. Stable but oxidized when heated to Ni_2O_3. Ni_2O_3. Not basic but forms salts corresponding to NiO.
	Fe_3O_4. Mixed base.	Co_3O_4. Mixed base.	
Sulphates MSO_4	← All isomorphous. With ammonium sulphate they form double salts. →		
Sulphates $M_2(SO_4)_3$	$Fe_2(SO_4)_3$.	None.	None.
Sulphides MS	← All black in colour and soluble in hot acids. →		
Chloride MCl_2	Hydrated salt oxidised and hydrolysed to iron(III) oxide on heating.	Hydrated salt pink and becomes blue (anhydrous) on heating.	Stable.
Hydroxide $M(OH)_2$	Green. Oxidized to brown $Fe(OH)_3$ in air.	Pink. Oxidized to brown $Co(OH)_3$ in air.	Stable.

Experiments with iron salts have probably been done in an earlier course, and the following brief instructions are included for revision purposes.

IRON

Experiment 95 Revisionary experiments with iron(II) and iron(III) salts

Make a stock solution of an iron(II) salt and one of an iron(III) salt. Take about 5 cm^3 of the required solution for each of the following tests:

(*a*) Add sodium hydroxide solution.
(*b*) Add ammonia solution.
(*c*) Add acidified potassium permanganate solution.
(*d*) Add potassium hexacyanoferrate(II) solution.
(*e*) Add potassium hexacyanoferrate(III) solution.
(*f*) Add potassium (or ammonium) thiocyanate solution. If the iron(II) sulphate is not fresh, a trace of iron(III) salt will give a result here; ammonium iron(II) sulphate should show a negative result.

The following table summarizes the results:

	Iron(II) salt	*Iron(III) salt*
(*a*) Sodium hydroxide solution	Green precipitate of iron(II) hydroxide, $Fe(OH)_2$	Red-brown precipitate of iron(III) hydroxide, $Fe(OH)_3$.
(*b*) Ammonia solution	As for sodium hydroxide solution.	
(*c*) Acidified potassium permanganate	Decolorizes the permanganate. Iron(II) salts are reducers (see p. 35).	Does not react.
(*d*) Potassium hexacyanoferrate(II) $K_4[Fe(CN)_6]$	Light blue precipitate of no special importance.	Deep blue precipitate 'Prussian blue'.
(*e*) Potassium hexacyanoferrate(III) $K_3[Fe(CN)_6]$	Deep blue precipitate. 'Turnbull's blue'.	Brown coloration of no special importance.
(*f*) Potassium thiocyanate KSCN	No action.	Blood-red coloration due to $[FeSCN]^{2+}$

Experiment 96 Conversion of iron(II) salt to iron(III) salt and vice versa

Oxidation of iron(II) salt. Take a test-tube filled to a depth of about 2 cm with a solution of an iron(II) salt (e.g. iron(II) sulphate). And excess (about an equal volume) of dilute sulphuric acid, and two or three drops of concentrated nitric acid, and boil. Cool, and add sodium hydroxide solution until a red-brown precipitate of iron(III) hydroxide is formed. This indicates that oxidation has taken place.

$$3Fe^{2+} + 4H^+ + NO_3^- \rightarrow 3Fe^{3+} + 2H_2O + NO$$

Other oxidizing materials such as chlorine, bromine, potassium permanganate or hydrogen peroxide may be substituted for nitric acid in the above experiment.

In all these cases, iron(II) ions are oxidized to iron(III) ions by electron loss.

$$Fe^{2+} \rightarrow Fe^{3+} + e^-$$

Reduction of iron(III) salt. Fill a test-tube to a depth of about 2 cm with an iron(III) salt solution (e.g. iron(III) chloride). Pass hydrogen sulphide into it until there is no further precipitate of sulphur, and on filtering a pale green solution is obtained.

Test the filtrate with potassium hexacyanoferrate(III) for proof of iron(II) salt.

$$2Fe^{3+} + H_2S \rightarrow 2Fe^{2+} + 2H^+ + S\downarrow$$

Iron(III) ions are reduced to iron(II) ions by electron gain.

$$Fe^{3+} + e^- \rightarrow Fe^{2+}$$

Sulphide ions are oxidized by electron loss.

$$H_2S \rightleftharpoons 2H^+ + S^{2-}; \quad S^{2-} \rightarrow S + 2e^-$$

Take a test-tube one-quarter full of iron(III) salt solution. Add an equal volume of concentrated hydrochloric acid and a few pieces of granulated zinc. Leave for half an hour, then filter. Test the filtrate with excess of sodium hydroxide solution to show that reduction to the iron(II) condition is complete. In the presence of acid, zinc atoms ionize and the electrons are accepted by iron(III) ions which are reduced to iron(II) ions.

$$Zn \rightarrow Zn^{2+} + 2e^-; \quad 2Fe^{3+} + 2e^- \rightarrow 2Fe^{2+}$$

Experiment 97 Preparation of iron(II) oxide, FeO

Material: iron(II) oxalate (ethanedioate); glass-wool.

Take a dry test-tube a quarter full of iron(II) oxalate, and close it with a plug of glass-wool. Heat, gently at first, and finally strongly enough to convert all the yellow oxalate to black iron(II) oxide. Remove the glass-wool and pour the oxide in small portions so that it sprinkles into an evaporating basin. Note the spontaneous ignition of the iron(II) oxide as it oxidizes to red iron(III) oxide.

$$FeC_2O_4 \rightarrow FeO + CO + CO_2$$

Dissolve any particles still in the test-tube in hydrochloric acid. Test the solution for iron(II) ions. Iron(II) oxide is a base, but, for convenience, iron(II) salts are prepared from metallic iron and acid.

Experiment 98 Preparation of iron(III) oxide, Fe_2O_3

Material: solution of any iron salt.

If the salt is an iron(II) salt, oxidize to the iron(III) condition as described in Experiment 96. Add excess ammonia solution to the iron(III) salt and filter off the iron(III) hydroxide. Heat the paper and contents in a crucible to leave red iron(III) oxide. Boil a little of the oxide in concentrated hydrochloric acid and show that it is a base.

Experiment 99 To show that triiron tetraoxide is a mixed base

Material: triiron tetraoxide (iron(II) iron(III) oxide).

Take enough of the oxide to cover the bottom of a test-tube and fill the tube a quarter full of concentrated hydrochloric acid. Warm gently and then filter. Divide the filtrate and test one part for iron(III) ions and the other for iron(II) ions. Both ions will be found to be present.

$$Fe_3O_4 + 8HCl \rightarrow 2FeCl_3 + FeCl_2 + 4H_2O$$

Experiment 100 Action of heat on hydrated iron chlorides

Material: solid hydrated iron(III) chloride.

Prepare a solution of iron(II) chloride by dissolving iron (filings) in a small quantity of concentrated hydrochloric acid. Evaporate in the test-tube until crystals appear. Heat strongly and test the vapour for hydrogen chloride (silver nitrate on a glass rod). Note the residue of iron(III) oxide, due to the iron(II) oxide, which first formed, being oxidized by air.

$$FeCl_2 + H_2O \rightarrow FeO + 2HCl$$
$$4FeO + O_2 \text{ (air)} \rightarrow 2Fe_2O_3$$

Heat a little iron(III) chloride in a test-tube. Test the gas for hydrogen chloride and note the residue of iron(III) oxide. Again hydrolysis has occurred.

$$2FeCl_3 + 3H_2O \rightarrow Fe_2O_3 + 6HCl$$

Experiment 101 Preparation of hydrated ammonium iron(II) sulphate, $Fe(NH_4)_2(SO_4)_2 \cdot 6H_2O$

Apparatus: measuring cylinder.

Material: ammonium sulphate.

Take 30 cm^3 of distilled water in a conical flask and add 4 cm^3 of concentrated sulphuric acid from the measuring cylinder. Add 5 g of iron gradually and then heat to boiling. Add 10 g of ammonium sulphate and evaporate to about two-thirds of its original volume. Cork loosely and set aside to crystallize. This salt is not an alum, but a double salt. An interesting fact is that it contains one-seventh of its mass of iron.

COBALT

Experiment 102 Reactions of cobalt(II) salts

Material: cobalt(II) chloride and carbonate; bleaching powder.

(*a*) Heat a little cobalt(II) carbonate in a small hard-glass tube. The brown residue is cobalt(II) oxide.

$$CoCO_3 \rightarrow CoO + CO_2$$

Transfer the oxide to a crucible and heat to redness. The black residue is tricobalt tetraoxide (cobalt(II) cobalt(III) oxide, Co_3O_4. Compare the stability of cobalt(II) oxide with iron(II) oxide and also the products of oxidation of the lower oxides.

(*b*) Take a test-tube one-quarter full of a solution of a cobalt(II) salt. Add excess of sodium hydroxide solution. The pink precipitate is cobalt(II) hydroxide.

$$Co^{2+} + 2OH^- \rightarrow Co(OH)_2\downarrow$$

Note the change to brown cobalt(III) oxide, Co_2O_3, on exposure to air and compare with iron(II) hydroxide in this respect. Show that cobalt(II) hydroxide is soluble in ammonia solution due to the formation of the complex ion $[Co(NH_3)_6]^{2+}$.

(*c*) Add a suspension of bleaching powder to a test-tube containing a solution of a cobalt(II) salt. The black precipitate is cobalt(III) hydroxide (the bleaching powder having acted as an alkaline hydroxide and an oxidizer). Divide the precipitate into two parts. Add excess hydrochloric acid to one part to obtain a brown solution of the unstable cobalt(III) chloride, and heat the second part to obtain oxygen and a residue containing cobalt(II) oxide. Hence cobalt(III) hydroxide behaves as a weak base, and differs from iron(III) hydroxide in acting as a hydrated peroxide.

(*d*) Evaporate some cobalt(II) chloride solution to dryness. Note the blue colour of the anhydrous salt. Cobalt(II) chloride is stable. (Compare with hydrated iron(II) chloride.) The action of heat may be shown in a picturesque manner by writing on a piece of paper with cobalt(II) chloride solution and allowing to dry. If the solution is fairly dilute, the writing becomes invisible. On being warmed, the writing appears bright blue due to dehydration of the salt.

Experiment 103 Preparation of hexaamminecobalt(III) chloride

Material: cobalt(II) chloride.

Add 8 g of ammonium chloride and 12 g of cobalt(II) chloride-6-water to 15 cm^3 of water and bring the solution to the boil. Then add 2 spatula-loads of animal charcoal and cool the flask under the cold water tap. Next add 25 cm^3 of concentrated ammonia solution and cool the flask still further (to below 10 °C). Then add five 5 cm^3 portions of 20-volume hydrogen peroxide with constant shaking of the flask. When all the peroxide has been added, heat the mixture to 60 °C and maintain it at that temperature, but no more, for about thirty minutes. On cooling the flask in ice and water, crystals will be deposited of the crude product.

$$2Co^{2+} + 2NH_4^+ + 10NH_3 + H_2O_2 \rightarrow 2Co(NH_3)_6^{3+} + 2H_2O$$

After filtering off the crude solid, put it straight away in a mixture of 100 cm^3 of water, acidified with 3 cm^3 of concentrated hydrochloric acid, which is gently boiling. The cobalt salt dissolves, leaving the charcoal, so the solution is filtered while it is still hot. To the filtrate add 10 cm^3 of concentrated hydrochloric acid and then cool it in ice and water. Golden-brown crystals are produced which are remarkably stable to acids.

NICKEL

Experiment 104 Reactions of nickel(II) salts

Material: nickel(II) sulphate and carbonate.

(*a*) Heat some nickel(II) carbonate in a hard-glass test-tube. The greenish-brown residue is nickel(II) oxide, which, on transference to a crucible and further heating, goes to the black nickel(III) oxide, Ni_2O_3. This is a close resemblance to iron(II) oxide although the action is less rapid. Show that nickel(III) oxide dissolves in dilute sulphuric acid to give green nickel(II) salt.

$$2Ni_2O_3 + 4H_2SO_4 \rightarrow 4NiSO_4 + 4H_2O + O_2$$

(*b*) Add sodium hydroxide solution to a solution of nickel(II) sulphate. The light green precipitate is nickel(II) hydroxide.

$$Ni^{2+} + 2OH^- \rightarrow Ni(OH)_2\downarrow$$

It is stable in air. (Contrast with iron(II) and cobalt(II) hydroxides.) Show that the hydroxide is soluble in ammonia solution to give a blue solution (due to the complex ion $[Ni(NH_3)_6]^{2+}$) and compare this action with that of cobalt(II) hydroxide.

(*c*) Heat a solution of nickel(II) chloride, made by dissolving the carbonate in hydrochloric acid. Note that the chloride crystals are stable when heated. Neither nickel(II) chloride nor cobalt(II) chloride resembles iron(II) chloride in this respect.

Experiment 105 To prepare sulphides of iron, cobalt and nickel

Material: nickel(II) sulphate; aqua regia (3 volumes of concentrated hydrochloric acid and 1 volume of concentrated nitric acid).

Make solutions of iron(II), cobalt(II) and nickel(II) salts. Pass hydrogen sulphide into each. A slight precipitate of iron(II) sulphide may be seen as a dark coloration, but neither of the other salts shows a precipitate.

Now add a little ammonia solution, and pass more hydrogen sulphide. All three give a black precipitate of the metallic sulphide.

Show that iron(II) sulphide alone dissolves in dilute hydrochloric acid. The sulphides of cobalt and nickel will dissolve in concentrated hydrochloric acid in the presence of potassium chlorate or in aqua regia. Transfer the sulphides to evaporating basins, add concentrated hydrochloric acid and a crystal of potassium chlorate. Heat until dissolved. The cobalt salt is pink in solution, the nickel salt greenish-yellow.

COPPER AND SILVER

	Copper(II)	*Copper(I)*	*Silver*
Metal	← All metallic. →		
	← All form stable salts. →		
Oxide	Basic.	Basic.	Unstable. Soluble in ammonia solution.
Hydroxide	Basic.	Not formed.	Not formed.
Chloride	Soluble in water. With ammonia solution gives deep blue solution.	Insoluble in water. Soluble in concentrated hydrochloric acid. Soluble in concentrated ammonia solution.	Insoluble in water. Soluble in concentrated hydrochloric acid. Soluble in concentrated ammonia solution.
Iodide	Unstable; forms copper(I) iodide and iodine.	Stable. Insoluble in water.	Stable. Insoluble in water.
Flame coloration of compounds	Green with blue zone.	—	—

COPPER

Experiment 106 Reactions of solid copper compounds

Material: copper(II) oxide.

(*a*) Mix intimately a little copper(II) oxide with anhydrous sodium carbonate, and heat this mixture on a charcoal block in the reducing flame of the blowpipe. Brown scales of copper are formed.

(*b*) Moisten a little copper(II) oxide with concentrated hydrochloric acid on a watch glass, dip a nichrome wire in the mixture and perform the flame test. A green flame with a blue zone is obtained.

(*c*) Obtain a borax bead on a platinum wire, dust on a little copper(II) oxide and reheat the bead in the oxidizing flame. A blue bead is seen (green when hot).

Experiment 107 Reactions of copper(II) salts

(*a*) Pass hydrogen sulphide into copper(II) sulphate solution and obtain a dark brown precipitate of copper(II) sulphide.

$$Cu^{2+} + S^{2-} \rightarrow CuS\downarrow$$

Wash the precipitate by decantation, add excess of dilute nitric acid and boil in a dish. The copper(II) sulphide readily dissolves.

$$CuS + 2H^{+} \rightarrow Cu^{2+} + H_2S\uparrow$$

(*b*) Take some copper(II) sulphate solution in a test-tube and add a solution of potassium iodide. A white precipitate of copper(I) iodide is thrown down and iodine is liberated.

$$2Cu^{2+} + 4I^{-} \rightarrow 2CuI\downarrow + I_2\downarrow$$

Add sodium thiosulphate solution to dissolve the iodine, when the copper(I) iodide is observed as a white precipitate. See Experiment 317, p. 343, for use of this reaction in the estimation of copper.

(*c*) Add sodium hydroxide solution to a solution of copper(II) sulphate in a boiling-tube. Note the blue gelatinous precipitate of copper(II) hydroxide.

$$Cu^{2+} + 2OH^{-} \rightarrow Cu(OH)_2\downarrow$$

Pour some of the suspension into a test-tube. Boil the portion remaining in the boiling-tube. The black precipitate is copper(II) oxide.

$$Cu(OH)_2 \rightarrow CuO\downarrow + H_2O$$

Add ammonia solution to the portion in the test-tube and show that the blue precipitate dissolves to form a deep blue solution. This solution contains the tetraamminecopper(II) ion $[Cu(NH_3)_4]^{2+}$.

Note the similarity of the action of ammonia on silver, copper(I) and copper(II) compounds.

(*a*) Add a solution of potassium hexacyanoferrate(II) to a solution of copper sulphate to obtain a brown precipitate of copper(II) hexacyanoferrate(II).

$$2Cu^{2+} + [Fe(CN)_6]^{4-} \rightarrow Cu_2Fe(CN)_6\downarrow$$

Experiment 108 Preparation of hydrated ammonium copper(II) sulphate $(NH_4)_2Cu(SO_4)_2{\cdot}6H_2O$

Material: ammonium sulphate.

Weigh 5 g of copper(II) sulphate and dissolve it in 50 cm^3 of boiling water. Weigh 2.6 g of ammonium sulphate and dissolve it in 10 cm^3 of water. Mix the solutions and evaporate them until crystallization begins, then set them aside to cool. This is a 'double salt'.

Experiment 109 Preparation of hydrated tetraamminecopper(II) sulphate, $[Cu(NH_3)_4]\,SO_4{\cdot}H_2O$

Apparatus: corked flask.
Material: ethanol.

Dissolve 10 g of copper(II) sulphate by boiling it in 50 cm^3 of water in a 200 cm^3 flask, and cool the solution to room temperature. Carefully add concentrated ammonia solution until the precipitate which forms redissolves. Cool to room temperature again. Add 200 cm^3 of ethanol from a pipette so that it forms a layer on top of the blue solution. Cork the flask loosely and leave it undisturbed for a week. Filter off the crystals and transfer them at once to a stoppered bottle. This compound is fundamentally different from the double salt, ammonium copper(II) sulphate, which behaves in solution as would its constituent sulphates. Tetraamminecopper(II) sulphate is a 'complex salt' in which the copper ion and ammonia form a single divalent ion: $[Cu(NH_3)_4]^{2+}$

Experiment 110 Preparation of copper(I) oxide

Material: potassium sodium tartrate (2,3-dihydroxybutanedioate, Rochelle salt); glucose.

Fill a boiling-tube to a depth of about 3 cm with copper(II) sulphate solution and add a spatula-load of the tartrate. When the salt has dissolved, add sodium hydroxide solution followed by a spatula-load of glucose, and boil. An orange-red precipitate of copper(I) oxide is formed by the reducing action of glucose on the copper(II) in solution. See also Experiment 226(*b*).

$$2Cu(OH)_2 + C_6H_{12}O_6 \rightarrow Cu_2O\downarrow + 2H_2O + \underset{\text{Gluconic acid}}{C_6H_{12}O}$$

Show that the precipitate (which readily settles) is soluble in concentrated hydrochloric acid, but gives free copper and the copper(II) salt with dilute sulphuric or dilute nitric acid. (In the latter case the copper is acted upon by any excess nitric acid.)

$$Cu_2O + 2HCl \rightarrow 2CuCl + H_2O$$
$$Cu_2O + H_2SO_4 \rightarrow Cu\downarrow + CuSO_4 + H_2O$$

Experiment 111 Preparation of copper (I) chloride

Apparatus: Buchner funnel.
Material: glass-wool.

Take sufficient copper(II) oxide to cover the bottom of a test-tube and add five times that bulk of concentrated hydrochloric acid. Warm to obtain a clear green solution of copper(II) chloride.

$$CuO + 2HCl \rightarrow CuCl_2 + H_2O$$

Add copper filings of about equal bulk to the copper oxide used, and boil for about 2 minutes. Filter the mixture through glass-wool into a beaker full of water. The white precipitate is copper(I) chloride.

$$Cu + CuCl_2 \rightarrow 2CuCl$$

Decant the supernatant liquid and show that the copper(I) chloride is soluble:

(*a*) in concentrated ammonia solution, due to the formation of a complex ion

$$CuCl + 2NH_3 \rightarrow [Cu(NH_3)_2]^+ + Cl^-$$

(*b*) in concentrated hydrochloric acid, due to the formation of another complex ion, this time with the chloride ion.

$$CuCl + 2Cl^- \rightleftharpoons [CuCl_3]^{2-}$$

This ion is unstable and decomposes on dilution with water. If it is desired to prepare a pure dry specimen of the copper(I) chloride, filter off the white solid using a filter pump and Buchner funnel, wash the solid with sulphurous (sulphuric(IV)) acid followed by glacial acetic (ethanoic) acid, and finally dry quickly by heating the solid on a water-bath. The product is kept in a bottle out of contact with the air.

Experiment 112 Reactions of copper (I) compounds

Material: copper(I) oxide; copper(I) chloride in concentrated hydrochloric acid.

(*a*) Add a few drops of potassium iodide solution to the solution of copper(I) chloride. The white precipitate is copper(I) iodide.

$$CuCl + I^- \rightarrow CuI\downarrow + Cl^-$$

(*b*) Pour a little of the solution of copper(I) chloride in concentrated hydrochloric acid into water. The white precipitate is copper(I) chloride which, although soluble in a high concentration of chloride ions, is insoluble in water.

$$CuCl + 2Cl^- \rightleftharpoons [CuCl_3]^{2-}$$

(*c*) To a spatula-load of copper(I) oxide in a test-tube, add some dilute hydrochloric acid and warm. A white precipitate of copper(I) chloride is seen.

$$Cu_2O + 2HCl \rightarrow 2CuCl + H_2O$$

(*d*) To a spatula-load of copper(I) oxide in a test-tube, add some dilute sulphuric acid. The solution turns blue, and red metallic copper is thrown down.

$$Cu_2O + H_2SO_4 \rightarrow CuSO_4 + Cu\downarrow + H_2O$$

(*e*) Repeat (*d*), using dilute nitric acid. A similar reaction takes place, but the liberated copper reacts with any excess dilute nitric acid. The products are, therefore, a green or blue solution of copper(II) nitrate, nitrogen oxide (turning brown on exposure to air) and water.

$$Cu_2O + 2HNO_3 \rightarrow Cu(NO_3)_2 + H_2O + Cu\downarrow$$
$$3Cu + 8HNO_3 \rightarrow 3Cu(NO_3)_2 + 2NO + 4H_2O$$

SILVER

Experiment 113 Reactions of silver compounds

Material: silver nitrate crystals.

(*a*) Grind a little solid silver nitrate with twice its bulk of anhydrous sodium carbonate in a mortar and heat on a charcoal block in the reducing flame of the blowpipe. The white bead, which will not mark paper (cf. lead), but which will dissolve in dilute nitric acid (cf. tin), is metallic silver.

(*b*) Fill a test-tube half full of silver nitrate solution and add three or four drops of concentrated hydrochloric acid. The white precipitate is silver chloride. Shake the mixture to coagulate the silver chloride, then decant it, wash it with water and allow it to settle.

$$Ag^+ + Cl^- \rightarrow AgCl\downarrow$$

Pour off the water and divide the solid into three parts. Leave one part exposed to light and it will turn violet. To another, add concentrated ammonia solution, when the solid will readily dissolve. Warm the third portion with concentrated hydrochloric acid, which will dissolve the silver chloride. The explanations of the solubility of silver chloride in ammonia and concentrated hydrochloric acid are similar to the explanations for the solubility of copper(I) chloride in the same reagents (see Experiment 112). For the reaction of silver halides with sodium thiosulphate see Experiment 164.

(*c*) Add a few drops of potassium chromate solution to a solution of silver nitrate. A brick-red precipitate of silver chromate, soluble in both dilute nitric acid and ammonia solution, is obtained.

$$2Ag^+ + CrO_4^{2-} \rightarrow Ag_2CrO_4\downarrow$$

(*d*) Add disodium hydrogenphosphate solution to silver nitrate solution to obtain a yellow precipitate of silver phosphate.

$$3Ag^+ + PO_4^{3-} \rightarrow Ag_3PO_4\downarrow$$

(*e*) Add 0.5 M ammonia gradually to silver nitrate solution in a test-tube. A brown precipitate of silver oxide is first formed which dissolves in excess of ammonia to form a complex ion $[Ag(NH_3)_2]^+$.

$$2Ag^+ + 2OH^- \rightarrow Ag_2O\downarrow + H_2O$$

Similarly, silver oxide is precipitated by sodium hydroxide solution but is not soluble in excess of the reagent.

Experiment 114 Recovery of silver from silver chloride

Material: glucose; laboratory residues of silver chloride.

Method (*a*) Wash the residues with water several times by decantation, dry them and mix well with twice the bulk of anhydrous sodium carbonate. Transfer the mixture to a crucible and heat strongly in a furnace. On cooling, a button of silver is left in the bottom of the crucible.

Method (*b*) Transfer the residues after washing by decantation to a dish. Add sodium hydroxide solution and glucose, and warm the mixture. When a portion of the solid dissolves completely in dilute nitric acid, decant the liquid from the grey silver, which remains in a finely divided condition. After washing, it can easily be taken up in dilute nitric acid.

ZINC, CADMIUM AND MERCURY

	Zinc	*Cadmium*	*Mercury(II)*	*Mercury(I)*
Metal	Metallic.	Metallic.	Metallic.	Metallic.
	← All form stable salts →			
Oxide	Basic.	Basic.	Basic.	Basic.
Hydroxide	Amphoteric	Basic.	Not formed.	Not formed.
Chloride	Soluble in water. With ammonia solution gives white precipitate then colourless solution.	Soluble in water. With ammonia solution gives white precipitate then colourless solution.	Soluble in water. With ammonia solution gives a yellow precipitate	Insoluble in water. With ammonia solution gives a black precipitate

Table continued

	Zinc	*Cadmium*	*Mercury(II)*	*Mercury(I)*
Iodide	Soluble in water. Colourless.	Soluble in water. Colourless.	Insoluble. Red.	Insoluble. Green.
Flame coloration of compounds	—	—	—	—

The properties of mercury do not show any marked agreement with those of other members of the main group or sub-group. It forms two series of salts: mercury(I) and mercury(II).

Mercury is very low in the electrochemical series and is therefore precipitated from solution by nearly all metals. Its hydroxide, carbonate and oxide are unstable and if heat is applied, mercury is obtained.

Mercury(I) salts give, in solution with water, the unusual ion Hg_2^{2+}; mercury(II) salts, on the other hand, are ionized only to a very slight extent.

ZINC

Experiment 115 Reactions of zinc and its compounds

Material: zinc sulphate; zinc foil.

(*a*) Hold a piece of zinc foil by means of crucible tongs in the Bunsen flame. The powder, yellow when hot and white when cold, is zinc oxide.

(*b*) Add sodium carbonate solution to a solution of zinc sulphate. A white precipitate of basic zinc carbonate is thrown down: $ZnCO_3 \cdot 2Zn(OH)_2 \cdot H_2O$. (Sodium hydrogencarbonate gives the normal carbonate.)

(*c*) Add sodium hydroxide solution a drop at a time to a solution of zinc sulphate in water. A white precipitate of zinc hydroxide is formed which dissolves in excess of the alkali to form sodium zincate. Zinc hydroxide is amphoteric.

$$Zn^{2+} + 2OH^- \rightarrow Zn(OH)_2\downarrow$$
$$Zn(OH)_2 + 2OH^- \rightarrow [Zn(OH)_4]^{2-}$$

If hydrogen sulphide is bubbled through the solution of sodium zincate, zinc sulphide is precipitated as a white solid.

(*d*) Add a few drops of ammonium sulphide solution to a solution of zinc sulphate. A white precipitate of zinc sulphide is obtained. (This precipitate is often discoloured.)

$$Zn^{2+} + S^{2-} \rightarrow ZnS\downarrow$$

(*e*) Dip a rolled filter paper into a *concentrated* solution of zinc sulphate to which some cobalt(II) nitrate solution has been added. Burn the filter paper on gauze on a tripod. A green ash (Rinmann's green) remains.

(*f*) Add ammonia solution, drop by drop, to a solution of zinc sulphate. The precipitate of zinc hydroxide dissolves in excess, due to the formation of a complex ion, $[Zn(NH_3)_4]^{2+}$.

CADMIUM

Experiment 116 Reactions of cadmium salts

Material: cadmium chloride or sulphate.

(*a*) Pass hydrogen sulphide into a solution of cadmium sulphate. The bright yellow precipitate is cadmium sulphide.

$$Cd^{2+} + S^{2-} \rightarrow CdS\downarrow$$

(*b*) Add to a test-tube one-quarter full of cadmium sulphate solution an equal bulk of concentrated hydrochloric acid, and pass hydrogen sulphide. No precipitate appears in acid of this concentration (approximately 6 M). Dilute the solution until the yellow precipitate appears. Cadmium sulphide is sometimes incompletely precipitated in Group II of the analysis tables if the solution is too acidic. Filter off some of the yellow cadmium sulphide and show that it is soluble in dilute nitric acid.

$$CdS + 2H^{+} \rightarrow Cd^{2+} + H_2S\uparrow$$

(*c*) Add sodium hydroxide solution to a solution of cadmium sulphate. Cadmium hydroxide (insoluble in excess) appears as a white precipitate. (cf. Experiment 115 (c)).

$$Cd^{2+} + 2OH^{-} \rightarrow Cd(OH)_2\downarrow$$

(*d*) Add ammonia solution, drop by drop, to a solution of cadmium sulphate. The white cadmium hydroxide is precipitated but dissolves in excess (cf. Experiment 115(*f*)).

MERCURY

Experiment 117 General reaction of mercury compounds

Material: mercury(II) chloride.

Grind a spatula-load of mercury(II) chloride in a mortar with two or three times its bulk of anhydrous sodium carbonate. Introduce the mixture into an ignition tube and heat gently. A grey deposit of mercury is obtained which, if scraped together, will give globules of mercury.

$$HgCl_2 + Na_2CO_3 \rightarrow HgCO_3 + 2NaCl$$

$$2HgCO_3 \rightarrow 2Hg + 2CO_2 + O_2$$

Experiment 118 Reactions of mercury(I) compounds

Material: mercury(I) nitrate solution (acidified with dilute nitric acid).

(*a*) Add a few drops of dilute hydrochloric acid to a little of the mercury(I) nitrate solution. A white precipitate of mercury(I) chloride (calomel) is seen.

$$Hg_2^{2+} + 2Cl^- \rightarrow Hg_2Cl_2\downarrow$$

Filter off the precipitate and pour ammonia solution on to the filter paper. A black precipitate of a mixture of mercury and mercury(II) aminochloride is obtained.

$$Hg_2Cl_2 + 2NH_3 \rightarrow NH_2HgCl + Hg + NH_4Cl$$

(*b*) Add sodium hydroxide solution (excess) to a little of the mercury(I) nitrate solution. A black precipitate of mercury(I) oxide mixed with mercury is produced.

$$Hg_2^{2+} + 2OH^- \rightarrow Hg_2O\downarrow + H_2O$$

(*c*) Add a few drops of potassium iodide solution to a little of the mercury(I) nitrate solution. A yellowish-green precipitate of mercury(I) iodide is obtained. The mercury(I) iodide (yellow) decomposes readily into mercury(II) iodide and mercury. This mixture is green.

$$Hg_2^{2+} + 2I^- \rightarrow Hg_2I_2\downarrow$$
$$Hg_2I_2 \rightarrow Hg\downarrow + HgI_2\downarrow$$

(*d*) Shake two or three crystals of mercury(I) nitrate with 3 cm^3 of water and then boil it. In common with many oxysalts of mercury, hydrolysis occurs and the white precipitate is the basic nitrate.

(*e*) Add a solution of tin(II) chloride to a little of the mercury(I) nitrate. A grey precipitate of mercury is obtained by reduction and the tin(II) salt is oxidized to a tin(IV) salt (see Experiment 119 (c)).

Experiment 119 Reactions of mercury(II) compounds

(*a*) Add sodium hydroxide solution to a little of the mercury(II) chloride solution. A yellow precipitate of mercury(II) oxide is obtained.

$$Hg^{2+} + 2OH^- \rightarrow HgO\downarrow + H_2O$$

(*b*) Add potassium iodide solution drop by drop to a little of the mercury(II) chloride solution. A precipitate (yellow then red) of mercury(II) iodide is obtained, soluble in excess. The latter solution when made alkaline with sodium hydroxide solution is Nessler's solution (see p. 107).

$$Hg^{2+} + 2I^- \rightarrow HgI_2\downarrow$$
$$HgI_2 + 2I^- \rightarrow HgI_4^{2-}$$

(*c*) Pour a little tin(II) chloride solution into a solution of mercury(II) chloride. The latter is reduced to mercury(I) chloride which comes down as a characteristic 'silky' white precipitate. If excess tin(II) chloride is added, the precipitate turns grey due to a further reduction to metallic mercury.

$$2Hg^{2+} + Sn^{2+} + 2Cl^- \rightarrow Hg_2Cl_2\downarrow + Sn^{4+}$$
$$Hg_2Cl_2 + Sn^{2+} \rightarrow 2Hg\downarrow + Sn^{4+} + 2Cl^-$$

(*d*) Bubble hydrogen sulphide slowly into a solution of mercury(II) chloride. Finally a black precipitate of mercury(II) sulphide is obtained. (The colour sequence may be white—yellow—brown—black. The initial precipitate is $HgCl_2{\cdot}2HgS$).

$$Hg^{2+} + S^{2-} \rightarrow HgS\downarrow$$

(*e*) Add a little dilute hydrochloric acid to a solution of mercury(II) chloride in a test-tube. Drop in a piece of bright copper foil and warm the tube. The foil is soon coated with a bright deposit of mercury.

$$Hg^{2+} + Cu \rightarrow Hg\downarrow + Cu^{2+}$$

Experiment 120 To illustrate the relation between mercury(I) and mercury(II) compounds

Apparatus: pestle and mortar.

Material: mercury; iodine; ethanol (methylated spirits).

Generally speaking, mercury(I) compounds are produced with mercury in excess, and mercury(II) compounds are produced if the non-metal is in excess.

Grind a little mercury (excess) with a small crystal of iodine in a mortar to which has been added a drop or two of ethanol. The yellowish-green mercury(I) iodide is formed. Add more iodine and grind again, when finally the red mercury(II) iodide is produced.

$$2Hg + I_2 \rightarrow Hg_2I_2 \quad \text{mercury in excess}$$
$$2Hg + 2I_2 \rightarrow 2HgI_2 \quad \text{iodine in excess}$$

Experiment 121 To show the relation between the two forms of mercury(II) iodide

Material: mercury(II) iodide.

Heat a few of the red crystals gently in an ignition tube; yellow crystals form as a sublimate in the cooler parts of the tube. Set the tube aside for some time—the red variety is gradually formed.

$$\underset{\substack{\text{Stable below} \\ 126\,^\circ\text{C}}}{\text{Red}} \rightleftharpoons \underset{\substack{\text{Stable above} \\ 126\,^\circ\text{C}}}{\text{Yellow}}$$

Alternatively put two spatula-loads of the red variety into a dish, cover with a watch-glass and heat gently until all has sublimed. Allow to cool. Remove the watch-glass and notice the yellow crystals adhering to it. On touching any of them with a glass rod, they immediately revert to the red variety.

Experiment 122 Preparation of mercury(II) iodide and Nessler's solution

Apparatus: measuring cylinder.

Material: mercury(II) chloride solution (5% m/V); potassium iodide solution (10% m/V).

Take 20 cm^3 of mercury(II) chloride solution in a measuring cylinder. Pour 10 cm^3 of this into a beaker and add potassium iodide solution until the red precipitate which first forms just redissolves, then add the second 10 cm^3 of mercury(II) chloride. Filter off the mercury(II) iodide.

Take a little of the mercury(II) iodide in a test-tube and add potassium iodide solution until the solution is clear. For equations, see Experiment 119(*b*).

Make alkaline with sodium hydroxide solution; this is *Nessler's solution.* Make a very dilute solution of an ammonium salt and add a drop of Nessler's solution to obtain a brown precipitate. Repeat with a still more dilute solution of ammonium salt and demonstrate the sensitivity of the test.

14
Group IIIB of the Periodic Table

B —	Al —	*Ga* —	*In* —	*Tl*
Boron	Aluminium	*Gallium*	*Indium*	*Thallium*

Elements in italics are not usually studied in an elementary course.

We see that this group is somewhat unsatisfactory from the point of view of comparison with other elements in the group. It will be remembered that the properties of gallium were predicted by Mendeléeff before the element was discovered.

ALUMINIUM

Experiment 123 The reactions of aluminium

Material: aluminium foil.

(*a*) Show that aluminium readily dissolves in warm dilute hydrochloric acid with the production of hydrogen.

$$2Al + 6H^+ \rightarrow 2Al^{3+} + 3H_2\uparrow$$

Hot concentrated sulphuric acid will attack it with the evolution of sulphur dioxide whilst nitric acid, dilute or concentrated, acts only very slowly on the metal.

(*b*) Add a little sodium hydroxide solution to aluminium powder in a test-tube. The reaction is extremely rapid, hydrogen being evolved.

$$2Al + 6OH^- + 6H_2O \rightarrow \underset{\text{aluminate ion}}{2[Al(OH)_6]^{3-}} + 3H_2\uparrow$$

(*c*) For the preparation of anhydrous aluminium chloride, see Experiment 167; hydrated aluminium chloride behaves similarly to hydrated iron chlorides (see Experiment 100).

Experiment 124 The reactions of aluminium salts

Material: aluminium sulphate solution; aluminium powder; flowers of sulphur.

(*a*) Add ammonia solution to a solution of aluminium sulphate. A white precipitate of aluminium hydroxide is thrown down.

$$Al^{3+} + 3OH^- \rightarrow Al(OH)_3\downarrow$$

Show that the precipitate is insoluble in excess ammonia.

(*b*) Repeat (*a*) using a solution of sodium hydroxide and adding the alkali drop by drop. The white precipitate is obtained but is readily dissolved in excess forming sodium aluminate. Aluminium hydroxide is amphoteric.

$$Al(OH)_3 + 3OH^- \rightarrow [Al(OH)_6]^{3-}$$

(*c*) Add blue litmus solution to a solution of aluminium sulphate in water. The litmus turns red. Add sodium carbonate solution and there is a rapid evolution of carbon dioxide. Aluminium salts in solution frequently act as acids due to hydrolysis:

$$Al^{3+} + 3H_2O \rightarrow Al(OH)_3 + 3H^+$$

(*d*) Make a mixture of dry aluminium powder with twice its bulk of dry flowers of sulphur and introduce into a test-tube enough of the mixture to cover the bottom. (*Care! Larger quantities may explode.*) Clamp the tube vertically, place a Bunsen burner under it and stand aside while the tube is heated. There is a vigorous action in which aluminium sulphide is synthesized. Add a few drops of water when the contents of the tube have cooled. Hydrogen sulphide is at once evolved due to the hydrolysis of the sulphide.

$$2Al + 3S \rightarrow Al_2S_3$$
$$Al_2S_3 + 6H_2O \rightarrow 2Al(OH)_3\downarrow + 3H_2S\uparrow$$

Similarly, the effect of bubbling hydrogen sulphide through a solution of aluminium sulphate is to produce the hydroxide, not the sulphide.

(*e*) Mix a little aluminium sulphate with twice its bulk of anhydrous sodium carbonate and heat on the charcoal block. A white infusible mass is left, which, if moistened with cobalt(II) nitrate solution and again heated, forms a bright blue solid.

Experiment 125 Oxidation of aluminium when in the form of an amalgam

Material: aluminium foil.

Aluminium is high in the electro-chemical series and therefore should be extremely reactive. Many of its actions, e.g. effect on exposure to air, suggest that there is a protective layer of oxide on its surface. The following experiment shows that aluminium is readily oxidized. Dip a piece of aluminium foil in a solution of mercury(II) chloride and rub the foil with a finger three or four times, then wash your hands thoroughly. Take the foil and expose it to the air. In a few moments it will be too hot to hold, and a feathery growth of aluminium oxide will be observed.

$$4Al + 3O_2 \rightarrow 2Al_2O_3$$

ALUMS

Aluminium sulphate has the property of forming a double salt with the sulphate of a monovalent metal. Thus with potassium sulphate it forms aluminium potassium sulphate, $KAl(SO_4)_2 \cdot 12H_2O$ (or $K_2SO_4 \cdot Al_2(SO_4)_3 \cdot 24H_2O$, known as potash alum or just as alum), and, where sodium or ammonium is denoted by X, similar compounds of formula $XAl(SO_4)_2 \cdot 12H_2O$. All these alums crystallize as octahedra. The general method of preparation is to mix equimolar quantities of the constituent sulphates in solution and allow them to crystallize. An extension of the term 'alum' includes similar compounds in which the sulphate of other trivalent metals have been substituted for aluminium sulphate, and, if Y denotes one of these metals, the general formula of an alum becomes $XY(SO_4)_2 \cdot 12H_2O$. For the above reasons, preparations of various alums are given at this point, and the important consequence of isomorphism —the formation of overgrowths—is appropriately included here.

Experiment 126 Preparation of hydrated aluminium potassium sulphate, $KAl(SO_4)_2 \cdot 12H_2O$, from its constituent salts

Material: potassium sulphate; aluminium sulphate crystals.

The relative molecular mass of potassium sulphate is 174: and of hydrated aluminium sulphate, $Al_2(SO_4)_3 \cdot 18H_2O$, it is 666. Equimolar masses of the sulphates will be as 174:666, say 1.45 g to 5.6 g. Weigh 1.45 g of potassium sulphate and dissolve it in 15 cm^3 of distilled water. Weigh 5.6 g of aluminium sulphate crystals and dissolve 20 cm^3 of water, warming to dissolve. Mix the solutions and leave overnight. From the crystals choose a well-formed specimen and allow it to grow in the solution. Dissolve the rest of the crystals in the minimum of water and recrystallize.

Experiment 127 Preparation of hydrated aluminium potassium sulphate from aluminium foil

Materials: aluminium foil; solid potassium hydroxide.

Dissolve approximately 1 g of potassium hydroxide pellets in about 40 cm^3 of water in a beaker and add upwards of 0.5 g of aluminium foil. When action ceases, the metal must be left in excess.

$$2Al + 6OH^- + 6H_2O \rightarrow 2[Al(OH)_6]^{3-} + 3H_2\uparrow$$

Decant off the liquid and add dilute sulphuric acid, with stirring, until a piece of litmus paper shows the solution to be slightly acidic. Heat the solution, filter it if necessary, evaporate it to small bulk and allow to cool. Filter off the crystals, wash and dry them.

$$[Al(OH)_6]^{3-} + 6H^+ \rightarrow Al^{3+} + 6H_2O$$

Experiment 128 Preparation of hydrated ammonium iron(III) sulphate, $NH_4Fe(SO_4)_2 \cdot 12H_2O$

Material: ammonium sulphate.

Find the relative molecular mass of this alum by substitution of relative atomic masses, and the relative masses of (*a*) ammonium sulphate (which is anhydrous) and (*b*) hydrated iron(II) sulphate ($FeSO_4 \cdot 7H_2O$) needed to make the corresponding mass of iron(III) sulphate. Then calculate the masses of these needed to make 20 g of the alum.

Take 11.5 g of iron(II) sulphate, dissolve in 30 cm^3 of dilute sulphuric acid, add 5 cm^3 of concentrated nitric acid and evaporate the solution to about half of its original volume. Meanwhile dissolve 2.7 g of ammonium sulphate in 10 cm^3 of water. Mix the solutions and set them aside to crystallize. Choose a good crystal and allow it to grow in the solution. Iron(III) alum crystals have an amethyst colour but deteriorate on standing in air for a long period due to the formation of basic iron(III) sulphate.

Experiment 129 Preparation of hydrated chromium(III) potassium sulphate, $KCr(SO_4)_2 \cdot 12H_2O$

Apparatus: thermometer.

Material: potassium dichromate; ethanol.

Instead of weighing equimolar masses of potassium sulphate and chromium sulphate, advantage is taken of the reducing action of ethanol on potassium dichromate in acid solution.

$$K_2Cr_2O_7 + 4H_2SO_4 + 3C_2H_5OH \rightarrow K_2SO_4 + Cr_2(SO_4)_3 + 7H_2O + \underset{\text{(acetaldehyde)}}{3CH_3CHO}$$

Heat 7.5 g of potassium dichromate in 50 cm^3 of water and cool the solution to room temperature. Add 6 cm^3 of concentrated sulphuric acid, stir with a thermometer and cool to 35 °C. Add, drop by drop,

5 cm^3 of alcohol, keeping the temperature below 50 °C, and when all has been added, set the solution aside to cool overnight. Choose a good crystal to grow, and place another crystal in the solution of alum from Experiment 126 to form an overgrowth.

Alternatively, sulphur dioxide may be used to reduce the dichromate.

15
Group IVB of the Periodic Table

C	Si	*Ge*	Sn	Pb
Carbon	Silicon	*Germanium*	Tin	Lead

The element in italics is not usually studied in an elementary course.

The properties of the series carbon, silicon, tin and lead, show a rising value in the metallic character at the end of the group and an increasing importance of the lower oxide. Thus carbon and silicon are non-metallic and exhibit a fairly close analogy, whereas tin and lead are metallic and therefore resemble one another more closely than they resemble the typical elements.

	Carbon	*Silicon*	*Tin*	*Lead*
Metallic character	Non-metal.	Non-metal.	Metal.	Metal.
Oxides	Acidic.	Acidic.	Acidic and basic.	Acidic and basic.
Hydrides	Many hydrides known.	Several hydrides known.	Very unstable	Extremely unstable.
Halides of form XCl_4	Stable, not ionized.	Unstable, not ionized.	Unstable, not ionized.	Very unstable, not ionized.
Halides of form XCl_2	—	—	Ionized and reducing agent.	Ionized, non-reducer.

SILICON

Experiment 130 Preparation and properties of silicon dioxide and silicon

Material: sodium silicate; magnesium powder; pure silicon dioxide (silicon(IV) oxide); silica tubing.

(*a*) To a dilute solution of sodium silicate in water add a little dilute hydrochloric acid and warm. A white precipitate of hydrated silicon dioxide is thrown down.

$$SiO_3^{2-} + 2H^+ \rightarrow SiO_2\downarrow + H_2O$$

cf. $$CO_3^{2-} + 2H^+ \rightarrow CO_2\uparrow + H_2O$$

(*b*) Heat a short piece of silicon dioxide tubing to red heat; drop it into cold water. It does not crack because its coefficient of expansion is very low and strains set up during expansion and contraction are too small to cause fracture.

(*c*) Remove some of the precipitated silicon dioxide[1] obtained in (*a*) to a test-tube, add sodium hydroxide solution and warm the tube. The solid dissolves, forming sodium silicate in solution.

$$SiO_2 + 2OH^- \rightarrow SiO_3^{2-} + H_2O$$

cf. $$CO_2 + 2OH^- \rightarrow CO_3^{2-} + H_2O$$

(*d*) Mix together approximately 3 g of dry silicon dioxide and 1 g of dry magnesium powder. Introduce the mixture into a *dry* test-tube, clamp at an angle, place a Bunsen underneath and stand aside during heating. A violent reaction ensues and, after the tube has cooled, brown pieces of silicon may be picked out.

$$SiO_2 + 2Mg \rightarrow 2MgO + Si$$

cf. $$CO_2 + 2Mg \rightarrow 2MgO + C$$

(*e*) Put one or two pieces of silicon on a crucible lid and play on the solid from above with a Bunsen burner. Silicon oxidizes to form silicon dioxide.

$$Si + O_2 \rightarrow SiO_2 \qquad \text{(compare carbon)}$$

(*f*) Add sodium hydroxide solution to amorphous silicon in a test-tube and warm. Hydrogen is evolved and sodium silicate remains in solution.

$$Si + H_2O + 2OH^- \rightarrow SiO_3^{2-} + 2H_2\uparrow$$

Experiment 131 Preparation of silane

Material: silicon dioxide; magnesium powder.

Repeat Experiment 130(*c*) using equal volumes of magnesium powder and silicon dioxide to fill a test-tube to the depth of about 2 cm. When cold transfer the contents of the tube to a beaker and add dilute hydrochloric acid. Silane is formed, which ignites to

[1] Anydrous silicon dioxide may not easily dissolve in alkalis. Its acidic properties may be shown by heating with a small piece of solid sodium hydroxide in a crucible. Agitation with water, when cool, will give a solution of sodium silicate.

form a fine fog of silica. Compare the preparation and combustion of methane.

$$2Mg + Si \rightarrow Mg_2Si$$
$$Mg_2Si + 4H^+ \rightarrow SiH_4 + 2Mg^{2+}$$
$$SiH_4 + 2O_2 \rightarrow SiO_2 + 2H_2O$$

TIN

Experiment 132 Reactions of tin(II) compounds

(*a*) Bubble hydrogen sulphide through a solution of tin(II) chloride. A dark precipitate is observed which is insoluble in dilute hydrochloric acid.

$$Sn^{2+} + S^{2-} \rightarrow SnS\downarrow$$

Filter off the precipitate and wash with distilled water. Transfer it to a dish and digest with a little yellow ammonium sulphide solution. The precipitate readily dissolves.

$$(NH_4)_2S + \underset{\text{from ammonium sulphide}}{SnS + S} \rightarrow \underset{\text{ammonium thiostannate}}{(NH_4)_2SnS_3}$$

Note: There is oxidation by the free sulphur in the ammonium sulphide. On the addition of a dilute acid to the thiostannate(IV), tin(IV) sulphide SnS_2 is precipitated.

(*b*) Add sodium hydroxide solution drop by drop to a solution of tin(II) chloride. A white precipitate of tin(II) hydroxide appears but dissolves in excess to form sodium stannite (stannate(II)).

$$Sn^{2+} + 2OH^- \rightarrow Sn(OH)_2\downarrow$$
$$Sn(OH)_2 + 2OH^- \rightarrow SnO_2^{2-} + 2H_2O$$

An exactly parallel reaction occurs if lead(II) nitrate is used in place of tin(II) chloride.

(*c*) Repeat (*b*) with tin(II) chloride using ammonia solution in place of sodium hydroxide solution. A white precipitate of the hydroxide is obtained but is not soluble in excess of ammonia solution.

(*d*) Tin(II) chloride is a powerful reducing agent. Add tin(II) chloride solution to solutions of the following reagents. Reduction occurs readily in each case.

Iron(III) chloride	Pale green iron(II) ions formed
Potassium permanganate	Manganese(II) ions formed
Potassium dichromate	Green chromium(III) ions formed

Mercury(II) chloride	White mercury(I) chloride or grey mercury precipitate formed.

See also Experiments 119(*c*) and 306.

Experiment 133 Preparation and properties of tin(IV) chloride

Apparatus: as in Figure 26; supply of dry chlorine.
Material: tin.

Into the retort put a little sand (to protect the glass during heating) followed by about six pieces of granulated tin. Insert the delivery tube and connect to a chlorine apparatus. The experiment should be carried out in a fume chamber. Warm the retort while chlorine passes over and note: (i) ignition of the tin; (ii) fine white crystals in the upper part of the retort, and (iii) the yellow distillate of tin(IV) chloride. The white crystals are $SnCl_4 \cdot 5H_2O$, due to traces of moisture in the apparatus.

$$Sn + 2Cl_2 \rightarrow SnCl_4$$

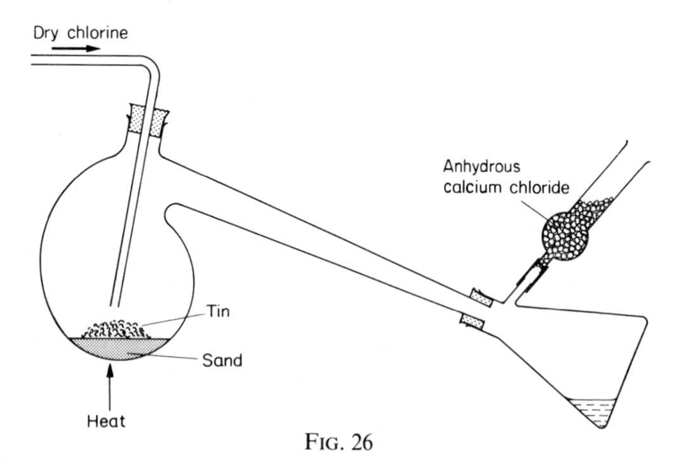

FIG. 26

Perform the following reactions with small portions of the liquid:

(*a*) Add a few drops of water and warm the mixture. Test for hydrogen chloride evolved. The white precipitate is hydrated tin(IV) oxide.

$$SnCl_4 + (2+x)H_2O \rightarrow SnO_2 \cdot xH_2O + 4HCl$$

(*b*) Add a little ammonia solution and warm. Divide the suspension of hydrated tin(IV) oxide into two parts and show its amphoteric nature by dissolving one part in sodium hydroxide solution and the other part in hydrochloric acid.

$$SnO_2 \cdot xH_2O + 4HCl \rightarrow SnCl_4 + (2+x)H_2O$$
$$SnO_2 \cdot xH_2O + 2NaOH \rightarrow \underset{\text{sodium stannate(IV)}}{Na_2SnO_3} + (1+x)H_2O$$

Experiment 134 The preparation of tin(IV) iodide

Material: solution of iodine in carbon tetrachloride (tetrachloromethane); tin.

Take 0.5 g of tin in a test-tube and add 5 cm^3 of the iodine solution, then agitate the solution periodically. Meanwhile proceed with another experiment because this reaction, marked by the solution becoming colourless, takes a while to reach completion. At the end of the reaction, pour the solution on to a watch-glass and allow the carbon tetrachloride to evaporate; this step should be done in a fume chamber because the solvent vapour is toxic. Bright orange crystals are formed: they are stable in air, unlike those of the chloride and bromide.

$$Sn + 2I_2 \rightarrow SnI_4$$

Experiment 135 Relative atomic mass of tin by preparation of tin(IV) oxide, SnO_2

Material: tin.

Weigh an evaporating basin, add about 1.5 g of tin and weigh again. Add not more than 5 cm^3 of concentrated nitric acid and warm the basin very gently on a gauze until action is proceeding moderately, then remove the flame. From time to time add a few drops of concentrated nitric acid until there is no further evolution of nitrogen dioxide. The pale yellow mass is hydrated tin(IV) oxide of formula $SnO_2 \cdot xH_2O$, the value of x being uncertain. Heat the hydrated tin(IV) oxide, gently while it appears to be moist, and later to red heat, to convert it to tin(IV) oxide. When cool, weigh the basin and oxide and from the results calculate the relative atomic mass of tin. As in all gravimetric estimations, reheating and reweighing should continue until the final mass is constant. The experiment is best performed in a fume chamber.

LEAD

Experiment 136 Reactions of lead(II) compounds

(*a*) To two portions of a solution of lead(II) acetate add respectively dilute hydrochloric acid and dilute sulphuric acid. In each case a white precipitate is thrown down.

$$Pb^{2+} + 2Cl^- \rightarrow PbCl_2\downarrow$$
$$Pb^{2+} + SO_4^{2-} \rightarrow PbSO_4\downarrow$$

Wash the lead(II) chloride by decantation, add about four times its bulk of water and heat. The solid dissolves and is reprecipitated on cooling.

Wash the lead(II) sulphate by decantation, add a concentrated solution of ammonium acetate and warm. The lead(II) sulphate dissolves.

(*b*) Add a little potassium chromate solution to 2 or 3 cm^3 of lead(II) acetate solution in a test-tube and observe the curdy yellow precipitate of lead(II) chromate thrown down.

$$Pb^{2+} + CrO_4^{2-} \rightarrow PbCrO_4\downarrow$$

Repeat, using a little potassium iodide solution in the place of potassium chromate. The yellow precipitate is lead(II) iodide (soluble in hot water).

$$Pb^{2+} + 2I^- \rightarrow PbI_2\downarrow$$

(*c*) Add sodium hydroxide solution a drop at a time to a solution of lead(II) acetate in water. The white precipitate is lead(II) hydroxide. This precipitate is soluble if excess of the alkali is added.

$$Pb^{2+} + 2OH^- \rightarrow Pb(OH)_2\downarrow$$
$$Pb(OH)_2 + OH^- \rightarrow \underset{\text{plumbite ion (plumbate(II))}}{[Pb(OH)_3]^-}$$

(*d*) Bubble hydrogen sulphide through a solution of lead(II) acetate in water. A black precipitate of lead(II) sulphide is thrown down.

$$Pb^{2+} + S^{2-} \rightarrow PbS\downarrow$$

Wash the precipitate by decantation, transfer it to a dish, add dilute nitric acid and boil. Some of the sulphide dissolves forming lead(II) nitrate solution, whilst some is oxidized to lead(II) sulphate.

(*e*) Add dilute sodium hydroxide solution to a solution of lead(II) acetate in a test-tube until the precipitate at first formed disappears. Add hydrogen peroxide solution and warm. A brown precipitate of lead(IV) oxide is formed.

(*f*) Add a solution of sodium carbonate to a solution of lead(II) acetate. The white precipitate is basic lead carbonate, $Pb(OH)_2 \cdot 2PbCO_3$.

$$3Pb^{2+} + 3CO_3^{2-} + H_2O \rightarrow Pb(OH)_2 \cdot 2PbCO_3\downarrow + CO_2$$

Sodium hydrogencarbonate solution precipitates the normal carbonate of lead.

$$Pb^{2+} + 2HCO_3^- \rightarrow PbCO_3\downarrow + CO_2 + H_2O$$

Experiment 137 Reactions of lead(IV) compounds

(*a*) Add a spatula-load of trilead tetraoxide to a test-tube filled to a depth of 2 cm with glacial acetic acid. Warm the mixture and the oxide will dissolve. (If a brown precipitate occurs at this stage, commence again using less oxide.) On cooling under the tap, white crystals of lead(IV) acetate come down.

$$Pb_3O_4 + 8CH_3COOH \rightarrow Pb(CH_3COO)_4 + 2Pb(CH_3COO)_2 + 4H_2O$$

Add two or three times the bulk of water to the mixture and warm. The lead(IV) acetate is hydrolysed and a brown precipitate of lead(IV) oxide is formed.

$$Pb(CH_3CO_2)_4 + 2H_2O \rightarrow PbO_2\downarrow + 4CH_3COOH$$

(*b*) Add a spatula-load of lead(IV) dioxide to a test-tube filled to a depth of 2 cm with concentrated hydrochloric acid and cool the tube under the tap. There is little evolution of chlorine at this temperature, and on filtering a golden yellow liquid containing the hexachloroplumbate(IV) ion is obtained.

$$PbO_2 + 4HCl \rightarrow PbCl_4 + 2H_2O$$
$$PbCl_4 + 2Cl^- \rightarrow [PbCl_6]^{2-}$$

Warm a portion of the yellow solution and test for chlorine. Cool the remaining solution under the tap, when white crystals of lead(II) chloride are obtained.

$$PbCl_4 \rightarrow PbCl_2 + Cl_2\uparrow$$

To another portion of the tetrachloroplumbate(IV) solution, add a concentrated solution of ammonia drop by drop. Fine yellow crystals of ammonium hexachloroplumbate(IV) are readily obtained.

$$PbCl_4 + 2NH_3 + 2HCl \rightarrow (NH_4)_2PbCl_6$$

To a third portion, add a few drops of sodium hydroxide solution. A red gelatinous precipitate (very similar to iron(III) hydroxide in appearance) is obtained, which, if heated, forms the well-known brown powder, lead(IV) oxide.

$$PbCl_4 + 2H_2O \rightarrow PbO_2\downarrow + 4HCl$$

(*c*) *Preparation of lead(IV) oxide and lead(II) nitrate.* Add 20 g of trilead tetraoxide carefully to 50 cm^3 of dilute nitric acid and boil for about a minute. Filter whilst hot and allow the filtrate to cool. Crystals of lead(II) nitrate separate out.

The residue of lead(IV) oxide is washed several times with hot water and dried by gentle heating in an evaporating dish.

$$Pb_3O_4 + 4HNO_3 \rightarrow PbO_2 + 2Pb(NO_3)_2 + 2H_2O$$

See also Experiment 153(*f*).

16
Group VB of the Periodic Table

N	P	*As*	Sb	Bi
Nitrogen	Phosphorus	*Arsenic*	Antimony	Bismuth

The element in italics is not usually studied in an elementary course.

The elements in Group VB form a good example of the gradation of properties to be found in the elements of the groups. The following table sums up these properties.

Property	*Nitrogen*	*Phosphorus*	*Antimony*	*Bismuth*
Character	Non-metal.	Non-metal.	Semi-metal.	Typical metal.
Hydrides	Very stable. Alkaline.	Fairly stable, Feebly alkaline.	More unstable.	Very unstable.
Oxides	N_2O_3, N_2O_5 both acidic.	P_4O_6, P_4O_{10} both acidic.	Sb_4O_6 amphoteric.	Bi_2O_3 basic. Bi_2O_5 acidic.
Halides	Explosive.	Not true salts.	Some Sb^{3+} ions. Hydrolysed to some extent.	Salts hydrolysed a little.
Acids	Nitrous acid reducing agent, also oxidizing agent. Nitric acid vigorous oxidizing agent.	Phosphorous acid reducing agent. Phosphoric acid few oxidizing properties.	Antimonites easily oxidized to antimonates.	Bismuthites hard to oxidize to bismuthates.
Sulphides	N_4S_4	P_4S_3	Sb_2S_3, Sb_2S_5	Bi_2S_3

NITROGEN

AMMONIA AND THE AMMONIUM ION

The properties of ammonia and the ammonium ion are summarized in the following, together with a reference to experimental work illustrating these properties:

(*a*) Ammonia gas when hot is a reducing agent, e.g. hot copper(II) oxide is reduced to copper by a stream of the gas.

(*b*) Ammonia tends to form complex ions with a number of metallic ions (see pp. 94–104).

(*c*) Ammonia with water gives a solution which is mainly undissociated, but which does contain sufficient hydroxide ions to behave as a weak alkali (see p. 288).

(*d*) The ammonium ion, NH_4^+, is a stable ion which is metallic in character. When warmed with a solution containing a high concentration of hydroxide ions, it yields ammonia gas (see pp. 309–11).

$$NH_4^+ + OH^- \rightarrow NH_3\uparrow + H_2O$$

By hydrolysis, ammonium salts may react acidic (*see* p. 291).

$$NH_4^+ \rightarrow NH_3 + H^+$$

(*e*) Ammonium salts sublime when heated (see p. 72).

Experiment 138 Reactions of the nitrites

Material: sodium nitrite (1 M) solution (nitrate(III)).

Nitrous (nitric(III)) acid, HNO_2 (which gives rise to the nitrites), may be considered as the acid formed from the lower oxide N_2O_3.

$$2HNO_2 \equiv H_2O \cdot N_2O_3$$
$$2NaNO_2 \equiv Na_2O \cdot N_2O_3$$

The oxide itself is unstable at temperatures above −21 °C, decomposing into nitrogen oxide and nitrogen dioxide.

$$N_2O_3 \rightarrow NO + NO_2$$

Note: Fumes of nitrogen dioxide are poisonous, and preferably the actions given below should be performed in a fume chamber. If it is necessary to perform them on the open bench, add water and pour away as soon as the observations have been made.

In reactions (*c*), (*d*) and (*e*) the nitrous acid is an oxidizing agent according to the equation

$$2NO_2^- + 4H^+ + 2e^- \rightarrow 2H_2O + 2NO \qquad \text{(electron gain)}$$

In reactions (*f*) and (*g*) the nitrous acid acts as a reducing agent being oxidized to nitric(V) acid.

$$H_2O + NO_2^- \rightarrow NO_3^- + 2H^+ + 2e^- \qquad \text{(electron loss)}$$

When nitrous acid acts as a reducing agent there is no evolution of gas. Care must be taken not to have the free nitrous acid in excess, otherwise decomposition occurs as in (*a*). See also Experiment 40.

(*a*) Add some dilute hydrochloric acid to a little sodium nitrite solution in a test-tube. The liquid turns pale blue, and effervescence and brown fumes are observed. The explanation is that with acids, nitrites liberate nitrous acid which is unstable and decomposes into nitrogen oxide and oxygen. The latter oxidizes nitrous acid to nitric acid.

$$3NO_2^- + 2H^+ \rightarrow NO_3^- + H_2O + 2NO$$

This nitrogen oxide finally reacts with the oxygen of the air to form nitrogen dioxide.

$$2NO + O_2 \rightarrow 2NO_2$$

(*b*) Make solutions of iron(II) sulphate and sodium nitrite and mix them. Add a few drops of dilute sulphuric acid, when the whole solution turns brown due to the formation of the unstable compound which iron(II) sulphate makes with nitrogen oxide (see p. 125). This test serves to distinguish between a nit*rite* and a nit*rate*.

(*c*) Add a few drops of concentrated hydrochloric acid to a little potassium iodide solution and pour the mixture into a solution of sodium nitrite. Iodine forms as a brown coloration or black precipitate due to the oxidation of potassium iodide to iodine.

In electronic terms, iodide ion is oxidized by electron loss; nitrous acid is reduced by electron gain.

$$2I^- \rightarrow I_2 + 2e^-$$
$$2HNO_2 + 2H^+ + 2e^- \rightarrow 2H_2O + 2NO$$

Adding,

$$2HNO_2 + 2H^+ + 2I^- \rightarrow 2H_2O + I_2 + 2NO$$

Nitrogen oxide then forms brown fumes with the oxygen of the air.

(*d*) Pass hydrogen sulphide into a solution of sodium nitrite acidified with dilute hydrochloric acid. There is a rapid action, sulphur being deposited. Brown fumes are observed.

Sulphide ions from the partially ionized hydrogen sulphide lose electrons and are oxidized; nitrite ions accept these electrons and are reduced.

$$S^{2-} \rightarrow S + 2e^-$$
$$2NO_2^- + 4H^+ + 2e^- \rightarrow 2H_2O + 2NO$$

Adding,

$$2NO_2^- + 4H^+ + S^{2-} \rightarrow 2H_2O + 2NO + S$$

(*e*) Add a piece of copper to the solution of sodium nitrite acidified with dilute sulphuric acid. The copper is rapidly attacked (much more rapidly than with nitric acid of similar concentration) giving a blue solution.

(*f*) Add a solution of sodium nitrite to bromine water acidified with dilute sulphuric acid. The bromine water is decolorized due to its reduction to hydrobromic acid.

The bromine is reduced to bromide ions by accepting electrons; the nitrite ion is oxidized (and so acts as a reducing agent) by supplying the electrons.

$$Br_2 + 2e^- \rightarrow 2Br^-$$
$$NO_2^- + H_2O \rightarrow NO_3^- + 2H^+ + 2e^-$$

Adding,

$$NO_2^- + H_2O + Br_2 \rightarrow NO_3^- + 2H^+ + 2Br^-$$

(*g*) Add a solution of sodium nitrite to a solution of potassium permanganate acidified with dilute sulphuric acid. The colour of the permanganate is discharged.

Permanganate ion oxidizes; nitrite ion reduces and is itself oxidized.

$$MnO_4^- + 8H^+ + 5e^- \rightarrow Mn^{2+} + 4H_2O \qquad \text{(electron gain)}$$
$$H_2O + NO_2^- \rightarrow NO_3^- + 2H^+ + 2e^- \qquad \text{(electron loss)}$$

Adding,

$$5NO_2^- + 2MnO_4^- + 6H^+ \rightarrow 5NO_3^- + 2Mn^{2+} + 3H_2O$$

(*h*) Add ammonium chloride solution to a solution of sodium nitrite and warm. Effervescence takes place and a colourless odourless gas is given off which gives negative results on testing with calcium hydroxide solution, litmus and a lighted splint. The gas is nitrogen.

$$NH_4^+ + NO_2^- \rightarrow 2H_2O + N_2\uparrow$$

See also Experiment 82(*f*).

Experiment 139 Reactions of the nitrates

Material: sodium nitrate; aluminium powder.

Nitric(V) acid and the nitrates correspond to the higher acidic oxide of nitrogen, N_2O_5.

$$2HNO_3 \equiv H_2O \cdot N_2O_5$$
$$2NaNO_3 \equiv Na_2O \cdot N_2O_5$$

The oxide is difficult to prepare and is unstable at temperatures above 0°C, forming nitrogen dioxide and oxygen.

(*a*) To a little sodium nitrate in a test-tube add concentrated sulphuric acid just to cover it (Care!), and warm gently. Nitric acid vapours are evolved with some decomposition, giving rise to brown fumes of nitrogen dioxide. The nitric acid condenses as oily drops on the cooler parts of the tube.

$$NaNO_3 + H_2SO_4 \rightarrow NaHSO_4 + HNO_3$$
$$4HNO_3 \rightarrow 2H_2O + 4NO_2 + O_2$$

For the action of heat on nitrates consult an elementary text-book. See also p. 382.

(*b*) Add two or three small pieces of copper to a little sodium nitrate in a test-tube and just cover with concentrated sulphuric acid (*Care!*), and warm gently. Brown fumes of nitrogen dioxide are readily evolved.

$$Cu + 2NaNO_3 + 3H_2SO_4 \rightarrow 2NaHSO_4 + CuSO_4 + 2H_2O + 2NO_2$$

(*c*) *Brown ring test.* Make a *cold* solution of iron(II) sulphate by shaking a few crystals with a little dilute sulphuric acid in a boiling-tube. Add a little sodium nitrate and again shake until it dissolves. The depth of liquid should be about 2 cm. Pour concentrated sulphuric acid carefully down the side until it has formed a layer about 1 cm deep under the iron(II) sulphate solution. A brown ring is observed at the junction of the two liquids (see p. 387).

(*d*) *Reduction of nitrate to ammonia.* Take two or three small crystals of sodium nitrate in a test-tube. Fill the tube about one-quarter full of concentrated sodium hydroxide solution. When dissolved, add a small quantity of either aluminium or zinc powder. Warm gently and test for ammonia. If Devarda's alloy[1] is available, use this in preference to the metal powder.

For the oxidizing action of nitric acid, see Experiment 38.

PHOSPHORUS

Experiment 140 Reactions of the phosphites

Material: disodium hydrogenphosphite (phosphonate).

Phosphorous (phosphonic) acid, H_3PO_3, behaves as a dibasic acid and may be regarded as the acid corresponding to the lower oxide.

$$2H_3PO_3 \equiv 3H_2O \cdot P_2O_3 \quad \underset{\substack{\uparrow \\ \text{non-acidic hydrogen}}}{H_2(HPO_3)}$$

[1] Devarda's alloy is 45% Al, 50% Cu, 5% Fe.

Hence sodium phosphite is Na_2HPO_3, i.e. disodium hydrogenphosphite.

(*a*) Add silver nitrate solution to a neutral solution of sodium phosphite. A white precipitate of silver phosphite is formed which, if warmed or allowed to stand, darkens due to reduction to metallic silver.

$$HPO_3^{2-} + 2Ag^+ + H_2O \rightarrow 2Ag\downarrow + HPO_4^{2-} + 2H^+$$

(*b*) Add mercury(II) chloride solution to a neutral solution of sodium phosphite. A white precipitate of mercury(I) chloride is formed which, if warmed, darkens due to further reduction to metallic mercury.

$$Hg^{2+} + HPO_3^{2-} + H_2O \rightarrow Hg\downarrow + HPO_4^{2-} + 2H^+$$

Experiment 141 Reactions of phosphorus and the phosphates

Apparatus: carbon dioxide apparatus; measuring cylinder; corks.

Material: disodium hydrogenphosphate-12-water; white and red phosphorus; iodine; phosphoric acid.

Phosphoric acid, H_3PO_4, behaves as a tribasic acid, although the normal salts are considerably hydrolysed in solution.

$$2H_3PO_4 \equiv 3H_2O \cdot P_2O_5$$

(*a*) Add two or three drops of disodium hydrogenphosphate solution to a test-tube half full of ammonium molybdate acidified with concentrated nitric acid. (The ammonium molybdate must be in considerable excess.) Gently warm the tube and a yellow precipitate of ammonium 12-molybdophosphate(3-), $(NH_4)_3\,[PMo_{12}O_{40}]$, is produced.

(*b*) Add a few drops of disodium hydrogenphosphate solution to a neutral solution of silver nitrate. A canary yellow precipitate of silver phosphate is produced. This precipitate is soluble in dilute nitric acid and also in ammonia solution.

$$3Ag^+ + PO_4^{3-} \rightarrow Ag_3PO_4\downarrow$$

(*c*) Add a few drops of disodium hydrogenphosphate solution to a solution containing magnesia mixture (magnesium sulphate, ammonia and ammonium chloride—the latter to prevent the precipitation of magnesium hydroxide) when a white crystalline precipitate of ammonium magnesium phosphate-6-water is produced.

$$Mg^{2+} + NH_4^+ + PO_4^{3-} + 6H_2O \rightarrow MgNH_4PO_4 \cdot 6H_2O\downarrow$$

(*d*) Add iron(III) chloride solution a drop at a time to the solution of disodium hydrogenphosphate. A buff precipitate of iron(III) phosphate is formed.

$$Fe^{3+} + PO_4^{3-} \rightarrow FePO_4\downarrow$$

This precipitate is soluble in dilute mineral acids and also in excess of iron(III) chloride.

(*e*) *Conversion of orthophosphate to pyrophosphate.* Heat about a quarter of a test-tube full of disodium hydrogenphosphate-12-water to red heat.

$$2Na_2HPO_4 \rightarrow Na_4P_2O_7 + H_2O$$

Make a solution of the residual sodium pyrophosphate (heptaoxo-diphosphate(V)) and compare its action on silver nitrate solution (white precipitate) with that of disodium hydrogenphosphate (yellow precipitate).

(*f*) *Preparation of orthophosphoric acid.* Take a little red phosphorus in an evaporating basin and add a little concentrated nitric acid. Warm gently (in a fume chamber preferably) and note the vigorous evolution of nitrogen dioxide. Add more nitric acid if some phosphorus remains undissolved and again heat. The liquid left is a solution of orthophosphoric acid, which may be evaporated to a thick syrup.

$$P_4 + 20HNO_3 \rightarrow 4H_3PO_4 + 20NO_2\uparrow + 4H_2O$$

(*g*) *Preparation of sodium salts of orthophosphoric acid.* Titrate a dilute solution of phosphoric acid against dilute sodium hydroxide solution using litmus as an indicator. Suppose a cm^3 of the acid neutralized 25 cm^3 of the alkali. Repeat the titration without litmus. This solution contains mainly disodium hydrogenphosphate from which crystals may be obtained by evaporating the solution to small bulk and allowing it to cool. Filter off the crystals, wash with a little cold distilled water and dry then between filter papers.

To obtain sodium dihydrogenphosphate add a cm^3 of the same phosphoric acid solution to 12.5 cm^3 of the caustic soda solution.

To obtain trisodium phosphate add a cm^3 of the same phosphoric acid solution to 37.5 cm^3 of the alkali.

Proceed in both cases to obtain crystals as indicated above.

$$NaOH + H_3PO_4 \rightarrow NaH_2PO_4 + H_2O$$
$$2NaOH + H_3PO_4 \rightarrow Na_2HPO_4 + 2H_2O$$
$$3NaOH + H_3PO_4 \rightarrow Na_3PO_4 + 3H_2O$$

(*h*) *Inter-conversion of red and white phosphorus.* Drop a very small piece of white phosphorus into a dry test-tube and sweep out the air by means of carbon dioxide. Insert a *loose* cork and heat the tube over a low flame. Allow to cool. The phosphorus vapour condenses to small droplets of white phosphorus once again. Repeat the experiment but insert a small crystal of iodine, placing it about 5 mm away from the white phosphorus. (See Figure 27.) On this occasion the red variety is obtained on the cooler side of the tube.

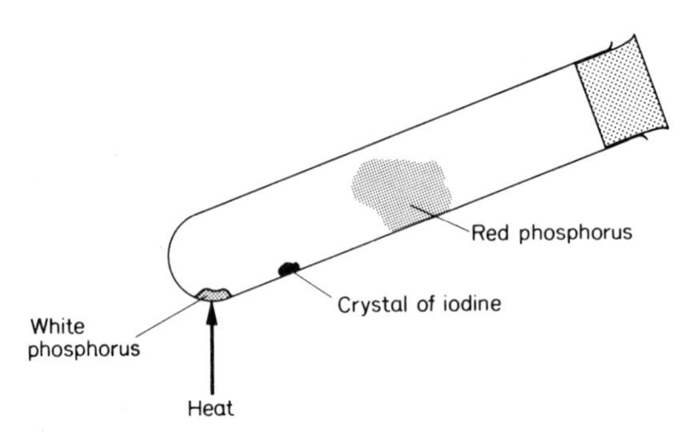

FIG. 27

Repeat the experiment using red phosphorus (no iodine) and the white variety is obtained on cooling.

Experiment 142 Preparation of phosphorus trichloride

Apparatus: retort, dry sand, lead-in tube, 100 cm^3 distilling flask and calcium chloride tube assembled as in Figure 26; condenser; water-bath; thermometer (100 °C).

Material: supplies of dry carbon dioxide and dry chlorine, and white phosphorus.

Arrange the apparatus in a fume chamber as in Figure 26, all parts being dry. Pass carbon dioxide to displace the air. Remove the lead-in tube temporarily and introduce a piece of dry phosphorus, repeating the addition till about 10 g have been introduced. (The dry sand is to protect the retort from cracking.) Pass dry chlorine through the lead-in tube. Spontaneous ignition occurs as the chlorine and phosphorus react.

$$2P + 3Cl_2 \rightarrow 2PCl_3$$

The yellow solid appearing in the retort is phosphorus pentachloride, formed by further chlorination.

$$PCl_3 + Cl_2 \rightarrow PCl_5$$

To purify, transfer the distillate to a distilling flask, fit it with a cork and fitted thermometer, attach to a sloping condenser and use another distilling flask with a calcium chloride guard tube as a receiver. Warm the liquid in the distilling flask on a water-bath and collect the product until the temperature is 76 °C.

Experiment 143 Preparation of phosphorus pentachloride

Apparatus: chlorine apparatus; apparatus as shown. (All parts carefully dried.)

Material: phosphorus trichloride; supply of chlorine.

Fit up the apparatus of Figure 28. Pass a stream of chlorine (well dried by passage through wash bottles containing concentrated sulphuric acid) into the flask, and allow the trichloride to drop slowly into the atmosphere of chlorine. The pentachloride collects as a yellow crystalline solid on the bottom of the flask.

$$PCl_3 + Cl_2 \rightarrow PCl_5$$

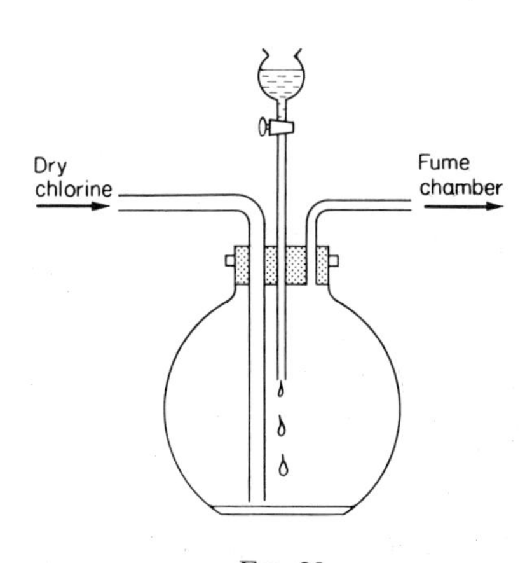

FIG. 28

The pentachloride should be transferred to a storage bottle.

Experiment 144 Action of water on the chlorides of phosphorus

Material: phosphorus tri- and penta-chlorides.

(*a*) Take a test-tube about one-eighth full of water and add one drop of phosphorus trichloride. Hold a rod moistened with silver nitrate near the mouth of the test-tube. The hydrolysis is vigorous and hydrogen chloride is evolved.

$$PCl_3 + 3H_2O \rightarrow 3HCl\uparrow + H_3PO_3$$

(*b*) Repeat as with (*a*) but use a small piece of solid phosphorus pentachloride. This reaction is also vigorous.

$$PCl_5 + 4H_2O \rightarrow 5HCl\uparrow + H_3PO_4$$

See also p. 208.

Experiment 145 Preparation of hydrated ammonium sodium hydrogenphosphate (microcosmic salt), $NaNH_4HPO_4 \cdot 4H_2O$

Material: disodium hydrogenphosphate-12-water.

Take 14 g of the phosphate crystals and 2.2 g of ammonium chloride in separate beakers and dissolve each in 10 cm^3 of water, heating to dissolve. Mix the solutions while hot and leave them to crystallize. Recrystallize with the minimum of water.

$$Na_2HPO_4 + NH_4Cl \rightarrow NaNH_4HPO_4 + NaCl$$

When heated, microcosmic salt decomposes into ammonia, water and sodium metaphosphate (polytrioxophosphate(V)).

$$NaNH_4HPO_4 \rightarrow NaPO_3 + NH_3\uparrow + H_2O$$

Make a loop on a platinum wire, make the wire red hot, and dip it into microcosmic salt. Heat to obtain a glassy bead of sodium metaphosphate, then dust the bead with manganese(IV) oxide and heat again. The amethyst colour is due (probably) to the formation of manganese orthophosphate.

ANTIMONY

Experiment 146 Reactions of antimony

Material: antimony(III) oxide (*Care! very poisonous*).

(*a*) *Marsh's test.* (Modified.) Fit up the apparatus shown in Figure 29 and pour dilute hydrochloric acid on to the zinc in the tube. After *testing that all air has been driven out* (the gases when collected burn smoothly without explosion), add a solution of the antimony compound in hydrochloric acid, and warm the delivery tube. A black deposit of antimony (insoluble in sodium hypochlorite solution or bleaching powder solution) forms on the cooler parts. The antimony in the compound has been reduced to stibine, SbH_3, which is readily decomposed by heat into the elements. (This test was originally devised for arsenic compounds, when the deposit is soluble.)

In Marsh's test proper, the gas must pass over lead(II) acetate paper to remove hydrogen sulphide and through anhydrous calcium chloride to dry it before reaching the hot part of the tube.

(*b*) *Reinsch's test*. Dissolve a little of the antimony compound by warming with dilute hydrochloric acid and immerse a clean strip of copper in the solution. A grey deposit of copper antimonide, Cu_3Sb_2, is formed.

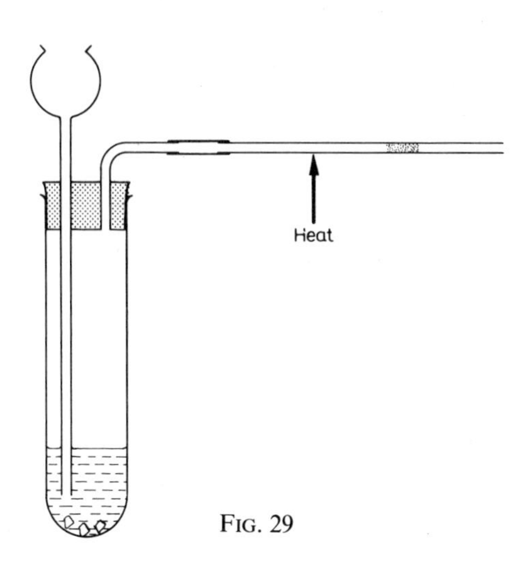

FIG. 29

Experiment 147 Reactions of antimony(III) compounds

Material: antimony(III) chloride; antimony(III) oxide; antimony potassium oxide tartrate (2,3-dihydroxybutanedioate).

(*a*) Antimony(III) oxide is insoluble in water but readily soluble in hydrochloric acid and in alkalis; it is therefore amphoteric.

(*b*) Pass hydrogen sulphide into a solution of antimony(III) chloride in hydrochloric acid. An orange precipitate of antimony(III) sulphide is obtained.

$$2Sb^{3+} + 3S^{2-} \rightarrow Sb_2S_3\downarrow$$

Filter this precipitate off and warm with sodium hydroxide solution or yellow ammonium sulphide. It is readily soluble in these solutions and, on acidifying, antimony sulphide is partially reprecipitated. Show that antimony sulphide will dissolve if warmed with concentrated hydrochloric acid but is not affected by ammonium carbonate solution. See also Experiment 31.

$$Sb_2S_3 + 6H^+ \rightarrow 2Sb^{3+} + 3H_2S\uparrow$$

(*c*) Add sodium hydroxide solution to a solution of antimony(III) chloride in dilute hydrochloric acid until the white precipitate just dissolves. On the addition of silver nitrate solution, a black precipitate of metallic silver is obtained, whilst the antimony(III) compound is oxidized to sodium antimonate.

(*d*) Dissolve a little antimony potassium oxide tartrate (tartar emetic, $K(SbO)C_4H_4O_6$) in water and add a little sodium hydrogencarbonate. Shake until the sodium hydrogencarbonate is dissolved and then add iodine solution in potassium iodide. The latter is decolorized, being reduced to hydrogen iodide, whilst the antimonite (antimonate(III)) becomes antimonate (antimonate(V)).

(*e*) Dissolve antimony(III) chloride in dilute hydrochloric acid and pour the clear solution into a boiling-tube full of water. A white precipitate of antimony chloride oxide is formed.

$$SbCl_3 + H_2O \rightleftharpoons SbOCl\downarrow + 2HCl$$

If concentrated hydrochloric acid is now added to a little of the white precipitate, it redissolves. The reaction is a good example of the reversibility of some reactions. (See Experiment 78(*b*).)

Experiment 148 Reactions of antimony(V) compounds

Material: potassium antimonate, $2KSbO_3 \equiv K_2O \cdot Sb_2O_5$.

(*a*) Pass hydrogen sulphide into a solution of a little potassium antimonate in dilute hydrochloric acid. An orange precipitate of antimony(V) sulphide, Sb_2S_5, is obtained. This precipitate is, like antimony(III) sulphide, soluble in solutions of either sodium hydroxide or yellow ammonium sulphide. The addition of acid to the solution brings about a reprecipitation of the sulphide.

(*b*) Add silver nitrate solution to a neutral solution of potassium antimonate. A white precipitate of silver antimonate (soluble in either ammonia or nitric acid) is formed.

$$SbO_3^- + 3Ag^+ + H_2O \rightarrow Ag_3SbO_4\downarrow + 2H^+$$

BISMUTH

Experiment 149 Reactions of bismuth compounds

Material: bismuth(III) nitrate; bismuth(III) chloride.

(*a*) Mix a little solid bismuth nitrate with anhydrous sodium carbonate and heat on the charcoal block with the mouth blowpipe. A pink globule of bismuth is obtained surrounded by a yellow incrustation of bismuth(III) oxide, Bi_2O_3.

(*b*) Pass hydrogen sulphide into a solution of bismuth(III) nitrate acidified with dilute hydrochloric acid. A dark brown precipitate of bismuth(III) sulphide (insoluble in yellow ammonium sulphide or in sodium hydroxide) is obtained.

$$2Bi^{3+} + 3S^{2-} \rightarrow Bi_2S_3\downarrow$$

Filter the precipitate off, wash a little into a dish with dilute nitric acid and warm. The precipitate readily dissolves.

(*c*) Dissolve a little bismuth chloride in dilute hydrochloric acid and then pour it into a boiling-tube full of water. A white precipitate of bismuth chloride oxide is formed. If a little of this is poured off into a test-tube and a few drops of concentrated hydrochloric acid added, the precipitate redissolves. (See Experiment 78(*a*).)

$$BiCl_3 + H_2O \rightleftharpoons BiOCl\downarrow + 2HCl$$

17
Group VIB of the Periodic Table

O	—	S	—	*Se*	—	*Te*	—	*Po*
Oxygen		Sulphur		*Selenium*		*Tellurium*		*Polonium*

Elements in italics are seldom studied in an elementary course.

Oxygen, sulphur, selenium and tellurium are more related to each other than to chromium etc., although sulphur has points of similarity with chromium in a few corresponding compounds. Chromium is considered as a transition element, see p. 84.

Generally, it may be said that the elements show a tendency to become more metallic with increase in relative atomic mass, and the maximum valency exhibited is six.

OXYGEN

Experiment 150 Preparation of oxides by direct oxidation

Apparatus: gas-jars; deflagrating spoons.

Material: supply of oxygen; sulphur; carbon; phosphorus; calcium; iron wire; magnesium.

Take six jars of the gas and introduce, respectively, a little of each element listed, in a deflagrating spoon, heating sufficiently to promote action; iron will need to be heated to red heat.

Note the nature of the oxide and test with litmus. Divide the oxides into (*a*) acidic; (*b*) alkaline, and (*c*) other oxides. Show that both (*b*) and (*c*) dissolve in hydrochloric acid and are therefore bases.

Experiment 151 Preparation of oxides by indirect oxidation

Material: copper foil; lead.

Take a small piece of copper foil in a test-tube and add concentrated nitric acid until the tube is about one-eighth full. Add more acid if the action dies down before the copper dissolves, to complete the solution of the metal. Transfer to a dish and cautiously evaporate the nitrate solution until crystals are formed, then heat strongly to complete the decomposition of the nitrate to oxide. When cool, show that the oxide is a base by dissolving it in dilute sulphuric acid.

Repeat the experiment, using lead, but finally dissolve the oxide in acetic acid instead of sulphuric acid.

Experiment 152 Preparation of hydrogen peroxide solution

Material: barium peroxide; phosphoric acid.

Dilute about 5 cm^3 of syrupy phosphoric acid with its own volume of water in a test-tube. Gradually add barium peroxide (cooling under the tap) until on filtering and testing by the reaction of Experiment 153(*a*), a rapid evolution of oxygen is obtained. Filter and use the solution for the tests described in Experiment 153.

$$2H_3PO_4 + 3BaO_2 \rightarrow Ba_3(PO_4)_2\downarrow + 3H_2O_2$$

Note: The solution may contain, in addition to hydrogen peroxide, a little excess phosphoric acid and some barium ions.

Experiment 153 Properties of hydrogen peroxide

Hydrogen peroxide reacts in the following ways:

(i) *as an oxidizing agent:*

$$H_2O_2 + 2H^+ + 2e^- \rightarrow 2H_2O \qquad \text{(electron gain)}$$

This property is shown in experiments (*b*) and (*d*).

(ii) *as a reducing agent:*

$$H_2O_2 \rightarrow 2H^+ + O_2 + 2e^- \qquad \text{(electron loss)}$$

This is illustrated in experiments (*c*), (e) and (*f*).

(iii) *being decomposed* catalytically into water and oxygen.

$$2H_2O_2 \rightarrow 2H_2O + O_2$$

This is a summation of the two previous processes and is illustrated in experiment (*a*) and the later part of (*e*).

Material: platinized asbestos; diethyl ether; lead(IV) oxide.

(*a*) To about 3 cm^3 of hydrogen peroxide in a test-tube, add a small piece of platinized asbestos or manganese(IV) oxide. Test the oxygen evolved. The action is catalytic. Refer to (iii) above.

(*b*) To about 3 cm^3 of lead(II) acetate solution, add hydrogen sulphide solution to precipitate lead sulphide (black). Decant the liquid. Add about 3 cm^3 of hydrogen peroxide and shake, adding more hydrogen peroxide, if needed, to oxidize the lead sulphide to white lead sulphate.

Refer to (i) above. Hydrogen peroxide, acting as an oxidizing agent, has accepted electrons.

$$4H_2O_2 + 8H^+ + 8e^- \rightarrow 8H_2O$$

Lead sulphide has been oxidized by loss of electrons.

$$PbS + 4H_2O \rightarrow PbSO_4 + 8H^+ + 8e^-$$

Adding,

$$PbS + 4H_2O_2 \rightarrow PbSO_4 + 4H_2O$$

(*c*) To a solution containing about 3 cm^3 of potassium permanganate solution acidified by dilute sulphuric acid, add a few drops of hydrogen peroxide. Observe the evolution of gas and the decoloration of the permanganate; retain the gas in the tube and test for oxygen.

Refer to (ii) above. Hydrogen peroxide, acting as a reducing agent, has lost electrons.

$$5H_2O_2 \rightarrow 10H^+ + 5O_2 + 10e^-$$

Permanganate has been reduced by gain of electrons.

$$2MnO_4^- + 16H^+ + 10e^- \rightarrow 2Mn^{2+} + 8H_2O$$

Adding,

$$2MnO_4^- + 6H^+ + 5H_2O_2 \rightarrow 2Mn^{2+} + 8H_2O + 5O_2$$

(*d*) To a test-tube containing about 3 cm^3 of potassium iodide solution acidified by dilute sulphuric acid, add a few drops of hydrogen peroxide. The brown coloration of iodine may then be tested with a drop of starch solution.

Refer to (i) above. As oxidizing agent, hydrogen peroxide has accepted electrons.

$$H_2O_2 + 2H^+ + 2e^- \rightarrow 2H_2O$$

Iodide ions have been oxidized by loss of electrons.

$$2I^- \rightarrow I_2 + 2e^-$$

Adding,

$$H_2O_2 + 2H^+ + 2I^- \rightarrow 2H_2O + I_2$$

(*e*) To a test-tube containing about 3 cm^3 of silver nitrate solution, add a slight excess of sodium hydroxide solution to form a brown precipitate of silver oxide; decant the supernatant liquid. Add a few drops of hydrogen peroxide to the silver oxide. Observe the evolution of oxygen as the silver oxide is reduced to black metallic silver.

Refer to (ii) above. Hydrogen peroxide, acting as a reducing agent, has lost electrons.

$$H_2O_2 \rightarrow 2H^+ + O_2 + 2e^-$$

Silver oxide has been reduced by electron gain.

$$Ag_2O + 2H^+ + 2e^- \rightarrow 2Ag + H_2O$$

Adding,

$$H_2O_2 + Ag_2O \rightarrow 2Ag + H_2O + O_2$$

Add a few more drops of hydrogen peroxide and show that the finely divided silver acts catalytically as did the platinized asbestos in (*a*).

(*f*) To a test-tube containing about 3 cm^3 of dilute nitric acid, add half a spatula-load of lead(IV) oxide. Add a few drops of hydrogen peroxide and observe the evolution of oxygen and the eventual solution of the lead(IV) oxide. Test the solution for lead(II) ions by adding a drop of potassium chromate solution. Hydrogen peroxide has acted as a reducing agent, as in (*e*). Lead(IV) oxide has been reduced by gain of electrons.

$$PbO_2 + 4H^+ + 2e^- \rightarrow Pb^{2+} + 2H_2O$$

(*g*) Fill a test-tube to a depth of about 2 cm with potassium dichromate solution and acidify with dilute sulphuric acid. Cover the solution with a layer of diethyl ether about 2 cm deep, and add a drop of well diluted hydrogen peroxide. The blue coloration in the ether layer is probably due to a chromium peroxide and this reaction serves as a delicate test for hydrogen peroxide.

OZONE, O_3

When oxygen is passed through a silent electric discharge, the issuing gas contains a small amount of ozone (trioxygen, 5–10%). Even this small percentage is higher than is stable with oxygen at room temperature.

$$3O_2 \rightarrow 2O_3; \ \Delta H = +289 \text{ kJ}$$

Consequently, when ozonized oxygen, obtained as above, is heated much of the ozone dissociates into oxygen until true equilibrium is reached for the given temperature. Chemically, ozone is an oxidizer, each molecule losing an atom of oxygen to the reducer and leaving a molecule of oxygen.

$$O_3 \rightarrow O_2 + O \text{ (combining with reducer)}$$

Experiment 154 Preparation and reactions of ozonized oxygen

Apparatus: as shown in Figure 30.

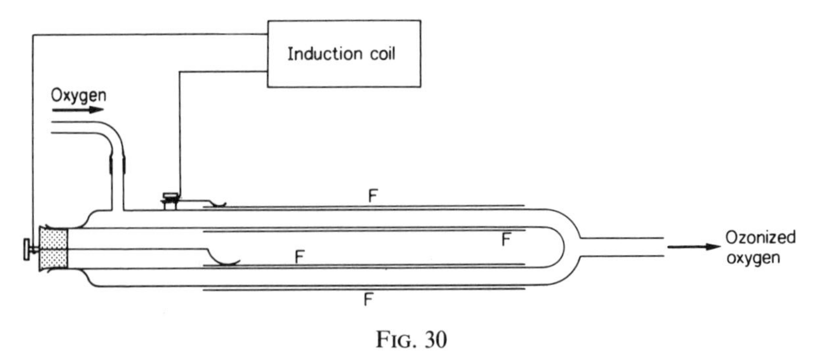

FIG. 30

Material: supply of oxygen; mercury; starch-potassium iodide solution.

The outside of the outer tube, and the inside of the inner tube, are coated with aluminium foil (marked F). Dry oxygen is passed through the space between the two tubes while the foil surfaces are kept at a high electrical tension by connecting them to an induction coil. The supply of oxygen is conveniently obtained from a cylinder and the flow is regulated to a moderate rate. Ozonized oxygen attacks rubber tubing, but connections can be made with pvc [poly(vinyl chloride) or poly(chloroethene)] tubing which is unaffected.

(*a*) Soak a piece of filter paper in a solution of starch-potassium iodide and hold in the issuing gas. The blue colour indicates free iodine.

$$2I^- + O_3 + H_2O \rightarrow I_2\downarrow + O_2 + 2OH^-$$

(*b*) Take a small drop of mercury in a clean, dry test-tube. Pass ozonized oxygen into the tube for a short time. Note the 'tailing' of the mercury. Heat the tube strongly and show that the mercury regains its usual mobility.

(*c*) Moisten a piece of filter paper with lead(II) acetate solution and hold it in a stream of hydrogen sulphide to form black lead sulphide. Put the piece in a test-tube and pass in ozonized oxygen for two or three minutes. Note the lighter colour as lead sulphide is oxidized to lead sulphate.

$$PbS + 4O_3 \rightarrow PbSO_4 + 4O_2$$

SULPHUR

The preparation of allotropes of sulphur will have been demonstrated in the elementary course. For revisionary purposes rhombic, plastic, and amorphous sulphur may readily be prepared as follows.

Experiment 155 Preparation of forms of sulphur

Material: crushed roll sulphur; carbon disulphide.

Fill a dry test-tube half full of sulphur and heat gradually to boiling (using a holder for safety). Pour the boiling sulphur into a beaker half full of water. Immerse any floating sulphur by stirring with a rod, and then remove and examine the plastic sulphur. Note the gradual loss of elasticity as the plastic form changes to rhombic.

Take enough sulphur to cover the bottom of a test-tube, add about 5 cm^3 of carbon disulphide, and stand the tube in hot water in a beaker. (*Carbon disulphide is inflammable and toxic—the experiment should be done in a fume chamber.*) When the liquid has boiled for two or three minutes, filter the liquid into a dish, cover with a filter paper held in position with an elastic band and pierce a few holes in the filter paper with a pin, and leave to crystallize. The residue in the original filter paper is amorphous sulphur and the crystals separating from the filtrate, rhombic sulphur.

Experiment 156 Preparation of sulphides

Material: flowers of sulphur; sheet copper.

(*a*) Take an ignition tube a third full of flowers of sulphur and heat until melted. Take a strip of copper and hook it over the rim of the tube so that its lower edge is a little above the surface of the sulphur. Heat to boil the sulphur and note the glow as the copper forms a mixture of copper(I) and copper(II) sulphides.

(*b*) Mix approximately equal parts of iron filings and sulphur and take a test-tube about half-full of the mixture. Heat until the action begins and note the glow of the mixture as combination to form iron(II) sulphide continues. (Compare with direct oxidation of metals.)

(*c*) Into a solution of copper(II) sulphate pass hydrogen sulphide. Filter off the precipitated copper(II) sulphide.

$$Cu^{2+} + S^{2-} \rightarrow CuS\downarrow$$

The precipitation of sulphides in both acid and alkaline solutions is important in the separation of metallic ions and is fully discussed in Chapter 44.

Experiment 157 Preparation of chlorides and oxychlorides of sulphur

Apparatus: distilling flasks (150 cm^3); condenser; water bath.

Material: siphon of sulphur dioxide; phosphorus pentachloride; supply of chlorine; natural camphor.

Care! These experiments should be done in a fume chamber.

(*a*) Take about ten spatula-loads of sulphur in a distilling flask. Fit the flask with a one-holed cork fitted with a tube to reach the level of the sulphur, and connect the tube to a supply of dry chlorine. Heat the sulphur on a gauze and pass in chlorine. Collect the liquid product (sulphur monochloride, disulphur dichloride, S_2Cl_2) in a dry test-tube. Warm a few drops of the product in a little water and test for sulphur dioxide and hydrochloric acid, and note the deposit of sulphur.

$$2S_2Cl_2 + 2H_2O \rightarrow 4HCl + 3S\downarrow + SO_2$$

(*b*) Take about ten spatula-loads of phosphorus pentachloride in a dry distilling flask which is attached to a sloping condenser (Figure 31). Fit the flask with a cork and delivery tube, the latter reaching well down into the flask. Stand the flask in a water-bath (cold water).

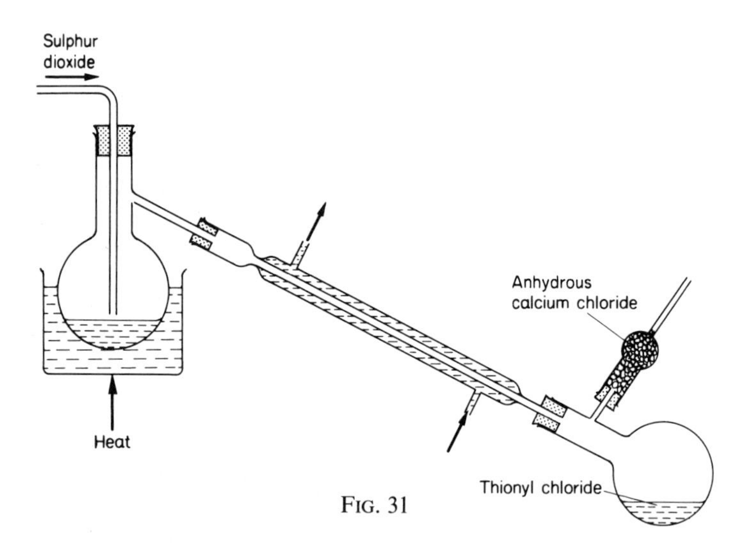

FIG. 31

Pass sulphur dioxide into the phosphorus pentachloride until it has completely liquefied. Heat the water-bath and collect the distillate of thionyl chloride (sulphur dichloride oxide), $SOCl_2$. The liquid still remaining in the flask is phosphorus trichloride oxide and this serves as a method of preparing the compound.

$$PCl_5 + SO_2 \rightarrow POCl_3 + SOCl_2$$

Add a few drops of the thionyl chloride to a little water and test for sulphurous acid and hydrochloric acid.

$$SOCl_2 + 2H_2O \rightarrow H_2SO_3 + 2HCl$$

(*c*) Fit up the apparatus as for (*b*). Cut 2 cm^3 of camphor into pieces and place in the dry flask. Pass sulphur dioxide until the camphor liquefies and then disconnect the sulphur dioxide siphon and substitute a supply of dry chlorine. Pass in chlorine until it is no longer absorbed. Heat the water-bath and collect the distillate of sulphuryl chloride (sulphur dichloride dioxide).

$$SO_2 + Cl_2 \rightarrow SO_2Cl_2$$

Add a few drops of the product to a little water and show that the resulting solution contains sulphuric and hydrochloric acids.

$$SO_2Cl_2 + 2H_2O \rightarrow H_2SO_4 + 2HCl$$

Experiment 158 Properties of sulphur dioxide and sulphites

Material: magenta; magnesium ribbon; sodium sulphite (sulphate(IV)); iodine in potassium iodide solution; siphon of sulphur dioxide.

As these properties have been studied at an earlier stage, it will be sufficient to give brief instructions:

(*a*) Pass the gas into a test-tube about half-full of water and use the stock solution to show that it is acid, decolorizes potassium permanganate, reduces potassium dichromate, decolorizes magenta, and gives a deposit of sulphur with hydrogen sulphide. Write the appropriate equations.

The above experiments may be performed in a simpler way by heating a little sulphur in an evaporating basin and testing the sulphur dioxide formed by drops of the various reagents on a glass rod.

(*b*) Prepare sodium sulphite and sodium hydrogensulphite (in solution) as follows.

Take two equal volumes of sodium hydroxide solution in test-tubes. Saturate one volume with sulphur dioxide to form sodium hydrogensulphite.

Add the second volume to the product to form sodium sulphite.

$$NaOH + SO_2 \rightarrow NaHSO_3$$
$$NaHSO_3 + NaOH \rightarrow Na_2SO_3 + H_2O$$

(*c*) Collect two gas-jars full of the gas. Hold a piece of magnesium ribbon in a pair of crucible tongs, heat until it is ignited and hold the burning metal in the gas. Note the combustion of the gas yielding sulphur and magnesium oxide.

$$2Mg + SO_2 \rightarrow 2MgO + S$$

Place a jar of hydrogen sulphide mouth to mouth with the second jar of sulphur dioxide. (See Experiment 66.)

(*d*) Make a hot solution of sodium sulphite and blow in air. Test the solution for sulphate.

$$2SO_3^{2-} + O_2 \rightarrow 2SO_4^{2-}$$

(*e*) Make a solution of about five spatula-loads of sodium sulphite crystals in 50 cm³ of water in a small flask. Add two spatula-loads of crushed roll sulphur and boil for at least an hour, adding water as required to replace that lost by evaporation. Transfer the solution to an evaporating basin and heat to a small bulk. Test the concentrated solution for sodium thiosulphate by (i) iodine solution, and (ii) addition of acid.

The absorption of oxygen in the first experiment and of sulphur in the second to form sodium sulphate and sodium *thio*sulphate respectively shows some resemblance between oxygen and sulphur.

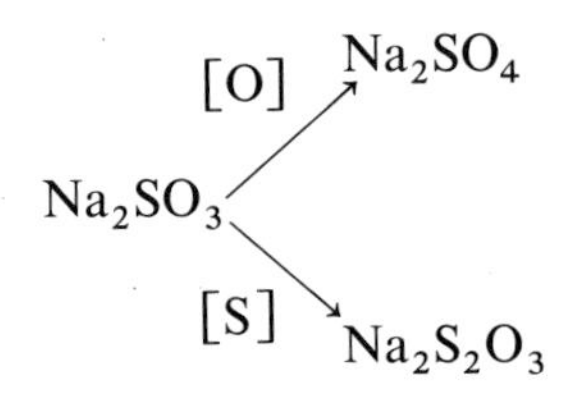

(*f*) Add dilute hydrochloric acid to a few crystals of sodium sulphite in a test-tube and warm. Sulphur dioxide is evolved which can be recognized by its smell and the decolorizing of a dilute solution of permanganate.

$$SO_3^{2-} + 2H^+ \rightarrow SO_2\uparrow + H_2O$$

(*g*) To a freshly made solution of sodium sulphite add barium chloride solution. A white precipitate is obtained which can be distinguished from barium sulphate by its solubility in dilute hydrochloric acid.

$$SO_3^{2-} + Ba^{2+} \rightarrow BaSO_3\downarrow$$

(*h*) Add a solution of iodine in potassium iodide drop by drop to a solution of sodium sulphite. The iodine is decolorized and the resulting solution may be tested for sulphate ions.

$$SO_3^{2-} + H_2O + I_2 \rightarrow SO_4^{2-} + 2I^- + 2H^+$$

Experiment 159 Reactions of sulphuric acid

Material: sucrose (cane sugar); formic (methanoic) acid; oxalic (ethanedioic) acid.

The reactions of **concentrated** sulphuric acid may be considered under the following headings:

(i) as a dehydrating agent, removing water, or the elements of water, from another substance. (See (*a*), (*b*), (*c*) and (*d*) below.)

(ii) as a displacer of acids from their salts, sulphuric acid being much less volatile than most other acids. (See (*e*) and (*f*).)

(iii) as an oxidizing agent. (See (*f*).)

For all the following experiments, a suitable quantity of the substance is sufficient to give a depth of about 5 mm in a test-tube and the concentrated acid added should be about twice that bulk.

(*a*) Leave copper(II) sulphate crystals and concentrated sulphuric acid standing for ten minutes. Observe the change from blue crystalline to white anhydrous copper sulphate.

$$CuSO_4 \cdot 5H_2O + (H_2SO_4) \rightarrow CuSO_4 + (H_2SO_4 \cdot 5H_2O)$$

(*b*) Warm the mixture of sucrose and concentrated sulphuric acid gently for a few seconds and then leave the tube to stand in a test-tube rack. Observe the vigorous action and the change of the white sugar to black carbon. Moistening the sucrose obviates warming.

$$C_{12}H_{22}O_{11} + (H_2SO_4) \rightarrow 12C + (H_2SO_4 \cdot 11H_2O)$$

(*c*) Observe the immediate effervescence when concentrated sulphuric acid is added to the formic acid. Warm the tube gently to drive the gas up the tube and light the gas as it emerges. Carbon monoxide burns with a blue flame.

$$HCOOH + (H_2SO_4) \rightarrow CO + (H_2SO_4 \cdot H_2O)$$

(*d*) Boil the mixture of oxalic acid and concentrated sulphuric acid and ignite carbon monoxide at the mouth of the tube.

$$H_2C_2O_4 \cdot 2H_2O + (H_2SO_4) \rightarrow CO + CO_2 + (H_2SO_4 \cdot 3H_2O)$$

(*e*) To crystals of sodium chloride add concentrated sulphuric acid, and test the fuming gas with silver nitrate solution on a glass rod; silver nitrate forms white silver chloride with hydrogen chloride.

$$NaCl + H_2SO_4 \rightarrow HCl + NaHSO_4$$

Hydrogen chloride has been displaced by the less volatile sulphuric acid.

Repeat the experiment with sodium acetate (ethanoate) and cautiously smell the acetic acid displaced. Repeat also with sodium formate (methanoate) and note that the displacement of formic acid is followed by dehydration as in (*c*). Refer to Experiment 139(*a*) for the displacement of nitric acid from nitrates.

(*f*) To crystals of potassium bromide add concentrated sulphuric acid and note that in this case a fuming gas is first formed and then a brown gas. Hydrogen bromide is displaced and then partially oxidized to bromine. Silver nitrate on a glass rod is turned to pale yellow silver bromide by hydrogen bromide. Sulphur dioxide (formed by reduction of sulphuric acid) may be shown by the decolorization of acidified potassium permanganate (manganate(VII) solution on a glass rod.

$$KBr + H_2SO_4 \rightarrow HBr + KHSO_4$$
$$2HBr + H_2SO_4 \rightarrow Br_2 + 2H_2O + SO_2$$

or

$$4H^+ + 2Br^- + SO_4^{2-} \rightarrow Br_2 + 2H_2O + SO_2$$

Repeat the experiment substituting potassium iodide and note the similar series of reactions but with the much greater extent of oxidation; much of the hydrogen iodide is oxidized to iodine. Warm the tube and observe the violet vapour of iodine.

$$4H^+ + 2I^- + SO_4^{2-} \rightarrow I_2 + 2H_2O + SO_2$$

Dilute sulphuric acid is highly ionized and serves as a source of hydrogen ions and sulphate ions; its reactions as an acid are demonstrated in (*g*) and its reaction as a sulphate in (*h*).

(*g*) Take two spatula-loads of zinc powder in a test-tube and add dilute sulphuric acid to a depth of about 2 cm. Close the tube with the thumb until sufficient hydrogen has been evolved to give a mild explosion when the mouth of the tube is held in a flame.

$$2H^+ + Zn \rightarrow Zn^{2+} + H_2$$

Repeat, using sodium carbonate in place of zinc; test for carbon dioxide by allowing the gas to pass into calcium hydroxide solution, when a turbidity is formed by precipitation of calcium carbonate.

$$2H^+ + CO_3^{2-} \rightarrow H_2O + CO_2$$
$$Ca(OH)_2 + CO_2 \rightarrow CaCO_3 + H_2O$$

(*h*) Take about a quarter of a test-tube full of dilute sulphuric acid and add an equal volume of barium chloride solution. The white precipitate is barium sulphate. It is insoluble in dilute hydrochloric acid.

$$SO_4^{2-} + Ba^{2+} \rightarrow BaSO_4\downarrow$$

Experiment 160 Preparation and reactions of sodium thiosulphate crystals, $Na_2S_2O_3 \cdot 5H_2O$

Apparatus: 250 cm^3 round bottomed flask fitted with reflux condenser.

Material: sodium sulphite; sodium thiosulphate crystals ('hypo'); iodine in potassium iodide; chlorine water; crushed roll sulphur.

Take a 250 cm^3 round bottomed flask and fit it with a reflux condenser. Into the flask pour 150 cm^3 of water and add 30 g of sodium sulphite and 15 g of crushed sulphur. Heat on a gauze for three hours. Filter the solution and evaporate to about 30 cm^3. Allow to cool and crystallize.

$$SO_3^{2-} + S \rightarrow S_2O_3^{2-}$$

(*a*) Heat a few crystals of sodium thiosulphate in a dry test-tube until they begin to melt. Observe water and sulphur as products. When cool, add dilute hydrochloric acid to the residue and recognize the evolution of hydrogen sulphide.

$$4Na_2S_2O_3 \rightarrow 3Na_2SO_4 + Na_2S_5$$
$$Na_2S_5 \rightarrow Na_2S + 4S$$

(*b*) Add iodine solution to a solution of sodium thiosulphate. The iodine is decolorized whilst the thiosulphate is converted into tetrathionate.

$$2S_2O_3^{2-} + I_2 \rightarrow S_4O_6^{2-} + 2I^-$$

See also Chapter 37.

(*c*) Add chlorine water in excess to a solution of sodium thiosulphate and test the resulting solution with barium chloride solution. The products are sodium sulphate and sulphur (which may be further oxidized to sulphuric acid).

$$S_2O_3^{2-} + Cl_2 + H_2O \rightarrow SO_4^{2-} + S\downarrow + 2H^+ + 2Cl^-$$

Bromine behaves similarly.

(*d*) Add concentrated hydrochloric acid to a solution of sodium thiosulphate in a test-tube. Sulphur is deposited and sulphur dioxide is evolved.

$$S_2O_3^{2-} + 2H^+ \rightarrow SO_2\uparrow + H_2O + S\downarrow$$

(See Experiment 164 for reaction between sodium thiosulphate and the silver halides. See Experiment 10 for reference to supersaturation.)

Experiment 161 Reactions of amidosulphuric (sulphamic) acid, NH_2SO_2OH

Material: sodium nitrite, amidosulphuric (sulphamic) acid.

(*a*) Make a solution of amidosulphuric acid, using a spatula-load in distilled water to a depth of about 2 cm in a test-tube. Note the high solubility of the acid. Add dilute hydrochloric acid and a few drops of barium chloride solution. Observe that at first there is little action, but, after standing, a white suspension of barium sulphate appears and becomes more evident on boiling as the amidosulphuric acid hydrolyses.

$$NH_2SO_2OH + H_2O \rightarrow NH_4HSO_4$$

(*b*) Make a solution of amidosulphuric acid as in (*a*). Make a solution of sodium nitrite in a similar way. Mix the solutions. The vigorous effervescence is due to the rapid evolution of nitrogen oxide, nitrogen dioxide and nitrogen. The acid is a strong (i.e. highly ionized) acid. It reacts with the nitrite to give oxides of nitrogen and its $-NH_2$ group also reacts with the nitrite to give nitrogen.

$$2H^+ + 2HNO_2 + 2e^- \rightarrow 2H_2O + 2NO$$
$$2NO + O_2 \rightarrow 2NO_2$$
$$NH_2SO_2OH + H^+ + NO_2^- \rightarrow N_2 + HSO_4^- + H_2O$$

Test the remaining solution for sulphate ion as in (*a*).

(*c*) Take a spatula-load of amidosulphuric acid in a test-tube and add sodium hydroxide solution until the tube is filled to a depth of about 2 cm, thereby ensuring an excess of sodium hydroxide. Heat and test the gas evolved for ammonia, using a piece of damp red litmus paper.

$$NH_2SO_2O^- + OH^- \rightarrow NH_3 + SO_4^{2-}$$

(*d*) Take two spatula-loads of amidosulphuric acid in a dry test-tube. Heat and test for (i) sulphur dioxide with a spot of potassium permanganate on a filter-paper; (ii) sulphur trioxide by allowing the white fumes to flow into a test-tube containing barium chloride solution acidified with hydrochloric acid. Note the crystalline sublimate; dissolve the crystals in sodium hydroxide solution (about 2 cm deep in a test-tube), heat and test for ammonia with damp red litmus paper. Acidify the remaining solution with hydrochloric acid and add barium chloride solution to test for sulphate ion. (See also p. 144.)

18
Group VIIB of the Periodic Table

F — Cl — Br — I — *At*
Fluorine Chlorine Bromine Iodine *Astatine*

The element in italics is not usually studied in an elementary course.

	Fluorine	*Chlorine*	*Bromine*	*Iodine*
Element	Yellow gas. Irritating smell.	Yellowish-green gas. Irritating smell.	Red liquid. Irritating smell.	Black solid. Irritating smell if gaseous.
Preparation	Electrolysis of fused KHF_2.	Heat on mixture of chloride, manganese(IV) oxide and concentrated sulphuric acid.	Heat on mixture of bromide, manganese(IV) oxide and concentrated sulphuric acid.	Heat on mixture of iodide, manganese(IV) oxide and concentrated sulphuric acid.
Activity	Very reactive. Combines with most metals and non-metals.	Very reactive. Combines with most metals and many non-metals.	Very reactive. Combines with most metals and many non-metals.	Reactive. Combines with most metals and a few non-metals.
Replacing action	Replaces all halogens from combination in simple salts.	Replaces bromine and iodine from bromides and iodides.	Replaces iodine from iodides.	—
Oxidizing action	Very powerful oxidizing agent.	Very powerful oxidizing agent.	Powerful oxidizing agent.	Weak oxidizing agent.
Action with alkalis	Forms the fluoride and displaces oxygen.	Forms hypochlorite with cold dilute, and chlorate with hot concentrated solution.	Similarly forms hypobromite and bromate.	Similarly forms hypoiodite which more readily yields iodate.
Halogen hydride	Fuming gas, very soluble forming weak acid. Reacts with silica. Does not dissociate into elements.	Fuming gas. Dissociates only at high temperature. Very soluble forming strong acid.	Fuming gas. Dissociation begins at red heat. Very soluble forming strong acid.	Fuming gas. Readily dissociates and is powerful reducing agent. Very soluble forming strong acid.

FLUORINE

The study of fluorine is complicated by its extreme activity—it combines with nearly all known elements.

Experiment 162 Preparation of hydrogen fluoride and its effect on glass

Apparatus: waxed microscope slide.

Material: calcium fluoride. *This experiment must be done in a fume chamber.*

Place a little powdered calcium fluoride in the bottom of a test-tube and pour over it some concentrated sulphuric acid. The microscope slide has been previously coated with wax and part of the wax removed by writing on the slide with a pin. Hold the slide face downwards over the tube and warm it gently. Whilst the reaction is in progress the similarity of hydrogen fluoride to the other halogens may be shown by (*a*) blowing across an ammonia bottle in the direction of the gas; (*b*) placing a piece of damp blue litmus paper in the gas. The action of silver nitrate on fluorides in solution is not typical as silver fluoride is soluble in water. *Steamy fumes* (*Care!*) of hydrogen fluoride are observed and if, after two or three minutes, the wax is removed from the slide, the writing will be seen to be etched on the glass. This is due to the formation of silicon fluoride.

$$CaF_2 + H_2SO_4 \rightarrow CaSO_4 + 2HF$$
$$\underset{\text{from glass}}{SiO_2} + 4HF \rightarrow SiF_4 + 2H_2O$$

After the experiment is completed, observe the greasy appearance of the test-tube due to etching.

Experiment 163 The reactions of fluorides

Material: sodium fluoride solution.

(*a*) Add silver nitrate solution: observe that the solution remains clear because silver fluoride, unlike the other halides, is soluble in water.

(*b*) Add calcium chloride solution: observe that a white precipitate forms because calcium fluoride, unlike the other halides, is insoluble in water.

$$Ca^{2+} + 2F^- \rightarrow CaF_2\downarrow$$

(*c*) Add a few drops of iron(III) chloride solution then test the solution for the presence of iron(III) ions with potassium thiocyanate: a negative result is obtained because of the formation of hexafluoroferrate(III) ions.

CHLORINE

Experiment 164 To compare and contrast the chloride, bromide and iodide of silver

Material: potassium chloride; potassium iodide; potassium bromide.

Fill each of three test-tubes in a rack, to a depth of about 3 cm, with solutions of potassium chloride, bromide and iodide respectively. (If the solids are supplied, dissolve a crystal of the size of a match head in about 3 cm^3 of water.) Add a solution of silver nitrate to each. The silver chloride is white, the silver bromide is slightly yellow, and the iodide is yellow. Divide each into three parts. Add dilute nitric acid to one portion and dilute ammonia solution drop by drop to the other in each case. All three precipitates are insoluble in nitric acid; silver chloride is soluble in ammonia, silver bromide slightly soluble, silver iodide insoluble.

$$Ag^+ + Cl^- \rightarrow AgCl\downarrow$$
$$Ag^+ + Br^- \rightarrow AgBr\downarrow$$
$$Ag^+ + I^- \rightarrow AgI\downarrow$$
$$AgCl + 2NH_3 \rightarrow \underset{\text{soluble complex ion}}{[Ag(NH_3)_2]^+} + Cl^-$$

Next show that all three precipitates are soluble in sodium thiosulphate solution.

$$AgCl + Na_2S_2O_3 \rightarrow \underset{\text{silver sodium thiosulphate}}{NaAgS_2O_3} + NaCl$$

Experiment 165 Some reactions of chlorides

Material: sodium chloride; solid potassium dichromate.

(*a*) Add 1 cm^3 of concentrated sulphuric acid to a spatula-load of sodium chloride in a test-tube. Hydrogen chloride is liberated.

$$NaCl + H_2SO_4 \rightarrow NaHSO_4 + HCl\uparrow$$

(*b*) Grind a spatula-load of sodium chloride with two of manganese(IV) oxide, transfer the mixture to a boiling-tube, add 2 or 3 cm^3 of concentrated sulphuric acid and warm. Chlorine is evolved (bleaches litmus) and some hydrogen chloride.

$$2NaCl + 2H_2SO_4 + MnO_2 \rightarrow Na_2SO_4 + MnSO_4 + 2H_2O + Cl_2\uparrow$$

(*c*) *Preparation of chromyl chloride, CrO_2Cl_2.* Heat a dry test-tube in the Bunsen flame to soften the glass in the region about a quarter to a third from the open end. Draw out the glass to reduce the diameter of the tube to about 5 mm at the heated part, and at the same time bend the open end slightly downwards. The apparatus will then serve as a small retort. When cold, introduce into the tube a mixture of not more than a spatula-load of finely ground potassium dichromate and half that amount of sodium chloride. Add just enough concentrated sulphuric acid to cover the mixture. Grasp the tube in one holder and in another hold a dry test-tube to act as a receiver. Heat the mixture *gently* and collect a few drops of the reddish-brown liquid, chromyl chloride (chromium(VI) dichloride dioxide).

$$K_2Cr_2O_7 + 4NaCl + 3H_2SO_4 \rightarrow K_2SO_4 + 2Na_2SO_4 + 2CrO_2Cl_2 + 3H_2O$$

Add a few drops of water to the compound. Test the gas (hydrogen chloride) with litmus paper and with silver nitrate solution on a glass rod. The yellow solution contains chromic acid. Add sodium hydroxide solution until the solution is neutral, then acidify with acetic (ethanoic) acid and add a solution of lead acetate. The yellow precipitate confirms the presence of chromate ion.

$$CrO_2Cl_2 + 2H_2O \rightarrow H_2CrO_4 + 2HCl$$

(*d*) For the action of silver nitrate see Experiment 164.

Experiment 166 Preparation of chlorine

Apparatus: as in Figure 32; Woulfe bottles.

Material: solid potassium permanganate (manganate(VII)).

The following is the most convenient method of preparing chlorine since it is easily controlled. Fit up the apparatus as shown in Figure

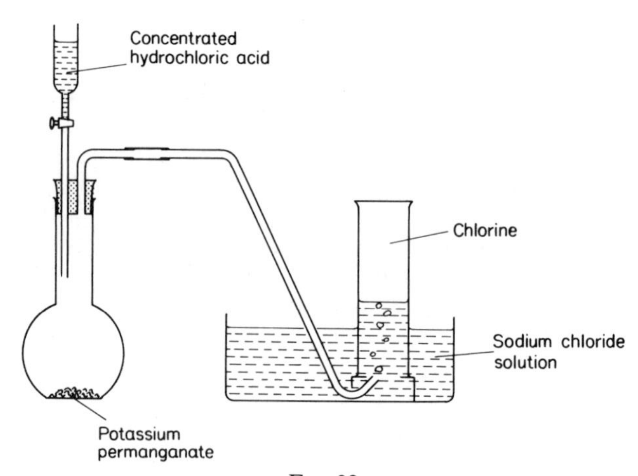

FIG. 32

32 and put ten spatula-loads of potassium permanganate into the flask. Fill the dropping funnel with concentrated hydrochloric acid (*Care! Make sure you have the correct acid.*) and allow the acid to run on to the permanganate to produce chlorine at the required speed.

$$2MnO_4^- + 16H^+ + 10Cl^- \rightarrow 2Mn^{2+} + 8H_2O + 5Cl_2\uparrow$$

If the gas is required dry, pass it through a little water to remove hydrogen chloride and then through concentrated sulphuric acid to dry it. Collect it by downward delivery.

Experiment 167 Preparation of anhydrous iron(III) chloride

Apparatus: as shown in Figure 33; chlorine generator.
Material: iron wire (steel wool); solid potassium permanganate.

Anhydrous iron(III) chloride (similar remarks apply to other chlorides, e.g. chromium(III), iron(II), magnesium, zinc, aluminium, tin(IV) etc.) cannot be prepared by evaporating a solution of the salt to dryness because hydrolysis takes place and the final product is iron(III) oxide. See Experiment 100.

Fit up the apparatus as shown in Figure 33; put steel wool in the combustion tube. Pass dry chlorine over for a minute to displace the air and then heat the tube by means of a Bunsen burner. As soon as

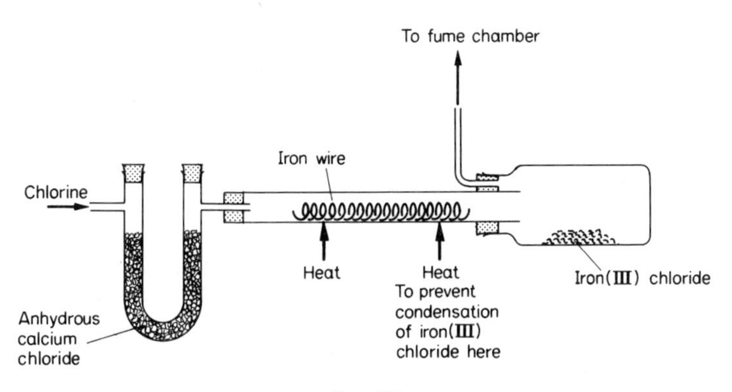

FIG. 33

the iron wire commences to burn, remove the flame and the wire will continue to burn if the supply of chlorine is sufficient. Most of the iron(III) chloride will condense as a mass of black crystals in the cool part of the tube. A trap (thistle funnel dipping into sodium hydroxide solution) may be used to absorb excess chlorine, or the gas may be led into the fume chamber.

$$2Fe + 3Cl_2 \rightarrow 2FeCl_3$$

Note: Anhydrous iron(II) chloride may be prepared in a similar apparatus. Dry hydrogen chloride is used in place of chlorine and the small colourless scales of iron(II) chloride are much less volatile and frequently adhere to the unattacked iron.

$$Fe + 2HCl \rightarrow FeCl_2 + H_2$$

BROMINE

(*Care! Liquid bromine will cause sores if it comes in contact with the flesh, and bromine vapour is painful to the eyes.*)

A gas-jar full of bromine may be obtained by warming a gas-jar by holding it upside down 30 cm above a Bunsen burner, then dropping 1 cm^3 of bromine into it, leaving the lid two-thirds on. After a little while the bromine will evaporate and fill the jar.

Experiment 168 Reactions of bromine

Apparatus: hydrogen generator.

Material: gas-jars of bromine vapour obtained from liquid bromine as described above; solution of fluorescein in ethanol.

(*a*) Invert a gas-jar of bromine over hydrogen, remove cover plates and mix. On applying a flame there is a feeble explosion.

$$H_2 + Br_2 \rightarrow 2HBr$$

(*b*) Invert a gas-jar of bromine over hydrogen sulphide. A deposit of sulphur is obtained and the colour of bromine is replaced by the misty fumes of hydrogen bromide.

$$Br_2 + H_2S \rightarrow 2HBr + S\downarrow$$

(*c*) Dip a filter paper in the solution of fluorescein and allow to dry. Place it in a gas-jar of bromine vapour, when the paper turns red due to the formation of eosin.

(*d*) See the end of Experiment 177 (b).

Experiment 169 Reactions of bromine water

Material: red phosphorus; sulphurous (sulphuric(IV)) acid.

Bromine dissolves to a slight extent in water forming a red solution of concentration approximately 4% *m*/*V*. Note the presence of the red vapour above the saturated solution.

(*a*) Add a spatula-load of iron filings to a test-tube one-quarter filled with bromine water. Shake. According to which element is in excess a pale green solution of iron(II) bromide (iron in excess), or a yellow solution of iron(III) bromide (bromine in excess) will be obtained.

$$Fe + Br_2 \rightarrow FeBr_2 \quad \text{or} \quad Fe + Br_2 \rightarrow Fe^{2+} + 2Br^-$$
$$2Fe + 3Br_2 \rightarrow 2FeBr_3 \quad \text{or} \quad 2Fe + 3Br_2 \rightarrow 2Fe^{3+} + 6Br^-$$

The presence of iron(II) or iron(III) ions may be shown by the addition of sodium hydroxide solution (a mixture of the two ions will give a black precipitate of Fe_3O_4).

(*b*) Hold a piece of blue litmus paper in the vapour above bromine water. It turns red and is bleached.

(*c*) Add a few drops of bromine water to a test-tube one-quarter filled with sulphurous acid. Test the solution for sulphate by adding dilute hydrochloric acid followed by barium chloride.

$$SO_3^{2-} + Br_2 + H_2O \rightarrow SO_4^{2-} + 2Br^- + 2H^+$$

(*d*) Add sodium hydroxide solution drop by drop to a test-tube one-quarter filled with bromine water until the colour is just discharged. The solution contains the hypobromite (bromate(I)) ion.

$$2OH^- + Br_2 \rightarrow OBr^- + Br^- + H_2O$$

This hypobromite solution will precipitate manganese(IV) oxide from a solution of manganese(II) sulphate and lead(IV) oxide from a solution of lead(II) nitrate.

(*e*) Add a spatula-load of red phosphorus[1] to a test-tube one-quarter filled with bromine water. Shake the tube and allow it to stand. The colour of the bromine water disappears. The bromine and phosphorus have combined and the resulting bromide of phosphorus has been decomposed to give phosphorous (phosphonic) or phosphoric acids.

$$P_4 + 6Br_2 \rightarrow 4PBr_3$$
$$P_4 + 10Br_2 \rightarrow 4PBr_5$$
$$2PBr_3 + 6H_2O \rightarrow 2H_3PO_3 + 6HBr$$
$$PBr_5 + 4H_2O \rightarrow H_3PO_4 + 5HBr$$

(*f*) Add a little bromine water to a solution of potassium iodide in water. Iodine is displaced.

$$2I^- + Br_2 \rightarrow I_2 + 2Br^-$$

Note: See Experiment 199 for the action of bromine on the unsaturated hydrocarbons.

[1] This reaction gives a suitable method of disposing of phosphorus residues.

Experiment 170 Preparation and properties of hydrogen bromide

Apparatus: boiling-tube fitted with cork and tube bent at right angles.

Material: potassium bromide; chlorine water.

For class purposes, hydrogen bromide may readily be prepared by the action of fairly concentrated sulphuric acid on potassium bromide. If the acid is concentrated there is considerable oxidation.

Dilute a quantity of sulphuric acid with water by pouring two volumes of concentrated sulphuric acid into one volume of water, adding the acid a little at a time. If each experimenter is preparing his own acid it will be sufficient if 2 cm^3 of acid are poured two to three drops at a time into 1 cm^3 of water in a boiling-tube. Into the diluted acid put a spatula-load of potassium bromide and warm gently (*Care!*).

$$KBr + H_2SO_4 \rightarrow KHSO_4 + HBr\uparrow$$

Test the misty fumes by carefully lowering a drop of each of the following reagents on the end of a glass rod into the gas.

(*a*) Silver nitrate solution. A pale yellow precipitate of silver bromide is formed.

$$Ag^+ + Br^- \rightarrow AgBr\downarrow$$

(*b*) Concentrated ammonia solution. Fumes of ammonium bromide are observed.

$$NH_3 + HBr \rightarrow NH_4Br$$

(*c*) Litmus solution. It is turned red.

(*d*) Chlorine water. Light brown or red coloration due to bromine.

$$2Br^- + Cl_2 \rightarrow Br_2 + 2Cl^-$$

(*e*) A drop of water. This may be removed and tested by dipping the rod into, e.g. a few drops of silver nitrate solution in a test-tube. The positive reaction shows the high solubility of the gas in water.

(*f*) Concentrated nitric acid. The hydrogen bromide is rapidly oxidized to bromine.

$$2HNO_3 + 2HBr \rightarrow 2H_2O + 2NO_2 + Br_2$$

(*g*) Attach a cork with tube bent at right angles and heat the tube strongly, having a piece of white paper ready to hold behind the tube. The hydrogen bromide is decomposed by strong heat into bromine and hydrogen. If the bromine is not easily visible its presence may be shown by the fluorescein test (Experiment 168(*c*)).

Experiment 171 Preparation of hydrogen bromide (demonstration)

Apparatus: as in Figure 34.
Material: bromine; red phosphorus.

Make a paste of 5–10 g of red phosphorus with water and sand (to moderate the action) and introduce this into the flask. Bromine is dropped in gradually from the tap funnel, the first few drops reacting with a flash of light. The bromine volatilized by the heat of the reaction is removed by passing the gases through a U-tube containing beads smeared with damp red phosphorus. (Do not have this material too wet or much of the hydrogen bromide will be absorbed here.) The hydrogen bromide, which is very similar in appearance to hydrogen chloride, is collected by displacement of air. Alternatively, if a solution is wanted, it may be collected by passing it through an inverted funnel into water (equations in Experiment 169(*e*)).

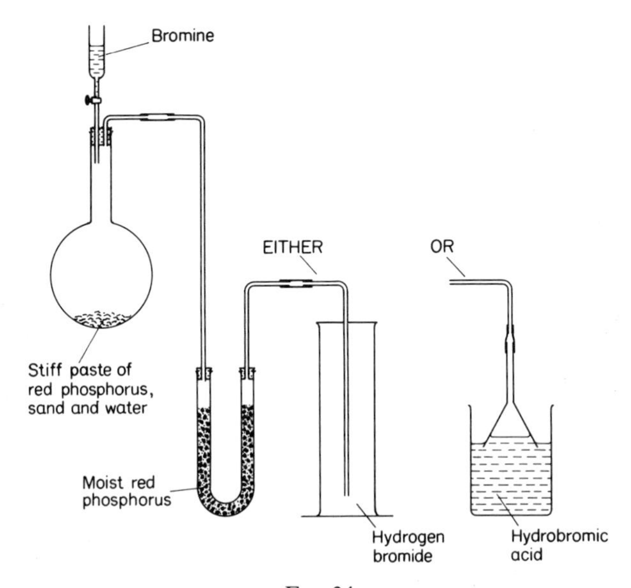

FIG. 34

The reactions of Experiment 170(*a*) to (*g*) may be repeated, using gas-jars of hydrogen bromide and correspondingly larger quantities of reactants. The action of heat on hydrogen bromide is best shown by filling a boiling-tube with hydrogen bromide, inserting a loose cork in the tube, and heating the tube strongly in the flame of a

Bunsen burner, holding a piece of white paper at the back as soon as decomposition begins.

$$2HBr \rightleftharpoons H_2 + Br_2$$

Experiment 172 Preparation of potassium bromide, KBr

Apparatus: 250 cm^3 flask; measuring cylinder; Buchner funnel; water-bath.

Material: bromine; potassium carbonate.

Take 100 cm^3 of distilled water in the flask. Measure 5 cm^3 of bromine in a measuring cylinder which already contains 2 to 3 cm^3 of water. Pour the bromine and water into the flask (and wash out the cylinder at once). Weigh 8 g of iron filings and add to the solution in portions of about 0.5 g, shaking well on each addition. If this operation of adding the iron is hurried, much heat is generated and some iron(III) bromide is formed and persists throughout the preparation. Heat the flask on a water-bath for ten minutes and filter quickly.

$$Fe + Br_2 \rightarrow FeBr_2$$

Prepare a solution of 20 g of potassium carbonate in 50 cm^3 of water. Add this solution to the green solution of iron(II) bromide, mix well and heat on the water-bath for ten minutes. The white (later green) precipitate is iron(II) carbonate.

$$FeBr_2 + K_2CO_3 \rightarrow 2KBr + FeCO_3$$

Filter quickly and evaporate the colourless solution to crystallization. Examine a drop of solution under a microscope for cubic crystals of the bromide.

Experiment 173 Reactions of bromides

Material: use one crystal the size of a match-head or 1 cm^3 of 1 M potassium bromide for each experiment; chlorine water; carbon tetrachloride (tetrachloromethane).

(*a*) Grind the bromide with a small quantity of manganese(IV) oxide, add 1 cm^3 of concentrated sulphuric acid to the mixture in a test-tube and warm gently. The red vapour of bromine may condense to small drops of liquid bromine on the sides of the test-tube.

$$MnO_2 + 2KBr + 2H_2SO_4 \rightarrow MnSO_4 + K_2SO_4 + 2H_2O + Br_2\uparrow$$

(*b*) Add a few drops of silver nitrate solution to a solution of potassium bromide. The pale yellow precipitate of silver bromide is insoluble in dilute nitric acid but sparingly soluble in excess ammonia solution.

$$Ag^+ + Br^- \rightarrow AgBr\downarrow$$

(*c*) Add a little chlorine water, a drop at a time, to a solution of potassium bromide. Bromine is liberated which turns the solution light brown or red. If two drops of carbon tetrachloride are added (cf. Experiment 176(*c*)) the bromine dissolves forming a brown or red solution.

$$Cl_2 + 2Br^- \rightarrow Br_2 + 2Cl^-$$

IODINE

Experiment 174 Preparation of hydrogen iodide

Apparatus: as in Figure 35; mortar and pestle.

Material: red phosphorus; iodine.

Grind a spatula-load each of dry red phosphorus and iodine in a mortar and introduce into a boiling-tube. Add three or four drops of water and fit the boiling-tube up as shown in Figure 35.

$$P_4 + 6I_2 \rightarrow 4PI_3$$
$$PI_3 + 3H_2O \rightarrow H_3PO_3 + 3HI\uparrow$$

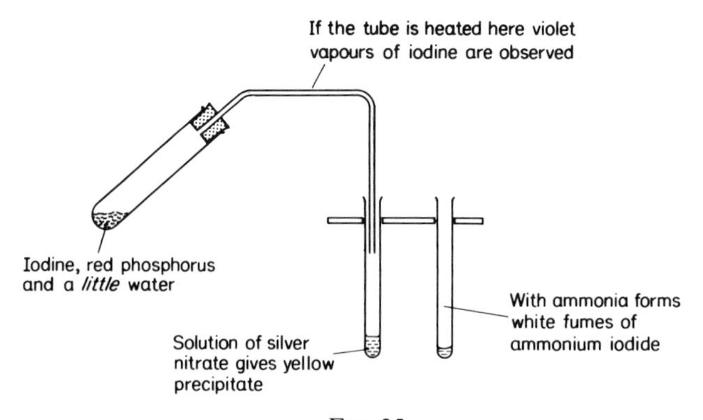

FIG. 35

The boiling-tube may be warmed to generate further quantities of hydrogen iodide. The following reactions of the gas may be shown.

(*a*) Pass the gas into silver nitrate solution in a test-tube. A yellow precipitate of silver iodide is seen.

(*b*) Pass the gas into a test-tube containing a few drops of concentrated ammonia solution. White fumes of ammonium iodide are produced.

$$HI + NH_3 \rightarrow NH_4I$$

(*c*) Pass the gas into a test-tube containing a few drops of concentrated nitric acid. The hydrogen iodide is easily oxidized to iodine and the nitric acid reduced to nitrogen dioxide.

$$2HNO_3 + 2HI \rightarrow 2H_2O + 2NO_2 + I_2$$

(*d*) Pass the gas for a short while into a test-tube containing a little concentrated sulphuric acid. The acid is reduced to sulphur dioxide, hydrogen sulphide or sulphur, showing that hydrogen iodide is a powerful reducing agent. See Experiment 159(f).

(*e*) Action of heat. Warm the delivery tube with a Bunsen burner. The violet vapours of iodine will be observed.

$$2HI \rightleftharpoons H_2 + I_2$$

Experiment 175 Preparation of hydrogen iodide (demonstration)

Apparatus: as in Figure 34; mortar and pestle.

Material: red phosphorus; iodine.

Grind 2 g of dry red phosphorus with 10 g of iodine and introduce the mixture into a dry flask fitted up as shown in the diagram (Figure 34). Run in water from the dropping funnel drop by drop until the flashing ceases. Be sure not to add too much water as the gas is very soluble. Ten to fifteen drops will be sufficient as a rule. Warm the flask gently and after a little while the gas comes off readily and may be collected.

The experiments of Experiment 174(*a*) to (*e*) may be repeated using larger quantities.

Experiment 176 Reaction of iodides

Material: Use one crystal of potassium iodide the size of a match-head or 1 cm^3 of a 1 M solution for each reaction; chlorine water; carbon tetrachloride (tetrachloromethane).

(*a*) Grind the iodide with a small quantity of manganese(IV) oxide and add 1 cm^3 of concentrated sulphuric acid to the mixture in a test-tube. Warm gently and observe the violet vapours of iodine.

$$MnO_2 + 2KI + 2H_2SO_4 \rightarrow MnSO_4 + K_2SO_4 + 2H_2O + I_2$$

With concentrated sulphuric acid alone iodine is also obtained on heating because hydrogen iodide is a powerful reducing agent.

(*b*) Add a little silver nitrate solution to a solution of potassium iodide. The yellow precipitate of silver iodide obtained is insoluble in both dilute nitric acid and ammonia solution.

$$Ag^+ + I^- \rightarrow AgI\downarrow$$

(*c*) Add a little chlorine water a drop at a time to a solution of potassium iodide. Iodine is liberated which turns the solution brown, whilst black crystals can usually be seen at the surface of the liquid by careful inspection.

$$Cl_2 + 2I^- \rightarrow I_2 + 2Cl^-$$

Add two drops of carbon tetrachloride. The iodine dissolves on shaking, giving an intense violet coloration.

(*d*) Add a few drops of lead(II) acetate solution to a solution of potassium iodide. A yellow precipitate of lead iodide is obtained.

$$Pb^{2+} + 2I^- \rightarrow PbI_2\downarrow$$

(*e*) Add potassium iodide solution, a drop at a time, to three or four drops of mercury(II) chloride solution. A red precipitate of mercury(II) iodide is obtained which will dissolve in excess of the iodide to form a complex ion. If sodium hydroxide solution is added at this stage, Nessler's solution is formed. This solution is used as a very delicate test for ammonia (see Experiment 122). (Equations for reactions are given in Experiment 119(b)).

Note: See Experiment 107(*b*) for reaction of potassium iodide solution with copper(II) ions.

Experiment 177 Preparation of iodic acid, HIO_3, and potassium iodate

Apparatus: retort; small flask; sand-tray; water-bath; measuring cylinder.

Material: iodine; fuming nitric acid; potassium hydroxide.

(*a*) *Iodic acid* (iodic(V) acid). *Care! This experiment should be performed in a fume chamber.* Take 5 g of iodine in a retort and add a measured 40 cm^3 of fuming nitric acid. Heat the retort on a sand-tray, keeping the temperature high enough to promote action. Collect any nitric acid which distils over and return it to the retort. When the iodine has all been oxidized to white crystals of iodic acid, pour the contents of the retort into an evaporating basin and heat almost to dryness on a water-bath. Collect the crystals and dry between filter paper.

$$I_2 + 10HNO_3 \rightarrow 2HIO_3 + 10NO_2 + 4H_2O$$

Heat some of the crystals in a dry test-tube, gently at first, and later strongly. Note the formation of moisture to leave iodine pentaoxide.

$$2HIO_3 \rightarrow I_2O_5 + H_2O$$

This is followed by decomposition to iodine and oxygen. (Test with glowing splint.)

$$2I_2O_5 \rightarrow 2I_2 + 5O_2$$

(*b*) *Potassium iodate* (potassium iodate(V)). Put a small piece (2 g) of potassium hydroxide into a boiling-tube and fill to a depth of 5 cm with water. When dissolved, add, a little at a time, ten spatula-loads (4.5 g) of iodine to the warm solution. Pour the solution into a watch-glass and allow to cool.

$$6OH^- + 3I_2 \rightarrow IO_3^- + 5I^- + 3H_2O$$

Decant the solution from the crystals, wash the latter with a *little* water and dry them on a filter paper. Heat a few crystals in a dry tube and show oxygen is evolved. (Potassium bromate may be made by a similar experiment using twenty drops (1 cm^3) of bromine in place of the iodine.)

PART III

Organic Chemistry

19
Introduction to Practical Organic Chemistry

So many organic compounds are important, either as essential ingredients of biological processes or as desirable supplements to civilized life, that a course of practical chemistry must include the study of some of them. Food, fuel, clothing, medicine, paper and dyes are, in the main, organic compounds; all are either necessary, or desirable, for civilized existence. Many organic compounds of biological importance are, however, of complex structure, and it would be unprofitable to attempt their study at this stage. The scientific method of approach is to study the structure and behaviour of the simpler compounds, and later to apply the knowledge so gained to a study of more complex compounds.

The aim of the present course is to select a comparatively few compounds, chiefly derived from either ethanol or benzene, and study their reactions. Many of these reactions will be carried out on a test-tube scale since the essential reactions can equally well be studied with a small quantity as with a large quantity of material, but in order to give opportunities for handling the types of apparatus used in preparative organic work, a number of preparations on a larger scale are included. The compounds which are to be considered contain one, or in some cases, two, **characteristic groups**. Thus, all primary alcohols contain the characteristic $-CH_2OH$ group which has specific properties independent of the rest of the molecule; consequently, as a class, alcohols have a large number of reactions common to all of them. Similarly the properties of an acid depend very largely on the presence of the $-COOH$ group in the molecule and many of the properties of acetic acid (ethanoic acid, CH_3COOH) are common properties of all such acids. The importance of a thorough knowledge of the properties associated with these characteristic groups cannot be exaggerated, since upon it depends the approach to advanced work in the subject.

GENERAL INSTRUCTIONS

1. Many organic compounds are easily oxidized and some are highly inflammable. Care should always be taken to guard against fire, especially when handling liquids. Remove Winchesters containing inflammable liquids to a safe place before lighting Bunsen burners. When distilling diethyl ether (ethoxyethane), lead away any vapour to the waste pipe of the sink. The distilling flask containing the ether should be heated over water brought to boiling point elsewhere, to avoid using a naked flame near the flask.

2. Apparatus must be dry as well as clean. The presence of moisture may lower the yield and cause unwanted products to be formed. For quick drying of apparatus, wash it first with ethanol, drain as well as possible, wash with the minimum of diethyl ether to dissolve traces of ethanol, and finally dry off the ether with a jet of air from the bellows.

3. Liquids containing moisture are usually dried by adding pieces of anhydrous calcium chloride.

Exceptions:

alcohols (use calcium oxide).
amines (use potassium hydroxide, anhydrous sodium sulphate, or potassium carbonate).
phenols (use anhydrous sodium sulphate).

4. When using distillation apparatus see that corks are well fitted to avoid losses. Introduce liquids into the flask through a funnel to prevent them running down the side tube. Add fragments of porous pot or anti-bumping granules (sintered aluminium oxide) to minimize 'bumping' during boiling. See that the bulb of the thermometer is at the level of the side tube to ensure that the temperature of the *vapour* is being recorded.

5. It is often necessary to boil liquids for a considerable time to complete a reaction. To prevent loss by vaporization, the flask is fitted with an upright condenser, which, in the case of liquids of high boiling points, need not be water-jacketed. This method of retaining a volatile liquid is called *refluxing*. (See Experiment 194, Figure 48.)

6. Note that organic preparations differ from inorganic preparations in being slower: organic compounds are not usually ionized and reactions are therefore molecular instead of ionic. Note also that yields of products are usually less than the expected. This is not a contradiction of accepted laws, but is due to the simultaneous formation of products other than those required. Thus, ethanol and concentrated sulphuric acid give the gas ethene (ethylene), but also give diethyl ether and ethyl hydrogensulphate. The quantities and conditions given in preparations are chosen to favour the highest yield of the required product, and instructions should, therefore, be closely followed.

7. Approximate yields have been given. This will enable a decision to be made where considered advisable to use smaller apparatus and a fractional quantity of materials. For example, in the preparation of aniline (Experiment 267(*b*)), a suitable modification would be to arrange for the reduction of nitrobenzene to be performed using fractional quantities, merging the products at the steam distillation stage.

Where the term 'yield is small' is given, it is not advisable to attempt the experiment on a smaller scale.

8. Illustrative diagrams which accompany the instructions for preparing individual compounds are presented, in general, in the classical manner using cork-closure apparatus to cater for the continued use of this type of apparatus where necessary or desirable.

Standard-joint apparatus has the advantages of simplicity of assembly, efficiency of closures and overall elegance. Where this type of apparatus is available, little difficulty should occur in translating items of the diagrams into their standard-joint equivalents; the recommended capacities of flasks, as given in the text, remain unchanged. It will also be recognized that the procedure as given in the text will be modified slightly in conformity to the change of apparatus, e.g. the instruction to close the neck of the distilling flask with a bored cork and thermometer will be read as an instruction to fit the flask with a two-way adapter and thermometer. The code letters of the following frequently used items of standard-joint apparatus are included, as required, under the heading *Apparatus* for individual experiments.

In a few cases, where major differences occur between the arrangements needed for cork-closure and standard-joint apparatus, diagrams for both are given; where one type is considered preferable, reference to the other is omitted.

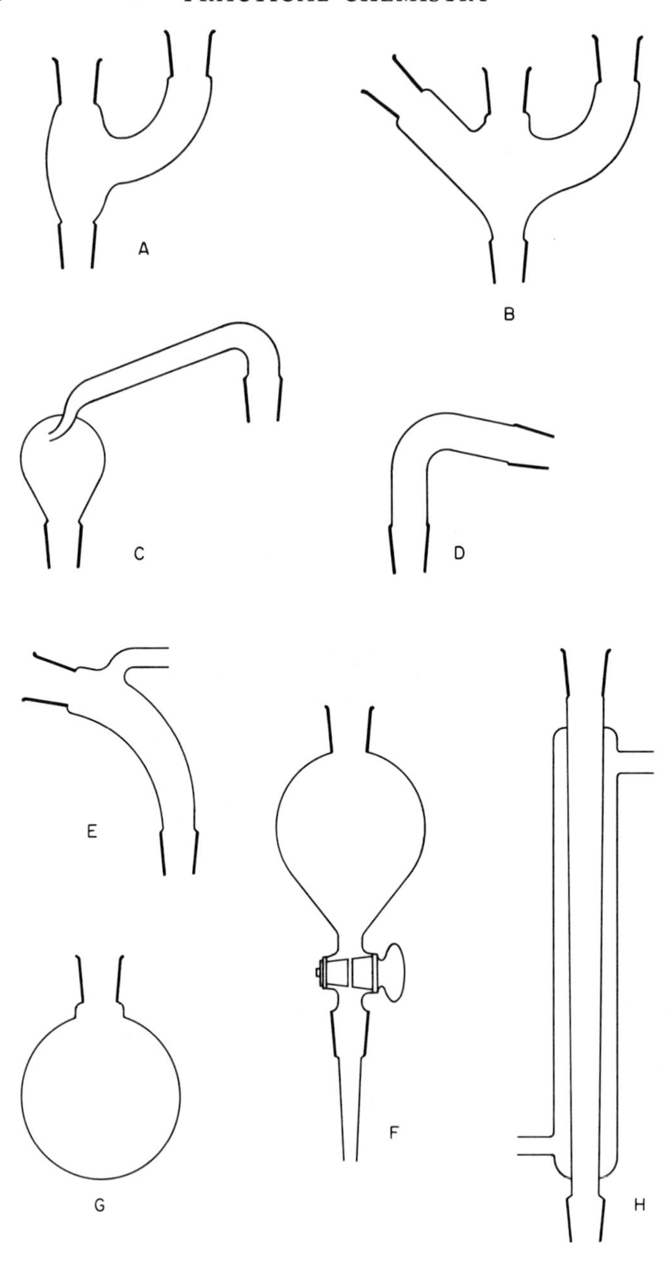

FIG. 36(a)

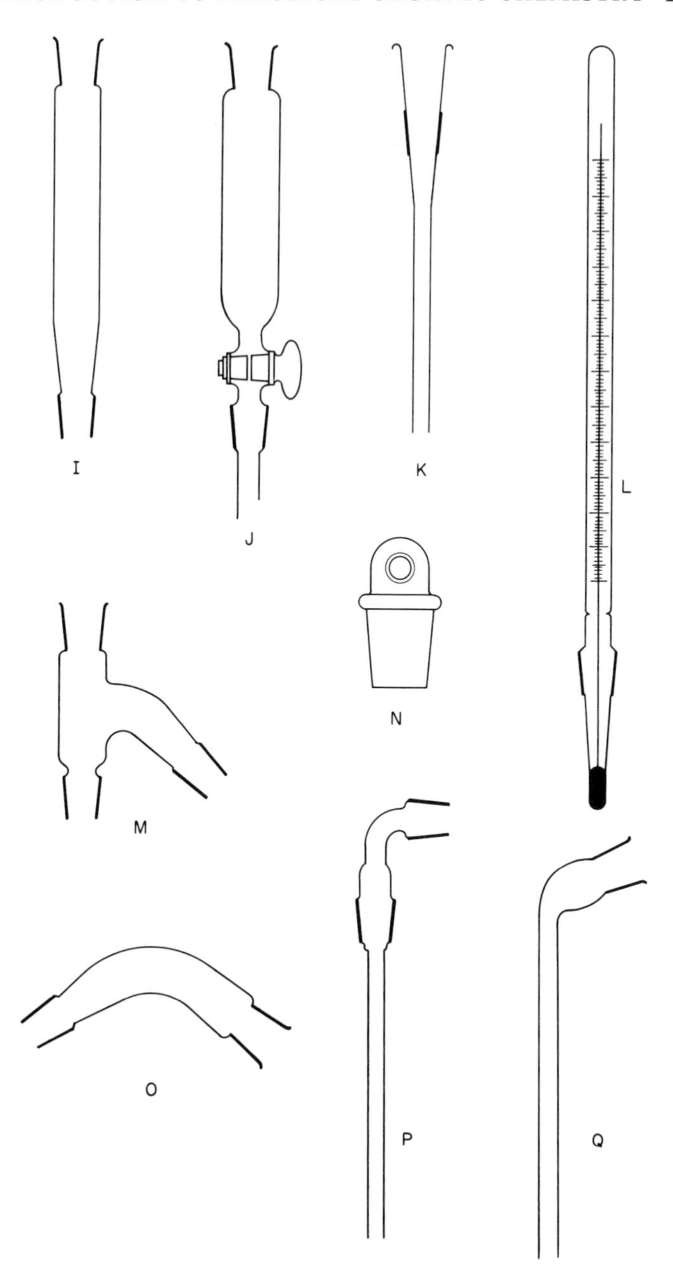

FIG. 36(b)

20
The Assessment of Purity

DETERMINATION OF THE MELTING POINT

If a compound is pure it will melt at a specific temperature and the change from solid to liquid will be sharp. If the temperature is raised gradually, this sudden change of state allows the melting point to be observed with accuracy.

The presence of an impurity has two effects on the melting point.

(*a*) it causes it to be lower than that of the pure compound, and

(*b*) it prolongs the period of melting, so that, instead of a sharp change, melting proceeds slowly over a range of temperature.

A sharp melting point is therefore a criterion of purity of a solid compound.

Experiment 178 To find the melting point of 1,3-dinitrobenzene

Apparatus: thermometer (100 °C or 360 °C); 10 cm length of glass tubing; boiling-tube; stirrer; rubber rings.

Material: 1,3-dinitrobenzene; olive oil (or glycerol (1,2,3-propanetriol) or medicinal paraffin).

Take a piece of glass tubing and rotate it in the hottest part of a Bunsen flame until the golden-yellow flame has persisted for some time and the glass has become very soft. Remove the tube from the flame and draw it out to the diameter of a match-stalk. Allow it to cool, break it at about the centre of the drawn portion, and seal off the small ends to form tubes as shown in Figure 37.

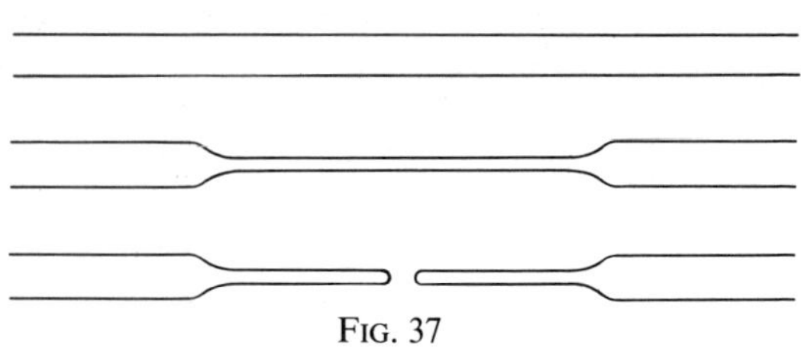

FIG. 37

Grind a small quantity of the given solid (with the flat side of a knife blade) and introduce a little into the narrow part of the tube. Set up the apparatus as shown and heat gradually, stirring the olive oil (or glycerol or medicinal paraffin) to keep its temperature uniform (see Figure 38). Watch the solid closely until it appears to collapse as it suddenly melts. Record the temperature. (Pure 1,3-dinitrobenzene melts at 90 °C.)

Note: (*i*) Do not read the temperature of solidification as some solids decompose at temperatures above their melting points.

(*ii*) Many compounds melt above 100 °C. If melting has not occurred at 95 °C, interrupt the experiment and replace the 100 °C thermometer by one graduated for higher temperatures.

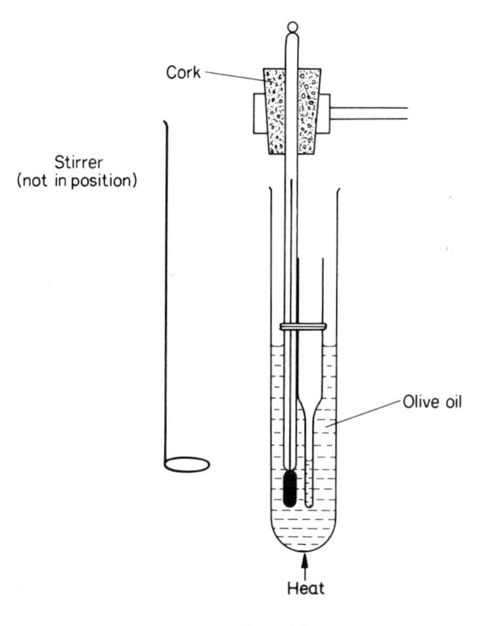

FIG. 38

(*iii*) Other suitable compounds for determination of the melting point are naphthalene, urea, and 3-nitrophenylamine.

(*iv*) The presence of an impurity always lowers the melting point (*see* Experiments 56—58.) If two substances have the same melting point they may or may not be identical. If they are identical a mixture of them will still have the same melting point, but if they are different each will act as an impurity to the other and the melting point will be lowered. This 'mixed melting-point' test is used as a final test when the identity of a substance has been established almost to the point of certainty.

DETERMINATION OF THE BOILING POINT

Experiment 179 To find the boiling point of acetone (propanone)

Apparatus: distilling flask (150 cm^3); condenser; thermometer (100 °C); flask (100 cm^3); fragments of porous pot. (Standard-joint apparatus EGGHLM.)

Material: acetone.

Arrange the apparatus as in Figure 39. Take about 50 cm^3 of acetone in the distilling flask and add a few fragments of porous pot. Attach the side tube of the flask to the condenser by a bored cork, and close the neck of the flask with a bored cork and thermometer, adjusting the thermometer so that its bulb is level with the side tube.

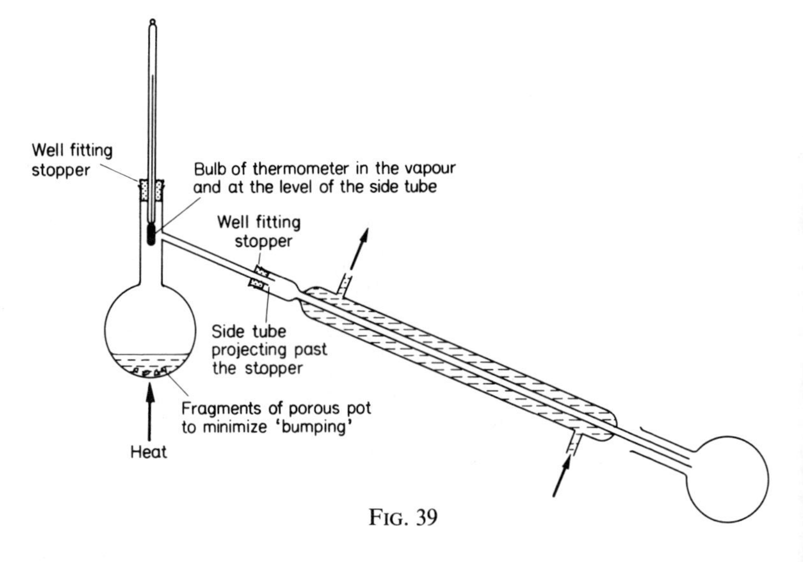

FIG. 39

Heat the flask and note the temperature when boiling begins. Record the temperature at half-minute intervals for a period of five minutes and take the average as the boiling point. The boiling point of acetone is 57 °C.

Experiment 180 Separation of a mixture of diethyl ether and aniline

Apparatus: As in Experiment 179, but with a distilling flask as a receiver as in Figure 40; thermometer (360 °C). (Standard-joint apparatus EGGHLM.)

Material: mixture of about 50 cm^3 each of diethyl ether (ethoxyethane) and aniline (phenylamine).

Set up the apparatus as described in Experiment 179, using the mixture in place of acetone. Ether is very inflammable and care must be taken to lead away any vapour which might otherwise be ignited.

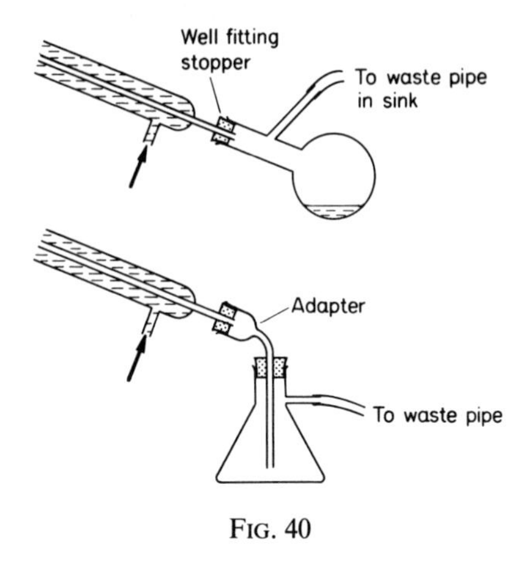

FIG. 40

Heat the flask on a water-bath containing water which has been brought to the boil elsewhere. When the temperature reaches 40 °C, detach the receiver and replace with a similar one. When no more liquid distils, remove the water-bath and heat the flask with a Bunsen flame. Discard the distillate collected between this temperature and 180 °C, and at 180 °C begin to collect the distillate of aniline in an ordinary flask (at about 130 °C allow the water to run out of the condenser). The receiver which was first used will contain almost pure ether (boiling point 35 °C) and the last receiver will contain almost pure aniline (boiling point 184 °C).

21
The Detection of Elements in an Organic Compound

Organic compounds contain carbon, and usually contain hydrogen. They may contain one or more of the elements oxygen, chlorine, bromine, iodine, nitrogen, sulphur, phosphorus; a metal may also be present. The detection of these elements is based on the following considerations.

When the compound is heated with copper(II) oxide, *carbon* is oxidized to carbon dioxide and *hydrogen* is oxidized to water.

When a compound containing a *halogen* is heated with copper, it gives a green flame; heated with calcium oxide, and then acidified with nitric acid, it gives a precipitate with silver nitrate; heated with sodium, and then acidified with nitric acid, it gives a precipitate with silver nitrate.

Many compounds containing *nitrogen* evolve ammonia when heated with soda lime. All organic compounds which contain nitrogen form sodium cyanide on fusion with sodium. If iron(II) sulphate solution is then added, sodium hexacyanoferrate(II) is left in solution.

$$2NaCN + FeSO_4 \rightarrow Fe(CN)_2 + Na_2SO_4$$
$$4NaCN + Fe(CN)_2 \rightarrow Na_4Fe(CN)_6$$

This solution with an iron(III) salt gives Prussian blue.

Compounds containing *sulphur*, when fused with sodium, give sodium sulphide, which, with acid, gives hydrogen sulphide.

Phosphorus is rarely present. When a compound containing phosphorus is heated with magnesium, the phosphorus combines to form magnesium phosphide, which, with water, gives phosphine. Phosphine (poisonous) is usually spontaneously inflammable in air, it gives white fumes with hydrogen iodide and is almost neutral to litmus.

$$Mg_3P_2 + 6H_2O \rightarrow 2PH_3 + 3Mg(OH)_2$$

If an organic compound contains a *metal*, the metal, metallic oxide or carbonate will be left as the final product on ignition, and the residue after solution in acid may be treated by the usual methods of qualitative analysis.

There is no general method of detecting *oxygen*. In some cases it will appear as water when the organic compound is heated alone, but in general its presence is inferred, its percentage being found by difference after the percentages of all other detected elements have been determined. Thus if the total percentages of elements known to be present were 80%, it would be assumed that oxygen accounted for the remaining 20%.

Experiment 181 Detection of carbon and hydrogen

Apparatus: cork fitted with a bent delivery tube.

Material: glucose (or other organic compound); anhydrous copper(II) sulphate.

Heat a crucible about half full of copper(II) oxide to redness to ensure its complete dryness and the absence of carbonate; cool in a desiccator. Fit a dry test-tube with a bent tube as shown (Figure 41).

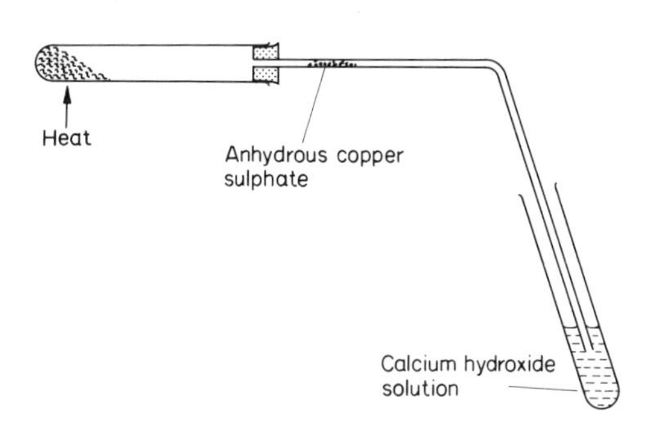

FIG. 41

Mix about a spatula-load of the given compound with about three times that quantity of dry copper(II) oxide and heat the mixture in the test-tube. Hold a test-tube half filled with calcium hydroxide solution so that any gas evolved may bubble through it. Note the cloudy suspension of calcium carbonate. Note also that the copper oxide has been reduced to copper.

$$2CuO + C \text{ (in the compound)} \rightarrow 2Cu + CO_2$$

Repeat the experiment, again with dry apparatus. Introduce a little anhydrous copper(II) sulphate into the horizontal part of the bent tube, and omit the calcium hydroxide solution. After heating, note any condensation in the cooler parts of the apparatus and the hydration of the anhydrous salt by the water formed.

$$CuO + 2H \text{ (in the compound)} \rightarrow Cu + H_2O$$

Experiment 182 Detection of nitrogen

Material: urea (or solid methyl orange, acetanilide (N-phenylethanamide or N-phenylacetamide)); soda lime; sodium. See also Experiment 185.

Method (*a*). Take enough of the given solid to cover the bottom of a test-tube and add about twice that quantity of soda lime. Heat the mixture and test the issuing gas for ammonia (red litmus paper). Many nitrogenous compounds do not give their nitrogen as ammonia, and a negative test must not be taken as a proof that nitrogen is absent. An amino-acid yields the corresponding amine.

Method (*b*). Take a small dry test-tube (about 100×16 mm). Take a piece of clean sodium no larger than a small pea and cut into fine shavings. Put half of the sodium into the tube, add about four times that quantity of the compound, and then add the rest of the sodium. Have ready a dish containing about 25 cm^3 of distilled water.

Hold a piece of wire gauze to guard yourself—**safety spectacles should also be worn**—and proceed as follows. Take the test-tube in a test-tube holder and heat gently until the first (sometimes vigorous) action has moderated, then heat strongly, concentrating the heat on the tip of the tube. The bottom of the test-tube must be kept red-hot for *a full minute*. Still guarding yourself with the gauze, plunge the red-hot end of the test-tube into the distilled water and stir with the broken part of the tube.

Filter the solution and take about 10 cm^3 in a test-tube. Add a few drops of a freshly prepared solution of iron(II) sulphate and boil to complete the formation of sodium hexacyanoferrate(II) (see earlier equations). Cool the solution under the tap, add a few drops of a solution of iron(III) chloride, and acidify with concentrated hydrochloric acid. A greenish-blue coloration, or precipitate, of Prussian blue indicates that the given compound contained nitrogen. After leaving to stand for a few minutes, filter, and examine the paper for particles of Prussian blue. These reactions comprise Lassaignés test.

Experiment 183 Detection of halogen

Material: chlorobenzene; bromoethane; iodoethane; bromobenzene; copper foil; sodium; calcium oxide.

Method (*a*). Hold a strip of copper in a holder and heat in a clear Bunsen flame until the flame is no longer coloured. Allow the copper to cool, then add a drop of the given compound. Heat again. A green flame indicates either chlorine or bromine or iodine in the compound.

Method (*b*). Take enough calcium oxide to cover the bottom of a test-tube and add a drop of the given compound. Heat the mixture until the tube gives a yellow coloration to the flame. Allow to cool and then add dilute nitric acid. Warm gently to obtain a clear solution and add a drop of silver nitrate solution.

White precipitate indicates chlorine in the given compound.
Pale yellow precipitate indicates bromine in the given compound.
Deep yellow precipitate indicates iodine in the given compound.

Method (*c*). Heat the compound with sodium[1] (see Experiment 182). After filtering, the solution (which contains sodium halide) is treated in one of the following ways, according to whether nitrogen is known to be present or absent.

(i) If nitrogen is known to be absent, acidify the filtrate with dilute nitric acid and add silver nitrate solution. Note the colour of the silver halide.

(ii) If nitrogen is known to be present, sodium cyanide (as well as the expected sodium halide) will be in the solution, and this, with acidified silver nitrate solution, will give a white precipitate of silver cyanide. The cyanide must, therefore, be removed before adding silver nitrate. Take an evaporating basin about a quarter full of the solution, add a slight excess of dilute nitric acid and evaporate (in a fume chamber) to about one-third of the total volume; the cyanide group is thus expelled as hydrogen cyanide. Cool the solution and test with silver nitrate as before.

Experiment 184 Detection of sulphur

Material: sodium benzenesulphonate (or 4-aminobenzenesulphonic acid or solid methyl orange); sodium; sodium nitroprusside (pentacyanonitrosylferrate(II)).

Fuse the compound with sodium as described in Experiment 182, method (*b*). Filter the solution and pour a drop of it into silver nitrate solution. A brown suspension of silver sulphide indicates sulphur present in the given compound. Divide the rest of the solution into two parts.

Part (*i*). Acidify with concentrated hydrochloric acid, heat, and test the vapour with lead(II) acetate solution on a filter paper. A brown stain of lead sulphide indicates sulphur.

Part (*ii*). Add a few drops of a freshly prepared (and very dilute) solution of sodium nitroprusside. A purple coloration indicates the presence of sulphur in the given compound.

---

[1] It is dangerous to use either trichloromethane (chloroform) or tetrachloromethane (carbon tetrachloride) in sodium fusion experiments.

Experiment 185 Middleton's method for detecting nitrogen, halogens and sulphur in organic compounds

Material: zinc dust.

In this method the use of sodium is avoided by using a mixture of the purest zinc dust and sodium carbonate. The reagent is made by mixing pure anhydrous sodium carbonate with twice its mass of zinc dust as intimately as possible in a mortar.

(*a*) If the given compound is a solid, take enough to give a depth of about 5 mm in the bottom of a *hard-glass* test-tube (100 × 16 mm) and add the reagent to bring the depth to 1 cm, mixing as well as possible. Now add more reagent, without mixing, until the depth is about 4 cm.

(*b*) If the given compound is a liquid, take enough of the reagent to give a depth of 1 cm, and add two or three drops of the compound. When the liquid has soaked the reagent, add more of the latter to give a total depth of 4 cm.

Use a test-tube holder and hold the tube *horizontally*. Heat the tube gently near the exposed surface of the mixture and gradually raise the temperature until the tube is red-hot there. *Very gradually* progress with the heating towards the closed end until the whole is red-hot, and finally turn the tube vertical and keep it red-hot for a full minute. Have ready an evaporating basin containing about 20 cm^3 of distilled water, and plunge the red-hot tube, closed end down, into it. Boil the contents of the basin, allow to cool, then decant through a filter. Retain the residue.

Test for nitrogen. Take about half of the filtrate, add 2 to 3 cm^3 of sodium hydroxide solution, and then iron(II) sulphate solution. Proceed as described in Experiment 182, remembering, however, that carbonate is present when acidifying. In Middleton's method, as in the sodium-fusion method, nitrogen becomes sodium cyanide during ignition.

Test for halogens. Take the remaining part of the filtrate and proceed as described in Experiment 183, method (*c*). If nitrogen is known to be present, the same procedure will be followed as in the sodium-fusion method.

Test for sulphur. Transfer the residue from the evaporating basin to a test-tube. Add about 10 cm^3 of dilute hydrochloric acid and test the issuing gas with lead(II) acetate paper.

Experiment 186 Estimation of nitrogen in urea (Kjeldahl's method)

Apparatus: long-necked boiling flask (250 cm^3); flask (500 cm^3);

conical flask; measuring cylinder; steam trap; tap-funnel. The 500 cm^3 flask and distilling apparatus are fitted as in Figure 42. (Standard-joint apparatus ACGHJ.)

Material: urea; potassium hydrogensulphate; standard (1 M) hydrochloric acid and sodium hydroxide solution.

The method depends upon the decomposition of organic nitrogen compounds by concentrated sulphuric acid to give ammonium sulphate. When the resulting solution is treated with excess of alkali the ammonia is expelled and absorbed in standard acid; the ammonia is estimated by back-titration of the standard acid. Some nitrogen compounds do not give satisfactory results, and among these, nitro-compounds are notable examples; they may, however, be reduced first to amines and then the estimation is satisfactory.

Weigh accurately about 0.5 g of urea and transfer to the long-necked flask. Add 10 cm^3 of concentrated sulphuric acid and add about a spatula-load of potassium hydrogensulphate (to raise the boiling point of the acid and make the decomposition of the nitrogen compound complete). Heat to boiling and continue to boil until the slight charring has cleared, then allow to cool to room temperature. Dilute by adding about 200 cm^3 of water. Transfer the solution to the distilling flask and add fragments of porous pot (to prevent bumping during boiling). Fit up the apparatus as in Figure 42. Take

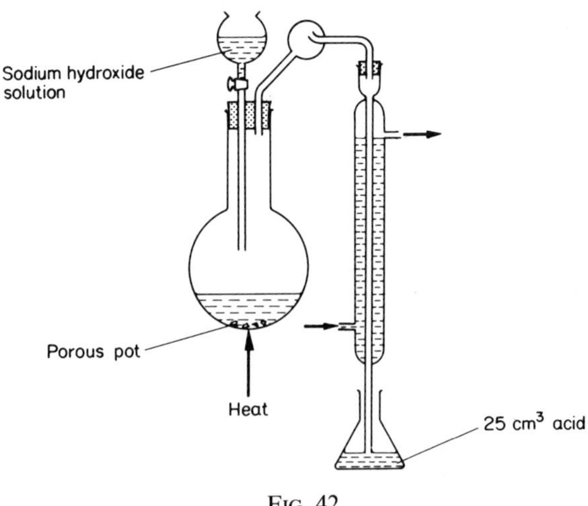

FIG. 42

exactly 25 cm^3 of the acid in the conical flask and arrange to keep the end of the condenser just below the surface of the acid throughout the period of distillation.

Dissolve a stick of sodium hydroxide (about 20 g) in 100 cm^3 of water and transfer to the tap funnel. Run this solution into the flask and heat to boiling. Ammonia is evolved and passes into the standard acid. Continue distillation until half the solution has distilled over, by which time all the ammonia has been expelled. Titrate the contents of the conical flask against standard alkali in the burette, using methyl orange as indicator.

Specimen results:

Mass of urea = 0.48 g

Volume of acid in receiver = 25 cm^3

Volume of alkali to neutralize acid after absorption of ammonia = 9.15 cm^3

Volume of acid used by ammonia = $(25-9.15)$ cm^3 = 15.85 cm^3

$$1000 \text{ cm}^3 \text{ 1 M HCl} \equiv 17 \text{ g of ammonia} \equiv 14 \text{ g of nitrogen}$$

$$15.85 \text{ cm}^3 \text{ 1 M HCl} \equiv \frac{14 \times 15.85}{1000} \text{ g nitrogen}$$

$$\% \text{ of nitrogen in urea} \equiv \frac{14 \times 15.85 \times 100}{1000 \times 0.48}$$

$$\equiv 46.2 \text{ (theoretical value 46.7)}$$

22
Hydrocarbons

ALIPHATIC HYDROCARBONS

Hydrocarbons of the aliphatic series are divided into three classes:

1. Alkanes or paraffins: examples methane CH_4, ethane C_2H_6.
2. Alkenes or olefines: examples ethene (ethylene) C_2H_4, propene C_3H_6.
3. Alkynes or acetylenes: example ethyne (acetylene) C_2H_2.

1. *Alkanes* are fully saturated hydrocarbons, i.e. each carbon atom is exerting its full tetravalency.

$$\begin{array}{ccc} & H & \\ & | & \\ H- & C & -H \\ & | & \\ & H & \end{array} \text{ methane} \qquad \begin{array}{cccc} & H & H & \\ & | & | & \\ H- & C- & C & -H \\ & | & | & \\ & H & H & \end{array} \text{ ethane} \qquad \begin{array}{ccccc} & H & H & H & \\ & | & | & | & \\ H- & C- & C- & C & -H \\ & | & | & | & \\ & H & H & H & \end{array} \text{ propane}$$

They are (*a*) unaffected by almost all reagents, and
(*b*) form compounds by *substitution*; a hydrogen atom must be displaced to allow another univalent atom or group to enter the molecule.

2. *Alkenes* are not saturated: the molecular formula shows two atoms of hydrogen less than would be required for full saturation:

$$\begin{array}{cccc} & H & H & \\ & | & | & \\ H- & C- & C & -H \\ & | & | & \end{array} \text{ ethene} \qquad \begin{array}{ccccc} & H & H & H & \\ & | & | & | & \\ H- & C- & C- & C & -H \\ & | & | & | & \\ & H & & & \end{array} \text{ propene}$$

The free valencies link to form a 'double bond' which readily breaks in the presence of a reagent with which the free valencies can unite.

$$\begin{array}{cccc} & H & H & \\ & | & | & \\ H- & C- & C & -H \\ & | & | & \end{array} \rightarrow \begin{array}{cccc} & H & H & \\ & | & | & \\ H- & C- & C & -H \\ & \smile & & \end{array} \rightarrow \begin{array}{cccc} & H & H & \\ & | & | & \\ H- & C- & C & -H \\ & | & | & \end{array} \rightarrow \begin{array}{cccc} & H & H & \\ & | & | & \\ H- & C- & C & -H \\ & | & | & \\ & Br & Br & \end{array}$$

(in presence of bromine)

This property of forming *addition* compounds is the chief characteristic of alkenes. The formation of addition compounds by other unsaturated groups will be noted later. (See Experiments 227 and 256(*b*).)

3. *Alkynes* are still less saturated: the molecular formula shows four hydrogen atoms less than required for saturation.

$$\begin{array}{ccccccc} & & | & & | & & \\ H & — & C & — & C & — & H \\ & & | & & | & & \end{array} \quad \text{ethyne (acetylene)}$$

The free valencies form a 'triple bond', which breaks to form addition compounds. Thus with bromine, ethyne forms tetrabromoethanes.

$$H—C—C—H + 2Br_2 \rightarrow \begin{array}{ccccccc} & & Br & & Br & & \\ & & | & & | & & \\ H & — & C & — & C & — & H \\ & & | & & | & & \\ & & Br & & Br & & \end{array}$$

Ethyne forms metallic derivatives which still retain the triple bond. They are very unstable and explosive.

ALKANES

Experiment 187 Preparation of the alkane, methane

Apparatus: as in Figure 43; mortar and pestle.

Material: fused sodium acetate (ethanoate);[1] soda lime; bromine.

About 20 g of sodium acetate (ethanoate) and an equal mass of soda lime are mixed intimately in a mortar and transferred to the hard glass tube (wrapped with gauze). The tube is heated, care being

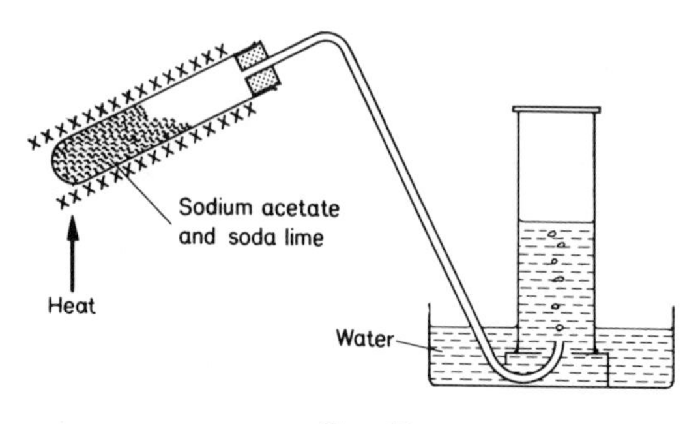

FIG. 43

[1] If fused sodium acetate is not available, the hydrated salt must first be heated in an evaporating basin on a water-bath, with constant stirring, until anhydrous.

taken that water is not sucked back because of variations of temperature. Collect three jars full of methane.

$$CH_3COONa + NaOH \rightarrow CH_4 + Na_2CO_3$$

Perform the following experiments.

(*a*) Apply a light to the last jar. Note colour of flame, and test the residual gas for carbon dioxide.

(*b*) Take the second jar, hold a dry jar mouth downward over it, and allow the methane to pass upward into the dry jar. While the upper jar is still inverted, light the gas in it. Note that moisture is formed. Tests (*a*) and (*b*) establish the presence of carbon and hydrogen in the hydrocarbon.

(*c*) Invert the third jar in a trough of water. Pour a drop or two of bromine into the water and adjust the jar to include the bromine in its mouth. Leave for twenty-four hours and then note that (i) methane is coloured by bromine, and (ii) the level of the water has not changed. Bromine has no appreciable action on methane (or on any members of the paraffin series) at room temperature.

By substituting the sodium salt of the next higher acid (propanoic acid), ethane may be prepared in a similar manner.

ALKENES

Experiment 188 Preparation of the alkene, ethene (ethylene)

Apparatus: large flask (1.5 dm^3) fitted with tap-funnel; delivery tubes; wash-bottle containing concentrated sodium hydroxide solution and fitted as in Figure 44.

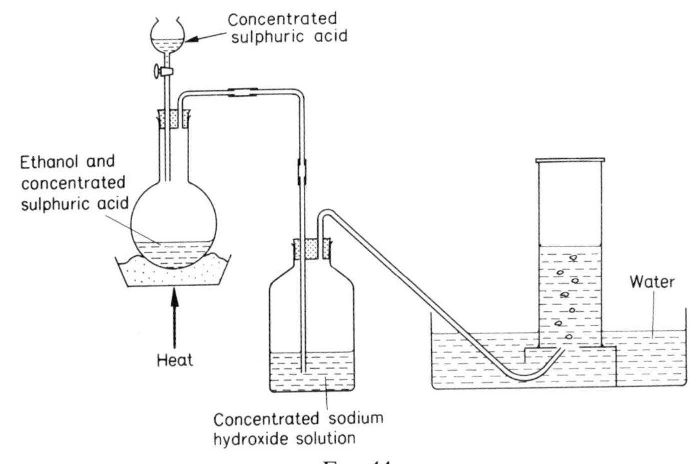

FIG. 44

Material: ethanol; anhydrous aluminium sulphate; bromine.

30 cm^3 of ethanol are poured into the flask and about five spatula-loads of aluminium sulphate are added to prevent frothing and charring during the subsequent heating. (Dry sand may be substituted but is less efficient.) 80 cm^3 of concentrated sulphuric acid is slowly run in from the tap-funnel. If a large yield of ethene is required a mixture of equal volumes of ethanol and acid is prepared and run in from the tap-funnel as required to maintain the flow of gas. The flask is heated on a sand-tray until action begins, after which the temperature is lowered a little. During the action some ethanol is oxidized at the expense of the acid, which is reduced to give a corresponding amount of sulphur dioxide, and it is to remove this that the gas is washed through sodium hydroxide solution.

$$C_2H_5OH + (H_2SO_4) \rightarrow C_2H_4 + (H_2SO_4 \cdot H_2O)$$

Collect four jars of ethene and carry out tests (*a*), (*b*) and (*c*) as given for methane (Experiment 187). Note points of similarity and of difference, and account for them knowing that ethene is a hydrocarbon but with a higher proportion of carbon than its corresponding paraffin, and that it is unsaturated.

To the fourth jar add a little dilute potassium permanganate (manganate(VII)) solution which has been made alkaline with a few drops of sodium hydroxide solution. On shaking, the purple colour changes to green potassium manganate (manganate(VI)) and later a brown suspension of manganese(IV) oxide is formed. The ethene has been oxidized to 1,2-ethanediol (ethylene glycol).

$$\begin{aligned} C_2H_4 + 2OH^- &\rightarrow C_2H_4(OH)_2 + 2e^- \\ MnO_4^- + e^- &\rightarrow MnO_4^{2-} \\ MnO_4^{2-} + 4H^+ + 2e^- &\rightarrow MnO_2\downarrow + 2H_2O \end{aligned}$$

Experiment 189 Preparation of cyclohexene

Apparatus: JAGIMLHEG.

Material: cyclohexanol; concentrated phosphoric acid (syrupy); Lissapol NX or as in Experiment 178; **ICE**.

Fill the tube I with glass beads so that it acts as a fractionating column. Into the tap-funnel put 25 cm^3 of cyclohexanol and into the distillation flask (G_1) 6 cm^3 of concentrated phosphoric acid. The receiver flask (G_2) should be cooled in ice. The distillation flask should be heated to 165 °C in an oil bath, e.g. Lissapol NX, for about an hour, during which time the cyclohexanol is dripped in. Then the temperature is raised to 190 °C for ten minutes. The temperature at the top of the fractionating column should not rise above 90 °C.

To the distillate add sodium chloride until no more will dissolve, separate the upper (organic) layer and then dry it with anhydrous magnesium sulphate. The crude product may be redistilled, if desired, in a small distillation apparatus (collect the fraction boiling at 80–83 °C).

Experiment 190 The reactions of alkanes and alkenes

Material: cyclohexane; 2-methylpropene or cyclohexene.

(*a*) Put a few drops of the substance on a watch-glass or in a crucible and set fire to them using a spill. Observe that the unsaturated substance burns with the more luminous and sooty flame.

(*b*) Put a few drops of potassium permanganate (manganate(VII)) into 2 cm^3 of sodium carbonate solution in a test-tube, and add it to a few drops of the substance in another tube. The purple colour is replaced by a green colour and then there may be a dark brown precipitate of manganese(IV) oxide in the case of the alkene. Acidified (sulphuric) permanganate may also be tried: the solution usually goes straight from purple to colourless in this case.

(*c*) Add bromine water slowly to a few drops of the substance: rapid decolorization occurs in the case of the alkene.

(*d*) Cool 5 cm^3 of the substance under the cold water tap and then add an equal volume of cold, moderately concentrated sulphuric acid (about 2.5 cm^3 of the concentrated acid poured into 2.5 cm^3 of water). Cool and shake the mixture for several minutes and observe that the alkene mixes (and reacts) with the acid. Wait a few more minutes and then pour the mixture into an equal volume of cold water: the alcohol corresponding to the alkene separates from the dilute sulphuric acid.

Experiment 191 The polymerization of an alkene

Material: styrene (phenylethene); paraffin.

Put 10 cm^3 of styrene into 20 cm^3 of paraffin in a small round-bottomed flask fitted with an air condenser. Boil the liquids for about thirty minutes and then allow them to cool. Pour the mixture into 100 cm^3 of methanol and filter off the polystyrene: the white solid hardens if it is pressed with a spatula under the surface of some more methanol.

ALKYNES

Experiment 192 Preparation of the alkyne, ethyne (acetylene)

Apparatus: as in Figure 45.
Material: calcium carbide (dicarbide); bromine.

The flask at the start should be quite dry. The bottom of the flask is covered with a thin layer of sand and a few pieces of calcium carbide are placed on this. Water is added drop by drop to obtain a steady flow of gas.

$$CaC_2+2H_2O \rightarrow C_2H_2+Ca(OH)_2$$

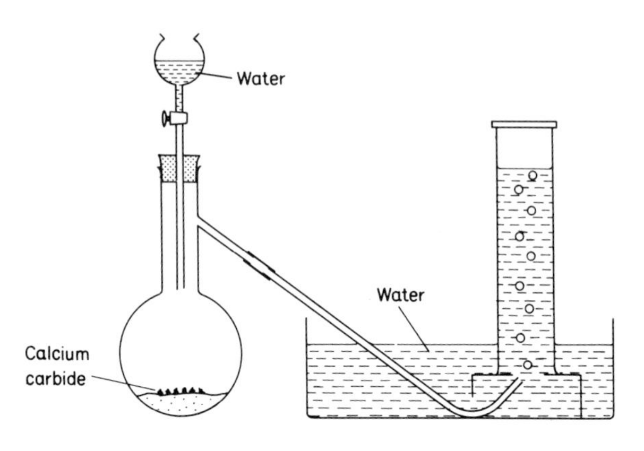

FIG. 45

Collect four jars of ethyne and carry out the following tests.

(*a*) Ignite the gas in the last jar. Note the smoky flame and the deposit of carbon. Compare with methane and ethene.

(*b*) Stand the second jar over bromine in water as was done with methane and ethene. The gas reacts to form tetrabromoethane.

$$C_2H_2+2Br_2 \rightarrow C_2H_2Br_4$$

(*c*) Prepare an ammoniacal solution of copper(I) chloride by dissolving a spatula-load of copper(II) carbonate in concentrated hydrochloric acid in a test-tube, adding a piece of copper and boiling the solution for a few minutes, then pouring it into another test-tube half full of water. Decant, and dissolve the white solid, copper(I) chloride, in ammonia solution. Now pour this solution into the third gas-jar. The red precipitate is copper(I) acetylide (dicarbide).

$$C_2H_2+2CuCl \rightarrow Cu_2C_2+2HCl \text{ (becoming } 2NH_4Cl)$$

(*d*) Prepare an ammoniacal solution of silver(I) oxide as described in Experiment 113(*e*). Pour it into the fourth gas-jar. A white precipitate of silver acetylide, Ag_2C_2, is obtained. Place a *small* portion on gauze and heat by means of a Bunsen burner. (*Care!*) The silver acetylide suddenly decomposes explosively.

AROMATIC HYDROCARBONS

Compounds derived from benezene or from homologues of benzene are called aromatic compounds. The structure of the benzene molecule is fundamentally different from that of any of the hydrocarbons already mentioned, and, since structure determines the properties of organic compounds, it would be expected that compounds derived from benzene would show marked differences from corresponding compounds derived from ethane or methane. At the same time it would be expected that the properties of an attached group would exert a similar influence on the properties of a compound, regardless of the particular radical to which it was attached. Aromatic compounds do, in fact, show the influence of both the radical and the group in the manner expected. Thus, acetic (ethanoic) acid and benzoic (benzenecarboxylic) acid (C_6H_5COOH) are similar in their acidic properties, but differ in those properties which are functions of the methyl (CH_3) or the phenyl (C_6H_5) radical. Further, the fundamentally different structures of the methyl and phenyl radicals influence the characteristics of the attached groups, and in some cases the resulting products are markedly dissimilar; thus, methylamine (CH_3NH_2) is unlike aniline (phenylamine $C_6H_5NH_2$) in many respects. It is because of these differences that it is advisable to divide organic compounds into two series, aliphatic and aromatic, and study them separately.

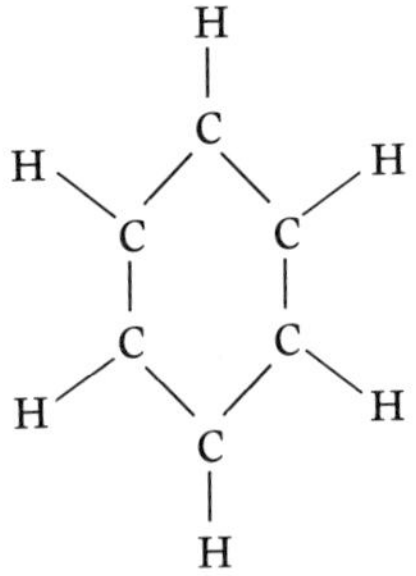

```
            H
            |
            C
   H      /   \      H
     \   /     \   /
       C         C
       |         |
       C         C
     /   \     /   \
   H       \ /       H
            C
            |
            H
```

FIG. 46

STRUCTURE OF BENZENE

Reference should be made to a thoeretical text-book for a full account of the work of Kekulé and others in elucidating the structure of the benzene molecule; it is sufficient here to state that the six carbon atoms are arranged in the form of a ring, with a hydrogen atom attached to each carbon atom (see Figure 46). In all attempts to assign a formula to benzene, the chief concern has been to overcome the difficulty arising from the apparent trivalency of carbon. For our present purpose this difficulty is met by accepting the view that in each of the six carbon atoms the fourth valency is usually 'dormant'. One example is given in Experiment 204 to show that, on rare occasions, the dormant valencies can become active, benzene thus behaving as an unsaturated compound. In the very great majority of cases, however, benzene compounds are made by substituting, either directly or indirectly, a univalent group for each hydrogen atom displaced; benzene behaves, normally, as a saturated compound. The toxic nature of benzene vapour must be borne in mind when performing experiments.

In the hexagon and circle descriptions of the benzene molecule there is, by convention, a carbon and a hydrogen atom at each of the six corners.

23
Halogen Compounds

ALKYL HALIDES

When the alkane, methane is mixed with chlorine and the mixture exposed to sunlight, hydrogen atoms are slowly, and progressively, displaced by chlorine.

$$\begin{array}{c}\text{H}\\ |\\ \text{H—C—H}\\ |\\ \text{H}\end{array} \rightarrow \begin{array}{c}\text{H}\\ |\\ \text{H—C—Cl}\\ |\\ \text{H}\end{array} \rightarrow \begin{array}{c}\text{Cl}\\ |\\ \text{H—C—Cl}\\ |\\ \text{H}\end{array} \rightarrow \begin{array}{c}\text{Cl}\\ |\\ \text{H—C—Cl}\\ |\\ \text{Cl}\end{array} \rightarrow \begin{array}{c}\text{Cl}\\ |\\ \text{Cl—C—Cl}\\ |\\ \text{Cl}\end{array}$$

methane	methyl chloride (chloro-methane)	methylene dichloride (dichloro-methane)	chloroform (trichloro-methane)	carbon tetrachloride (tetrachloro-methane)

These chloro-compounds are of considerable importance and, although none of them are actually made by direct chlorination of methane but by various alternative methods, the diagram shows how they are related.

Methyl chloride, the first product of chlorination, is an example of an alkyl halide. Other compounds of similar structure may be prepared, by various means, having bromine or iodine in place of chlorine, and having the ethyl, propyl or other alkyl radical in place of methyl. All such compounds are known as alkyl halides but are frequently named as haloalkanes.

There are two suitable general methods of preparation, both starting with the corresponding alcohol, and the essential reaction in both is the replacement of the hydroxyl (—OH) group of the alcohol by an atom of the halogen, e.g.

$$\begin{array}{c}\text{H}\;\;\text{H}\\ |\;\;\;|\\ \text{H—C—C—OH}\\ |\;\;\;|\\ \text{H}\;\;\text{H}\end{array} \rightarrow \begin{array}{c}\text{H}\;\;\text{H}\\ |\;\;\;|\\ \text{H—C—C—Br}\\ |\;\;\;|\\ \text{H}\;\;\text{H}\end{array}$$

Method (*i*). The hydracid of the halogen is prepared *in situ* with the appropriate alcohol. This is a suitable method of introducing —Cl or —Br, but not —I.

Method (*ii*). Phosphorus and the halogen (functioning as the phosphorus halide) react with the alcohol. This method is usually used for introducing —Br or —I.

Experiment 193 Preparation of bromoethane[1] (ethyl bromide)

Formula: C_2H_5Br.

Physical properties: colourless liquid, immiscible with water, boiling point 39 °C, density 1.5 g cm^{-3}.

Apparatus: distilling flask (500 cm^3); condenser; adapter; conical flask (200 cm^3); measuring cylinder; separating funnel; distilling flask (150 cm^3); thermometer (100 °C); water-bath; sand-tray. (Standard-joint apparatus GHMQL, EG.)

Material: ethanol; potassium bromide; anhydrous calcium chloride.

Arrange the apparatus as shown in Figure 47. Take 40 cm^3 of ethanol in the large distilling flask, and add 50 cm^3 of concentrated sulphuric acid, slowly and with shaking, from the measuring cylinder (using a thistle-funnel). Allow the mixture to cool to room temperature.

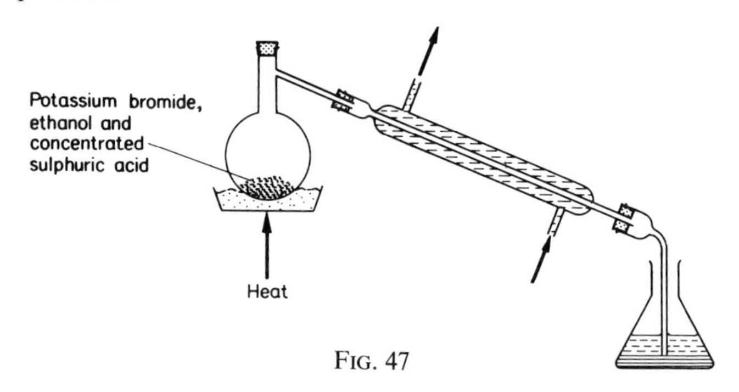

FIG. 47

Add 50 g of potassium bromide, cork the flask, and gently heat it on the sand-tray. The oily liquid which falls to the bottom of the water in the receiver is bromoethane. When no more oily drops can be seen in the condenser, remove the Bunsen and transfer the contents of the receiver into the separating funnel.

Run off the lower layer temporarily into a flask and discard the upper (aqueous) layer. Return the lower layer to the funnel and add an approximately equal volume of a sodium carbonate solution which has been prepared by dissolving about 5 g in 100 cm^3 of water. Shake the liquids, taking care to remove the stopper at frequent

[1] See also Experiment 210, p. 206, which is a small-scale version.

intervals to avoid a high pressure of carbon dioxide (formed by the action of acid impurity on the carbonate). Remove the lower layer as before, discard the aqueous layer and return the lower layer to the funnel. Wash with water and then run off the lower layer into a flask. Add enough granulated calcium chloride to cover the bottom of the flask, cork and leave overnight to dry.

Decant the clear liquid into a small distilling flask and attach the flask to a sloping condenser. Close the neck of the flask with a bored cork and thermometer, and adjust the latter so that its bulb is at the level of the side tube. Use a small dry flask as a receiver. Heat the liquid over a water-bath at a moderate temperature so that distillation proceeds steadily. Collect the fraction between 35 °C and 43 °C. If the product is to be kept for some time it should be transferred to an amber-coloured bottle and stored out of direct sunlight. Yield 35 g.

$$C_2H_5OH + KBr + H_2SO_4 \rightarrow C_2H_5Br + KHSO_4 + H_2O$$

Experiment 194 Preparation of 1-bromobutane (butyl bromide)

Apparatus: as in Figures 48 and 49.

Material: 1-butanol; sodium bromide; anhydrous magnesium sulphate.

To 20 cm³ of water in a small, round-bottomed flask, add 15 g of sodium bromide and 12 cm³ of butanol; agitate the mixture until a

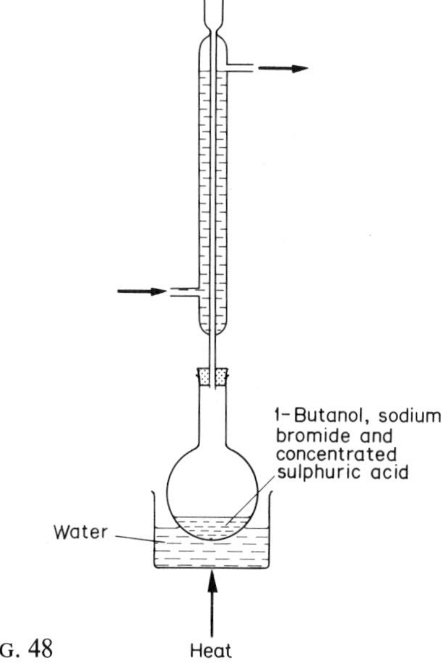

FIG. 48

solution is obtained. Then add, with constant shaking, and cooling if the temperature is likely to rise above 40 °C, 15 cm^3 of concentrated sulphuric acid. Add a few anti-bumping granules to the mixture and then boil it in a fume chamber under a reflux condenser (see Figure 48) for about an hour. The purpose of the condenser here is to condense vapour that otherwise would escape and return it, as liquid, to the reaction mixture in the flask; hence its name, reflux condenser.

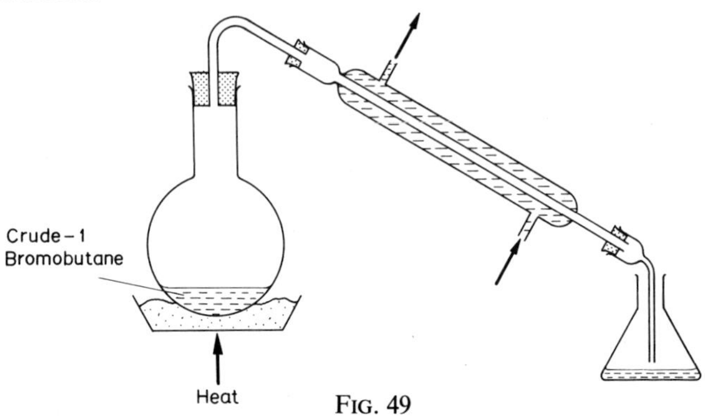

FIG. 49

Allow the flask and its contents to cool before re-arranging the apparatus for simple distillation (see Figure 49). Heat the flask until no more oily drops distil over. Put the distillate in a separating funnel, and after separating any water which forms the upper layer, rinse the crude product successively with 10 cm^3 of water, an equal volume of concentrated hydrochloric acid, 10 cm^3 of water again, 10 cm^3 of 0.5 M sodium hydrogencarbonate solution, and finally with another 10 cm^3 of water. Dry the bromobutane with 1 g of anhydrous magnesium sulphate for fifteen minutes before filtering it into a small flask, from which it may be redistilled if required (boiling point 102 °C).

Experiment 195 Preparation of iodomethane (methyl iodide)

Formula: CH_3I.

Physical properties: colourless immiscible liquid, boiling point 45 °C, density 2.3 g cm^{-3}.

Apparatus: flask (250 cm^3); condenser; water-bath; separating funnel; thermometer (100 °C); measuring cylinder; small corked flask; adapter; distilling flask (150 cm^3). (Standard-joint apparatus GHMLQ.)

Material: methanol; dry red phosphorus; iodine; anhydrous calcium chloride.

Measure 25 cm^3 of methanol into the flask and add 3 g of red phosphorus; fit a reflux condenser as shown in Figure 48.

Weigh 30 g of iodine in a small corked flask. Detach the condenser and add a small portion of the iodine, then attach the condenser again. Proceed in this way until the whole of the iodine has been added (half an hour).

Heat the flask on a water-bath for half an hour, then leave overnight.

Transfer the liquid to a distilling flask connected to a sloping condenser as in Figure 47. Distil over a water-bath and collect the distillate under water in the receiver.

Shake the distillate with sodium hydroxide solution until the lower layer is no longer coloured with excess iodine. Run off the lower layer into a flask, add pieces of granulated calcium chloride and leave to dry.

Distil the clear liquid in a small distilling flask fitted with a thermometer and attached to a sloping condenser. Collect the fraction distilling between 43 °C and 46 °C and store in a coloured bottle. Yield 20 g.

$$3CH_3OH + PI_3 \rightarrow 3CH_3I + H_3PO_3$$

The red phosphorus must be dry; fresh stock should be used.

Experiment 196 Preparation of iodoethane (ethyl iodide)

The procedure is the same as for iodomethane using 35 cm^3 of ethanol in place of the 25 cm^3 of methanol. The dried iodoethane is distilled between 68 °C and 73 °C.

$$3C_2H_5OH + PI_3 \rightarrow 3C_2H_5I + H_3PO_3$$

Iodoethane is a colourless liquid, immiscible with water, having a boiling point of 72 °C and a density of 2 g cm^{-3}.

Experiment 197 Reactions of alkyl halides

Apparatus: cork with bent tube.

Material: chloropropane; iodomethane; iodoethane; bromoethane; sodium ethoxide; a solution of potassium hydroxide in ethanol; zinc dust.

(*a*) *Reaction with silver nitrate.* Add about 1 cm^3 of silver nitrate solution to two drops of iodomethane. Note the precipitate of silver iodide (yellow). Iodomethane readily hydrolyses in aqueous silver nitrate. Repeat with iodoethane in place of the methyl compound. Compare the incomplete precipitation with the former case. Now add a few drops of dilute nitric acid and heat. The more complete precipitation of silver iodide is due to hydrolysis of the alkyl halide by the acid. Repeat using bromoethane in place of iodoethane.

$$C_2H_5I + H\cdot OH \rightarrow C_2H_5OH + HI$$

(*b*) *Reaction with sodium ethoxide.* Take a test-tube containing two spatula-loads of sodium ethoxide. Add about five drops of bromoethane and allow it to soak into the solid. Warm gently. Recognize the smell of diethyl ether. (See also Experiment 221.)

(*c*) *Reaction with potassium hydroxide.* Take ten drops of iodoethane in a test-tube. Add about 5 cm^3 of the solution of potassium hydroxide in ethanol and fit the tube with a cork and bent delivery tube. Arrange to collect any gas evolved over water in a test-tube. Warm the mixture. Show that the gas collected burns with a smoky flame. The gas is ethene, but contains some iodoethane as impurity.

$$C_2H_5I + KOH \rightarrow C_2H_4 + KI + H_2O$$

(*d*) *Reduction.* Prepare a zinc-copper couple by sprinkling about five spatula-loads of zinc dust into a beaker half full of copper(II) sulphate solution. Leave for a minute, then decant and wash by decantation. Wash with ethanol and filter. Take about 4 cm^3 of iodoethane in a test-tube, add an equal volume of ethanol, then the zinc-copper couple. Close the test-tube with a cork and delivery tube and collect the gas (ethane) over water. Burn the gas. Note the slightly luminous flame (tinged with green due to iodoethane as impurity).

$$Zn + C_2H_5OH + C_2H_5I \rightarrow C_2H_6 + Zn^{2+} + I^- + C_2H_5O^-$$

AROMATIC HALIDES

INTRODUCTION OF —Cl GROUP INTO A SIDE-CHAIN

Toluene consists essentially of the two radicals, phenyl (—C_6H_5) and methyl (—CH_3) linked together. Actions affecting the methyl group (the side-chain) follow the general course expected with aliphatic compounds. Thus the introduction of the —Cl group into the side-chain requires a different method from either of the two given for introducing this group into the benzene nucleus (see Experiments 271, 272, pp. 267–8). The experiment should be conducted in bright sunlight (a factor which is required for displacing hydrogen by chlorine in methane), and the toluene must be kept at boiling point.

Experiment 198 Preparation of benzyl chloride

Formula: $C_6H_5CH_2Cl$.

Physical properties: colourless fuming, irritating liquid, boiling point 176 °C.

Apparatus: retort (250 cm^3); condenser fitted as in Figure 50(*b*) or Standard-joint apparatus as in Figure 50(*a*); distilling flask; sand-tray; thermometer (360 °C).

Material: toluene (methylbenzene); steady supply of chlorine; concentrated sulphuric acid in a drying bottle; phosphorus trichloride (not essential).

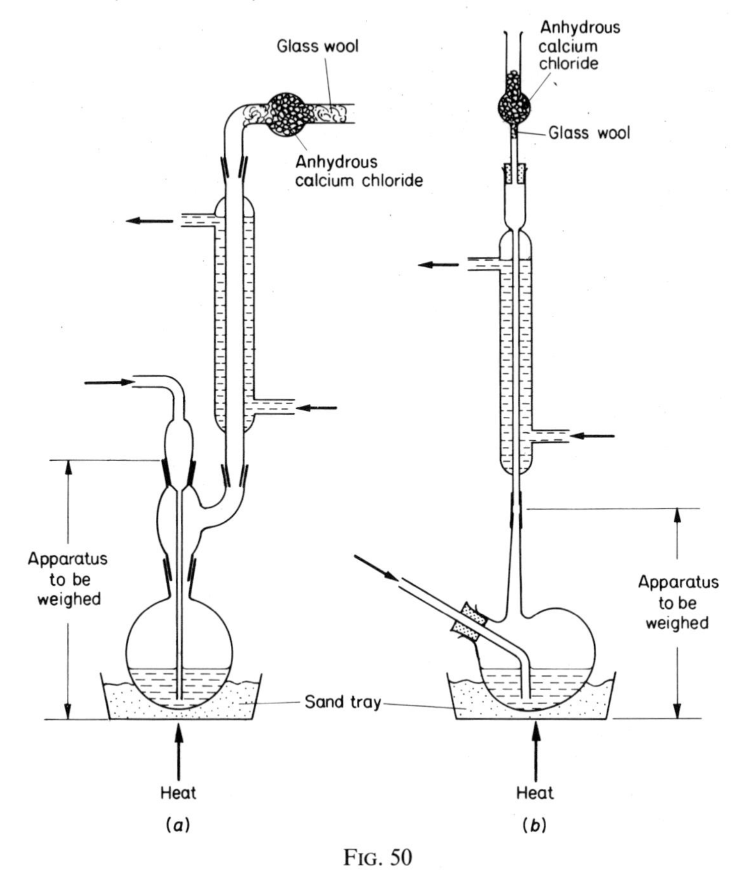

FIG. 50

Weigh the retort together with its leading-in tube. Add 50 cm^3 of toluene and weigh again. The addition of a little phosphorus trichloride helps the chlorination, but it is not essential. If it is decided to use it, add three or four drops and weigh again. This will be the total mass referred to later.

Connect the retort to the condenser either by a bored cork or by a piece of overlapping rubber tubing. Connect the flask in which the chlorine is to be generated to the drying bottle and that in turn to the tube leading into the retort.

It is advisable to close the calcium chloride tube with a cork and a delivery tube which leads to the window or to an inverted funnel dipping into sodium hydroxide solution.

Heat the retort on a sand-tray until the toluene boils, and then pass in dry chlorine. The liquid becomes progressively darker and hydrogen chloride is evolved. After three hours of steady chlorination allow the retort to cool. Since 1 mole of toluene (92 g) is required to lose 1 g of hydrogen and gain 35.5 g of chlorine, the required gain for the mass of toluene taken can be calculated. Find whether the original total mass has gained by the calculated amount, and if not, reassemble the apparatus and continue to chlorinate as before. When the action is complete, transfer the liquid to a distilling flask and distil. Unchanged toluene distils off at 110 °C and benzyl chloride ((chloromethyl)benzene) at 176°C. Collect the fraction distilling between 175 °C and 178 °C. Note the irritating smell and compare with its isomer, 2-chlorotoluene (chloro-2-methylbenzene). Yield 40 g.

DIHALIDES

Experiment 199 Preparation of 1,2-dibromoethane (ethylene dibromide)

Apparatus: as in Figure 51.
Material: as in Experiment 188.

Using initial quantities the same as those in Experiment 188, and having ready an additional quantity of a mixture of ethanol and acid (150 cm^3 of each), prepare the gas and pass it through the two bottles containing bromine covered with a little water. Use 25 cm^3 of bromine in the first bottle and 10 cm^3 in the second (measured in a cylinder).

The exit of the last bottle is connected to a U-tube filled with a loosely packed mixture of soda lime and glass-wool to prevent traces of the irritating bromine vapour from escaping.

Continue to pass the gas, adding more mixture from the tap-funnel as required, until the bromine is decolorized to an oily liquid, dibromoethane (total time about two hours)[1].

$$CH_2 = CH_2 + Br_2 \rightarrow CH_2BrCH_2Br$$

The product is washed in a separating funnel, first with sodium

[1] Note carefully that the wash-bottle containing sodium hydroxide solution must be disconnected *before* the Bunsen is removed at the end of the experiment.

hydroxide solution diluted with its own volume of water, then with water. It is run off into a flask, left to dry with calcium chloride overnight, and then distilled. The fraction collected at 130 °C–132 °C is retained. Yield 50 g.

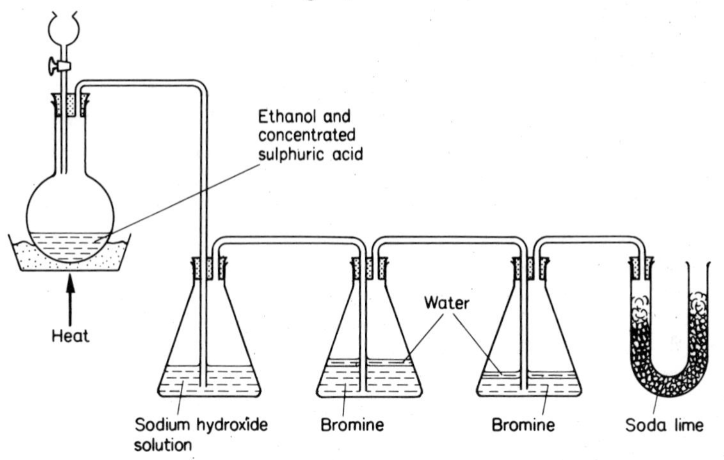

FIG. 51

TRIHALIDES

Chloroform ($CHCl_3$) and iodoform (CHI_3) are halogenated compounds of methane. They are both prepared from either ethanol or acetone (propanone) by the action of a hypochlorite (chlorate(I)) or hypoiodite (iodate(I)). The action in both cases probably takes place in three stages, involving: (*a*) oxidation (when the alcohol is used) by the halogen; (*b*) halogenation of the oxidized product; (*c*) alkaline hydrolysis of the compound so formed.

(*a*) $CH_3CH_2OH + Cl_2 \rightarrow CH_3CHO + 2HCl$
CH_3CHO (acetaldehyde, ethanal)

(*b*) $CH_3CHO + 3Cl_2 \rightarrow CCl_3CHO + 3HCl$
CCl_3CHO (chloral, trichoroethanal)

(*c*) $2\,CCl_3{\cdot}CHO + Ca(O{\cdot}H)_2 \rightarrow 2CHCl_3 + Ca^{2+}(HCOO^-)_2$
$CHCl_3$ (chloroform, trichloromethane); $Ca^{2+}(HCOO^-)_2$ (calcium formate)

When acetone is used, the intermediate compound in (*b*) is trichloroacetone (1,1,1-trichloropropanone), CH_3COCCl_3.

Experiment 200 Preparation of chloroform (trichloromethane)

Formula: $CHCl_3$

Physical properties: sweet-smelling liquid, boiling point 61 °C, density 1.5 g cm^{-3}.

(*a*) *Using bleaching powder*

Apparatus: large (1 dm^3) flask; condenser and adapter; mortar and pestle; sand-tray; separating funnel; distilling flask (100 cm^3); water-bath; thermometer (100 °C); measuring cylinder. (Standard-joint apparatus DGHQ.)

Material: ethanol or acetone[1]; bleaching powder; anhydrous calcium chloride.

Arrange the apparatus as shown in Figure 49. Make sure that all connections are well made.

Weigh 100 g of bleaching powder and grind it with water taken from a measured 300 cm^3 to make a cream. Decant the mixture into a large flask; use the rest of the water to rinse the mortar, and add the washings to the flask. Add either 35 cm^3 of ethanol or 40 cm^3 of acetone[1], then connect the flask to the condenser, and gently shake the flask to mix the contents thoroughly. The reaction usually begins spontaneously, and a bath of cold water should be ready to moderate the action if it is too vigorous. If, however, the action is slow in starting, warm the flask on a water-bath until action begins. When the first action has proceeded for about ten minutes, heat the flask on a sand-tray and distil until no more oily drops can be seen in the condenser.

Transfer the contents of the receiver to a separating funnel, allow to settle, then run off the lower layer temporarily into a flask. Discard the aqueous layer and return the chloroform layer to the funnel. Shake with an equal volume of sodium hydroxide solution to remove free chlorine or hydrogen chloride. Allow to settle and run off the lower layer into a flask. Add enough granulated calcium chloride to cover the bottom of the flask, cork the flask and leave until the liquid is clear.

Decant the dry liquid into a distilling flask connected to a sloping condenser. Close the neck of the flask with a bored cork and thermometer and distil over a water-bath, collecting the fraction between 60 °C and 62 °C. Yield 30 g.

(*b*) *Using sodium hypochlorite*

The declining use of bleaching powder commercially and the increased availability of sodium hypochlorite solution have favoured an alternative method using the latter material. The reactions occurring are

[1] Acetone gives a better yield.

essentially the same as when bleaching powder is used, the equations given above representing the stages in which the reactions take place, allowing for the substitution of two sodium ions for one calcium ion.

Apparatus: flask (1 dm^3); condenser; separating funnel; distilling flask (100 cm^3); water-bath; thermometer (100 °C); measuring cylinder. Standard-joint apparatus GHDJQ.

Material: acetone (propanone); sodium hypochlorite solution (12–15% of available chlorine); anhydrous calcium chloride.

Arrange the apparatus with the flask fitted with a reflux condenser. Measure 300 cm^3 of hypochlorite solution and pour it into the flask. Measure 40 cm^3 of acetone and add 1 cm^3 of it to the hypochlorite by pouring it down the condenser tube. Swirl the mixture and allow to cool. Repeat the additions of 1 cm^3 portions, taking care to mix thoroughly and to cool on each addition, the operation taking about forty-five minutes.

Direct separation by tapping off is difficult because of the tendency to form a three-phase system. Replace the reflux condenser by a sloping condenser, heat the mixture gently and collect the oily distillate of moist chloroform. With care, collection of unnecessary water can be avoided, but any surplus may be readily tapped off. Dry the product with calcium chloride and distil as described in the previous experiment.

Experiment 201 Reactions of chloroform

Material: potassium hydroxide dissolved in ethanol; iodine; diethyl ether (ethoxyethane); ethanol; 1,3-benzenediol (resorcinol); aniline (phenylamine).

(*a*) *Reaction with silver nitrate.* Add a few drops of silver nitrate solution to a test-tube containing about 2 cm^3 of chloroform. Shake well. Silver chloride is *not* precipitated, chloroform being un-ionized. Add a few drops of ethanol to give a solution, and show that even when aqueous silver nitrate is in solution with chloroform, silver chloride is not precipitated. Chloroform is a covalent compound.

(*b*) *Hydrolysis of chloroform.* Boil a few drops of chloroform with about ten times that volume of the potassium hydroxide solution. Acidify with dilute nitric acid and add silver nitrate solution. A white precipitate of silver chloride is obtained. Under these conditions, chloroform has been hydrolysed to potassium chloride and potassium formate.

$$CHCl_3 + 4OH^- \rightarrow HCOO^- + 3Cl^- + 2H_2O$$

(*c*) *Isocyanide test.* Mix in a test-tube, *one drop each* of chloroform and aniline. Add five drops of the potassium hydroxide solution and warm gently *in a fume chamber*. The sickly smell is due to phenyl isocyanide (isocyanobenzene), and similar products are obtained when any other primary amine is substituted for aniline. Pour the product into an excess of concentrated hydrochloric acid and then wash the test-tube thoroughly. This test is not recommended.

$$CHCl_3 + C_6H_5NH_2 + 3OH^- \rightarrow C_6H_5NC + 3Cl^- + 3H_2O$$

(*d*) *Solvent action.* Take a few drops of chloroform in a test-tube and add a crystal of iodine. Note the ease with which iodine dissolves. Add water to show the immiscibility of chloroform, then add a solution of iodine in ether to show (i) the relative densities of ether and chloroform, and (ii) the different colours of the two iodine solutions.

(*e*) *Reaction with resorcinol.* Dissolve a few crystals of resorcinol in the minimum of sodium hydroxide solution. Add a few drops of chloroform and warm the mixture gently. Note the red fluorescence in the aqueous layer.

Experiment 202 Preparation of iodoform (triiodomethane)

Formula: CHI_3.

Physical properties: yellow crystalline solid, melting point 119 °C.

Apparatus: small flask; water-bath; thermometer (100 °C); measuring cylinder.

Material: ethanol, iodine.

Make a solution of 15 g of anhydrous sodium carbonate in 50 cm^3 of distilled water. Keep about 5 cm^3 of the solution for use later, and to the remainder add 7 cm^3 of ethanol.

Warm the solution on a water-bath at about 60 °C and add 9.5 g of iodine gradually, with continual stirring until the solution is permanently brown. Discharge the colour by adding a few drops of the sodium carbonate solution kept back for this purpose.

Heat for a few minutes in the water-bath at 70 °C, then set aside to cool. Filter to isolate the crystals of iodoform. Yield 5 g.

$$C_2H_5OH + I_2 \rightarrow CH_3CHO + 2HI \quad \text{(oxidation)}$$
$$CH_3CHO + 3I_2 \rightarrow CI_3CHO + 3HI \quad \text{(substitution)}$$
$$2CI_3CHO + CO_3^{2-} + H_2O \rightarrow 2CHI_3 + 2HCOO^- + CO_2 \quad \text{(hydrolysis)}$$

Compare these equations with those given for the preparation of chloroform, p. 195.

Experiment 203 Reactions of iodoform

Material: iodoform; potassium hydroxide dissolved in ethanol; ethanol.

(*a*) *Reaction with silver nitrate.* Repeat Experiment 201(*a*), using a few crystals of iodoform in place of the chloroform. Show that iodoform does not react with silver nitrate solution.

(*b*) *Hydrolysis of iodoform.* Repeat Experiment 201(*b*), using a few crystals of iodoform in place of chloroform. Write the appropriate equation.

(*c*) *Decomposition of iodoform.* Heat a few crystals of iodoform in a dry test-tube. Note the violet vapour of iodine as the compound decomposes.

Experiment 204 Preparation of benzene hexabromide

Apparatus: wide-mouth bottle (about 500 cm^3); thistle funnel.
Material: benzene, bromine.

Pour sufficient benzene into the bottle to form a thin layer on the bottom, and stand a crucible on it. By means of a thistle funnel, run three or four drops of bromine into the crucible. Stopper the bottle and leave it where it will be in sunlight during a few days. The colourless crystals are the addition compound benzene hexabromide (1,2,3,4,5,6-hexabromocyclohexane).

[benzene ring] $+ 3Br_2 \rightarrow$ [cyclohexane ring with six Br substituents]

$$C_6H_6 + 3Br_2 \rightarrow C_6H_6Br_6$$

Cyclohexane, C_6H_{12}, is represented by a hexagon (without an inner circle) and the six bromine atoms are, in effect, replacements for six hydrogen atoms.

AROMATIC SUBSTITUTION

A group may enter the benzene molecule by displacing a hydrogen atom to form a compound of the type, C_6H_5X; this process is known as *direct* substitution. The groups —NO_2, —SO_3H, —Cl, —Br, —CH_3 (and similar alkyl groups) and —$COCH_3$ may be introduced directly; of these, the nitro group (—NO_2) is the most important because, once introduced, it may be converted, by stages, into other groups to give a large variety of compounds of technical importance. The process of introducing a group into the benzene molecule by converting another group already present is known as *indirect* substitution; examples of this process are given later (see Experiments 271–2).

Experiment 205 Preparation of bromobenzene

Formula: C_6H_5Br.

Physical properties: colourless liquid, boiling point 157 °C, density 1.5 g cm^{-3}.

Apparatus: four evaporating basins; measuring cylinder; separating funnel; thermometer (360 °C); distilling flask (150 cm^3); condenser; flask (250 cm^3). See Figure 52. (Standard-joint apparatus GHJMQ.)

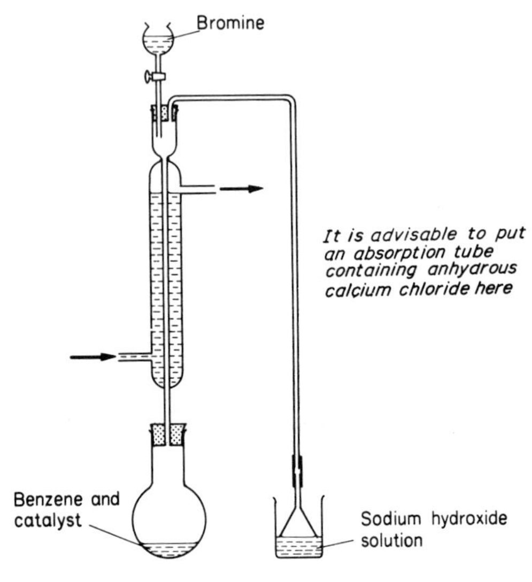

FIG. 52

Material: benzene; bromine; aluminium foil; ethanol; anhydrous calcium chloride. *Care! The experiment should be carried out in a fume chamber.*

The introduction of bromine by this method requires a catalyst. To prepare the catalyst, weigh 0.5 g of aluminium foil and cut it into six pieces. Take four evaporating basins, containing respectively a solution of mercury(II) chloride, water, ethanol and benzene. Place a piece of aluminium in the mercury(II) chloride solution for a minute, then wash it in turn in each of the other liquids, leaving it finally to stand in the benzene. Repeat with the other pieces.

Take 55 cm^3 of benzene in the flask (Figure 52), add the prepared catalyst, and attach the flask to the rest of the apparatus. Pour sodium hydroxide solution into the beaker until the rim of the funnel is just submerged. Pour a measured 20 cm^3 of bromine into the tap-funnel. *(Care! when using bromine.)* Allow the bromine to drip into the benzene at the rate of one drop a second. (The action is exothermic. Hydrogen bromide is evolved freely and is absorbed by the sodium hydroxide.) When all the bromine has been added, allow the mixture to cool, then transfer it to a separating funnel and shake with dilute sodium hydroxide solution. Run off the lower layer into a flask, add pieces of calcium chloride and leave to dry. When clear, decant into a distilling flask. Distil and collect the fraction between 150 °C and 160 °C. Yield 50 g.

$$C_6H_6 + Br_2 \rightarrow C_6H_5Br + HBr$$

Note: Iron and pyridine will also catalyse this reaction. This may be illustrated as follows.

Put six drops of benzene into each of three test-tubes followed by three drops of bromine. To one tube add iron filings and to another a few drops of pyridine. Warm if necessary to promote the action. The tube without a 'carrier' shows no evidence of reaction whilst hydrogen bromide is evolved in each of the other two.

Experiment 206 Preparation of iodobenzene

Formula: C_6H_5I.

Physical properties: dense yellow liquid, boiling point 190 °C, density 1.8 g cm^{-3}.

Apparatus: flask (250 cm^3); measuring cylinder; separating funnel; water-bath; condenser; distilling flask; apparatus for steam distillation; thermometer (360 °C).

Material: redistilled aniline (phenylamine); sodium nitrite; potassium iodide; diethyl ether; anhydrous calcium chloride; **ICE**.

The method follows the usual course of Sandmeyer reactions, but in this case the copper(I) salt may be replaced by the potassium salt.

Diazotize 20 cm^3 of aniline in exactly the same way as in the preparation of phenol from aniline (Experiment 217, p. 209). To the cold diazonium salt add a solution of 40 g of potassium iodide in 100 cm^3 of water. Mix well and allow to stand for about a quarter of an hour. Fit the flask with a reflux condenser and heat on a water-bath until effervescence (nitrogen) has almost ceased, then steam distil. See Figure 69.

Transfer the aqueous distillate to a separating funnel and extract three times with ether, using 30 cm^3 of ether for each extraction. Wash the ether-iodobenzene solution first with water, then with dilute sodium hydroxide solution to remove traces of phenol, and then with water again. Run off the ether layer into a flask and dry with calcium chloride.

Distil the dry liquid, first over hot water in a water-bath (no Bunsen) and, when the ether has distilled off, over a gauze. When the temperature has reached 130 °C, empty the condenser and continue to distil, collecting the fraction between 187 °C and 192 °C. Yield 30 g.

$$C_6H_5NH_2 + 2H^+ + NO_2^- \rightarrow C_6H_5N_2^+ + 2H_2O$$
$$C_6H_5N_2^+ + I^- \rightarrow C_6H_5I + N_2\uparrow$$

24

Hydroxy Compounds

ALCOHOLS

Experiment 207 Preparation of ethanol (ethyl alcohol) C_2H_5OH

Formula: C_2H_5OH

Physical properties: colourless liquid with wine-like odour, miscible with water in all proportions, boiling point 78.3 °C.

Apparatus: Winchester quart bottle (2.5 dm^3); distilling flasks (1000 cm^3 and 100 cm^3); fractionating column; thermometer; condenser; adapter; conical flask; a Bunsen valve (see Experiment 295); separating funnel. Standard joint apparatus GIMLHQ.

Material: glucose; yeast; diammonium hydrogenphosphate; potassium nitrate; calcium oxide; anhydrous potassium carbonate. The yeast nutrient can be obtained from Grey Owl Laboratories, Amondsbury, Glos.; Boots Ltd, etc.

Add 125 g of glucose to 500 cm^3 of water and boil till dissolved. Add either 1 g of ammonium phosphate and 1 g of potassium nitrate, or the juice of a lemon, or a yeast nutrient, then pour the hot solution into 1500 cm^3 of cold water contained in a Winchester quart bottle. Cream 30 g of yeast with a little of the warm diluted glucose solution and add to the main bulk of solution in the Winchester bottle. The *must*, as the solution is now called, is left at a temperature of 25–30 °C until fermentation ceases (usually three days). A Bunsen valve should be fitted to the bottle.

$$C_6H_{12}O_6 \rightarrow 2C_2H_5OH + 2CO_2$$

The whole volume of liquid is then distilled in portions, porous pot being added to prevent bumping. A fractionating column is used and the portion coming over below 95 °C is retained. The distillate is then redistilled in the same apparatus, the fraction boiling between 78 and 83 °C being retained. This second distillate is saturated with anhydrous potassium carbonate, when two layers separate, the upper one being ethanol with a little water. The lower layer is removed, using a separating funnel. The ethanol is transferred to a conical flask containing a few lumps of fresh calcium oxide and corked. After about twelve hours, the alcohol is distilled using a 100 cm^3 distilling flask and condenser. It should boil at 78.3 °C. (Yield about 35 cm^3.)

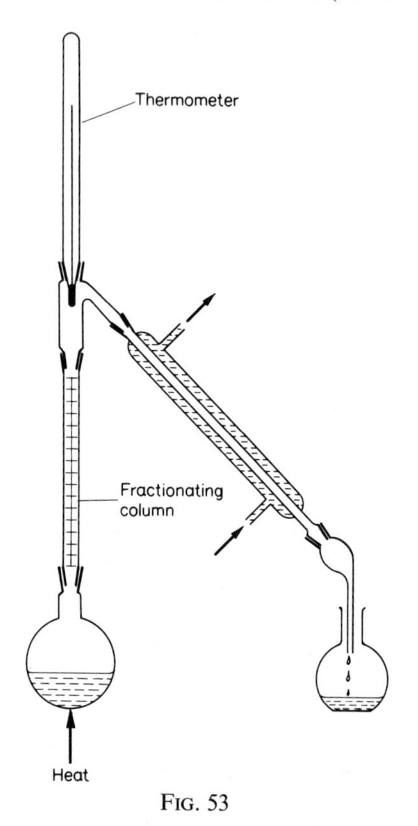

FIG. 53

Experiment 208 Preparation of sodium ethoxide

Primary alcohols contain the characteristic —CH_2OH group. Of all the hydrogen atoms present in a molecule of an alcohol, only the one directly attached to the oxygen atom can be replaced by sodium. Thus ethanol when treated with sodium gives a solid, sodium ethoxide, and gives off hydrogen.

$$2C_2H_5OH + 2Na \rightarrow 2C_2H_5ONa + H_2$$
$$(C_2H_5O^- Na^+)$$

Similarly, methanol gives sodium methoxide (CH_3ONa), and hydrogen.

These sodium compounds are readily hydrolysed, even by atmospheric moisture, to regenerate the alcohol.

$$C_2H_5O^- + H_2O \rightarrow C_2H_5OH + OH^-$$

Apparatus: water-bath.
Material: ethanol; sodium.

Take about 12 cm^3 of ethanol in an evaporating basin and add a piece of sodium about the size of a pea. Note the immediate effervescence (hydrogen). When the sodium has dissolved, evaporate the solution on a water-bath, stirring to break the crust of solid as it forms. Transfer the dry solid to a dry test-tube and seal it off to exclude moisture. If methylated spirits has been used, the product will contain sodium methoxide as an impurity. (Methylated spirits consists of 85% of ethanol, the remainder being impure methanol.) Yield about 1 g (almost theoretical).

Experiment 209 Determination of the relative atomic mass of sodium, using the sodium-ethanol reaction

Apparatus: siphon apparatus (as in Figure 54); measuring cylinder.
Material: ethanol; sodium.

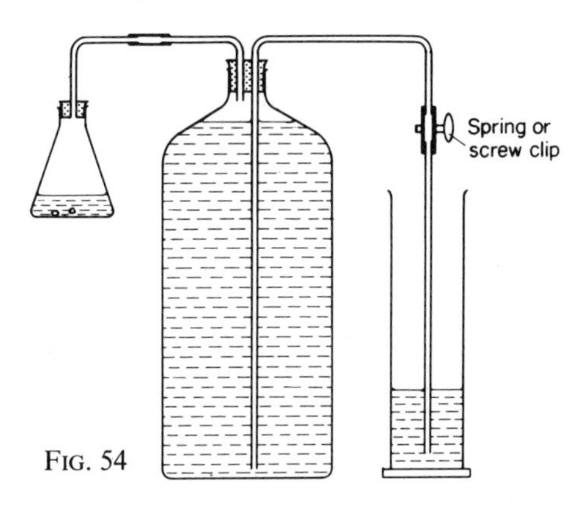

FIG. 54

Arrange the apparatus as shown and fill the siphon tube with water. Take about 15 cm^3 of ethanol in the (dry) flask. Weigh a piece of clean sodium (about the size of a pea) in a weighing bottle. Remove the flask, add the sodium, and replace the flask immediately. Open the clip on the siphon tube as quickly as possible and collect the water displaced by the hydrogen. Make the usual adjustments of levels after the gas has cooled and correct the volume of hydrogen to standard temperature and pressure.

Find the mass of sodium required to displace 11.2 dm^3 of hydrogen at s.t.p.

The remaining solution may bc evaporated to give solid sodium ethoxide as in the previous experiment.

REACTIONS OF ALCOHOL WITH ACIDS

Alcohols react in some respects as bases. When an alcohol reacts with an acid (organic or inorganic), water is eliminated between the —OH of the alcohol and the displaceable hydrogen of the acid, e.g.

$$C_2H_5OH + HBr \rightarrow \underset{\text{(bromoethane)}}{C_2H_5Br} + H_2O$$

Organic acids, with rare exceptions (see p. 255), contain the characteristic carboxyl (—COOH) group, and it is the hydrogen of this group which reacts with the hydroxyl of the alcohol, e.g.

$$C_2H_5OH + CH_3COOH \rightleftharpoons \underset{\text{(ethyl acetate)}}{CH_3COOC_2H_5} + H_2O$$

The products of both of the above reactions are called esters to distinguish them from the salts of inorganic chemistry, with which they have a superficial similarity. Salts and esters differ both in the speeds of their formation and in their properties.

Experiment 210 Reaction of ethanol with hydrogen bromide

Apparatus: boiling-tube fitted with cork and bent tube.
Material: ethanol; potassium bromide.

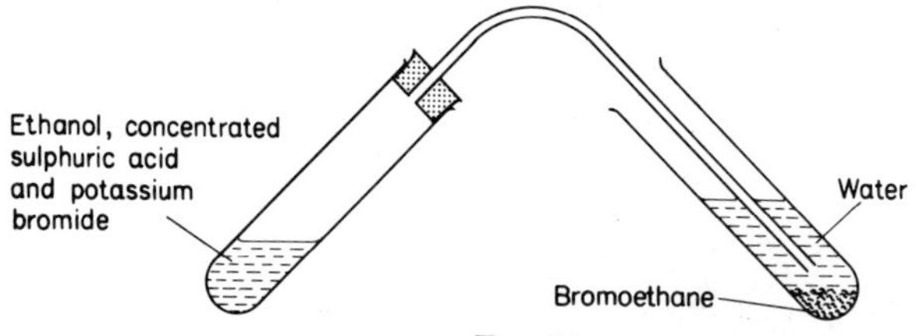

FIG. 55

Take enough ethanol to fill the boiling-tube to a depth of 3 cm. Add about one-third of that volume of concentrated sulphuric acid, cooling and shaking the mixture during the addition. Add crystals of potassium bromide until the total volume is about twice that of the liquid. Fit the delivery tube and immerse the end of it deeply in water in a test-tube (Figure 55). Warm the mixture in the boiling-tube and note the oily drops of a dense liquid (bromoethane) collecting below the water.

The preparation of this compound on a larger scale was described in Experiment 193, p. 188.

Experiment 211 Reaction of ethanol with acetic acid

Material: ethanol; glacial acetic acid (pure ethanoic acid).

Make a mixture of about ten drops of ethanol, five drops of glacial acetic acid and three drops of concentrated sulphuric acid. Warm gently. Pour the liquid into a test-tube half full of water (in which the ester is insoluble), and note the fruity smell of ethyl acetate.

The preparation of ethyl acetate on a larger scale is given in Experiment 247, p. 241.

OXIDATION OF ALCOHOLS

Primary alcohols are readily oxidized to give aldehydes. In every case the hydrogen atom of the —OH group and a hydrogen atom attached to the same terminal carbon atom are removed by the oxidizer, e.g.

$$\mathrm{H{-}\overset{\displaystyle H}{\underset{\displaystyle H}{\overset{|}{\underset{|}{C}}}}{-}\overset{\displaystyle H}{\underset{\displaystyle H}{\overset{|}{\underset{|}{C}}}}{-}OH} \rightarrow \mathrm{H{-}\overset{\displaystyle H}{\underset{\displaystyle H}{\overset{|}{\underset{|}{C}}}}{-}\overset{\displaystyle H}{\overset{|}{C}}{=}O} + 2H^+ + 2e^-$$

(acetaldehyde)

Experiment 212 Oxidation of ethanol

Material: a saturated solution of sodium (or potassium) dichromate; ethanol.

Take approximately 5 cm^3 of the dichromate solution. Add five drops of ethanol and five drops of concentrated sulphuric acid. Warm gently. Note the irritating smell of acetaldehyde (ethanal), and also the change in colour of the dichromate as it is reduced to chromium sulphate.

$$Na_2Cr_2O_7 + 4H_2SO_4 + 3C_2H_5OH \rightarrow Na_2SO_4 + Cr_2(SO_4)_3 + 7H_2O + 3CH_3CHO$$

or

$$Cr_2O_7^{2-} + 8H^+ + 3C_2H_5OH \rightarrow 2Cr^{3+} + 7H_2O + 3CH_3CHO$$

The preparation of acetaldehyde on a larger scale is given in Experiment 224, p. 219; and the oxidation of methanol by atmospheric oxygen is described in Experiment 223, p. 218.

Experiment 213 Distinguishing test for ethanol

Material: A saturated solution of iodine in aqueous potassium iodide; ethanol.

To about 5 cm^3 of the iodine solution add five drops of ethanol. Add sodium hydroxide solution carefully until the colour has almost gone. Stand the test-tube in water at about 70 °C for two or three minutes, then remove and leave to cool. The yellow crystals are iodoform, and the smell is reminiscent of antiseptics (see p. 198).

As this reaction does not occur with methanol, it serves as a distinguishing test between *ethyl* and *methyl* alcohols.

Note: Other compounds, e.g. acetaldehyde, acetone (propanone), and 2-propanol also give the reaction, the essential requirement being that either the compound itself, or the compound it becomes on oxidation with iodine, contains the group CH_3CO linked to an atom of hydrogen or to an alkyl radical.

Experiment 214 Reaction between ethanol and phosphorus pentachloride

Material: ethanol; phosphorus pentachloride.

To about 1 cm^3 of ethanol add a few crystals of phosphorus pentachloride. The action is vigorous and hydrogen chloride is evolved.

$$C_2H_5OH + PCl_5 \rightarrow C_2H_5Cl + HCl + POCl_3$$

This action is typical of a great many reactions between phosphorus pentachloride (or trichloride) and compounds containing an hydroxyl group.

Experiment 215 Tests for methanol

(*a*) Methanol is oxidized to formaldehyde (methanal) by: (i) catalytic oxidation by oxygen of the air (see pp. 61 and 218); (ii) a dichromate and sulphuric acid using the procedure of Experiment 212 but substituting methanol for ethanol.

Formaldehyde is readily recognized by its pungent odour.

(*b*) Methanol forms methyl salicylate with salicylic (2-hydroxybenzoic) acid and concentrated sulphuric acid. See Experiment 246 (*c*), p. 240. Methyl salicylate is recognized by its smell of oil of wintergreen.

Experiment 216 To distinguish between primary, secondary and tertiary alcohols (Lucas)

Material: ethanol; 2-propanol; 2-methylpropan-2-ol; zinc chloride dissolved in concentrated hydrochloric acid.

To a small quantity of the alcohol in a corked test-tube add an excess of the zinc chloride solution, shake the mixture and allow it to stand. With primary alcohols of low relative molecular mass the aqueous layer remains clear; with secondary alcohols, chlorides separate on standing; with tertiary alcohols, two layers appear immediately.

PHENOL

The hydroxyl group may be introduced by two methods.

1. By 'diazotizing' aniline (phenylamine) and then heating.
2. By heating sodium benzenesulphonate with sodium hydroxide.

Note: (*a*) the hydroxyl group cannot be introduced directly into the benzene nucleus,

(*b*) —Cl, —Br and —I cannot be displaced by —OH (compare the hydrolysis of alkyl halides, Experiment 197(*a*), p. 191. It is important to remember that once a halogen has been introduced into the benzene nucleus it is very difficult to displace it.

Care! Phenol is corrosive. When making or using it, take care that solid phenol or a concentrated solution of it does not fall on the skin.

Experiment 217 Preparation of phenol from aniline

The action of nitrous acid (sodium nitrite in acid solution) on an amino-group depends (*a*) on the position of the $—NH_2$ group and (*b*) on the temperature.

If the $—NH_2$ group is *in the nucleus* and the temperature is in the region of 0 °C, the nitrogen of the $—NH_2$ and the nitrogen of the nitrous acid remain in the molecule of the new compound to form a diazonium compound.

$$C_6H_5NH_3^+ + NO_2^- + H^+ \rightarrow C_6H_5N_2^+ + 2H_2O$$

If the $—NH_2$ group is in the nucleus and *the temperature is not kept low*, there is an evolution of nitrogen and the $—NH_2$ group is replaced by —OH to give a phenol.

If the $—NH_2$ group is in the side-chain, it behaves as it does in an aliphatic compound; *whether hot or cold* it is replaced by an —OH group.

Formula of phenol: C_6H_5OH.

Physical properties: colourless needle crystals, melting point 43 °C.

Apparatus: flask (250 cm^3); steam distillation apparatus as shown in Figure 69; separating funnel; measuring cylinder; thermometer (360 °C); freezing mixture (**ICE**-salt) in a large basin or trough.

Material: redistilled aniline; sodium nitrite (nitrate(III)); anhydrous sodium sulphate; diethyl ether.

Take 100 cm^3 of water in the 250 cm^3 flask, add a measured 24 cm^3 of concentrated sulphuric[1] acid, and while still hot add 20 cm^3 of aniline. Cool the solution to room temperature and then place in the freezing mixture.

Prepare a solution of sodium nitrite by dissolving 16.5 g in 40 cm^3 of water in the conical flask, and transfer it to the tap funnel.

When the anilinium hydrogensulphate solution has cooled to 0 °C, allow the sodium nitrite solution to drip in very slowly and then adjust the funnel so that its end is under the surface of the liquid in the flask. Keep the liquid agitated and see that the temperature does not rise above 2 °C during the addition. Shake the mixture and leave in the freezing mixture for a quarter of an hour, then transfer it to a water-bath and heat for half an hour. Arrange the apparatus for steam distillation (Figure 69). Pour the mixture into the large flask and heat almost to boiling before passing in the steam. When no more phenol distils over, transfer the aqueous distillate to a separating funnel and extract with ether, using 90 cm^3 of ether in three fairly equal portions. Dry the ether-phenol solution by adding about 10 g of anhydrous sodium sulphate. (To obtain this quantity, heat about 25 g of sodium sulphate-10-water in an evaporating basin until there is no further loss of moisture.)

When dry, distil the solution, first over hot water in a water-bath (no Bunsen) until the ether has all distilled and then over a gauze. Empty the condenser at 130 °C and continue with the distillation. Collect the fraction between 180 °C and 185 °C; this solidifies. Yield 7 g.

$$C_6H_5NH_3^+ + NO_2^- + H^+ \rightarrow C_6H_5N_2^+ + 2H_2O$$
$$C_6H_5N_2^+ + H_2O \rightarrow C_6H_5OH + N_2 + H^+$$

Experiment 218 Preparation of phenol from sodium benzenesulphonate

Apparatus: nickel basin (diameter about 75 mm); thermometer (360 °C); flask (250 cm^3); separating funnel; water-bath; distilling flask and condenser.

[1] Sulphuric acid is used in preference to hydrochloric acid because the intermediate compound benzenediazonium chloride is *slightly* decomposed to give chlorobenzene as an impurity. Benzenediazonium hydrogensulphate does not decompose in this manner.

Material: sodium benzenesulphonate; sodium hydroxide; diethyl ether; anhydrous sodium sulphate.

Care! The mixture used in the first stage is very corrosive.

Place 35 g of solid sodium hydroxide (sticks or pellets) in the basin, add 5 cm^3 of water and heat with a small flame until solution is complete. Fit a thermometer with a small test-tube to enclose the bulb. Add gradually 20 g of crushed sodium benzenesulphonate, using the fitted thermometer as a stirring rod, and keeping the temperature between 240 °C and 250 °C. After an hour, the mass is allowed to cool.

Add small quantities of water and heat to dissolve the solid and decant the solution into a flask. (The solution contains sodium phenate (phenoxide), C_6H_5ONa, sodium sulphite, and excess sodium hydroxide.) *Keeping the solution cool,* acidify with concentrated hydrochloric acid. Note the smell of sulphur dioxide from the sodium sulphite, and the brown oil (phenol).

Extract the phenol with ether, using 45 cm^3 in three portions. Dry the ether-phenol solution with anhydrous sodium sulphate and distil as in Experiment 217. The yield is small.

$$C_6H_5SO_3^- + 2OH^- \rightarrow C_6H_5O^- + SO_3^{2-} + H_2O$$
$$C_6H_5O^- + H^+ \rightarrow C_6H_5OH$$

Experiment 219 Reactions of phenol

Apparatus: thermometer (100 °C).

Material: phenol; Millon's reagent; sodium nitrite; zinc powder.

Make a stock solution of phenol by covering the bottom of a test-tube with the solid and filling it almost to the top with water. Use this solution in approximately equal parts for tests (*a*), (*b*) and (*c*).

(*a*) Add one drop of iron(III) chloride solution. Note the violet coloration.

(*b*) Add bromine water and shake. Add sufficient bromine water to leave the solution a definite brown colour. Note the white precipitate of 2,4,6-tribromophenol.

$$C_6H_5OH + 3Br_2 \rightarrow C_6H_2Br_3OH + 3HBr$$

(*c*) Add a drop of Millon's reagent (see p. 443). Stand the test-tube in a beaker of water and heat gradually. Note the red coloration.

(*d*) Take ten drops of concentrated sulphuric acid in a test-tube.

Add a quantity of sodium nitrite about the size of a match-head and warm gently until dissolved. Add one or two crystals of phenol. Pour the liquid into a beaker one-quarter full of water. Note the red coloration. Add sodium hydroxide solution and note the change of colour to blue. This is *Liebermann's Reaction.*

(*e*) Take enough phenol to cover the bottom of a test-tube and add water to a depth of 2 cm. Cork the test-tube and shake to obtain an emulsion. Remove the cork, place the test-tube in a beaker of water and warm. Stand a thermometer in the phenol solution and note the temperature when the emulsion clears. Allow the solution to cool and note the reappearance of cloudiness. (See Experiment 14, p. 18.)

(*f*) Take the emulsion from the previous experiment and carefully add sodium hydroxide solution until a clear solution of sodium phenate is obtained. Add a few drops of concentrated hydrochloric and note the reappearance of phenol.

$$C_6H_5OH + OH^- \rightarrow C_6H_5O^- + H_2O$$
$$C_6H_5O^- + H^+ \rightarrow C_6H_5OH$$

(*g*) Mix in a dry test-tube a full spatula of phenol and about three times that bulk of zinc powder. Heat the mixture and light the benzene vapour at the mouth of the test-tube. The zinc has reduced the phenol and become zinc oxide.

$$C_6H_5OH + Zn \rightarrow C_6H_6 + ZnO$$

See also Experiment 268 (*e*), p. 265.

25

Ethers

The best-known member of this group of compounds is the liquid diethyl ether (ethoxyethane), usually called simply 'ether'. A lower member, dimethyl ether (methoxymethane), is a gas. The structural formula of any ether contains two alkyl radicals attached to an oxygen atom:

$$C_2H_5{-}O{-}C_2H_5 \quad \text{diethyl ether}$$
$$CH_3{-}O{-}CH_3 \quad \text{dimethyl ether}$$

or, generally, $R_1{-}O{-}R_2$ where R_1, R_2 are alkyl radicals which may be either the same or different.

Two methods are available for preparing these compounds.

1. By distilling a mixture of an alcohol and concentrated sulphuric acid. When the distillation has begun it is (theoretically) only necessary to add more alcohol to obtain a continuous supply of ether. The process falls short of the ideal through the dilution of the acid by the water formed in the action, but it is still known as Williamson's Continuous Process. The action takes place in two stages as shown in the equations

$$C_2H_5OH + H_2SO_4 \rightarrow C_2H_5HSO_4 + H_2O$$
$$C_2H_5HSO_4 + C_2H_5OH \rightarrow (C_2H_5)_2O + H_2SO_4$$

The method could not be used to make a mixed ether, i.e. an ether containing two different alkyl radicals.

2. By the reaction of an alkyl halide with the sodium compound of an alcohol. Here the alkyl radical in the halide may be the same as, or different from, the alkyl radical of the alcohol, and the method could therefore be used for making any ether. For the preparation of ethers the first method is to be preferred where possible, both because of its simplicity and the higher yield obtained.

$$R_1O^- + R_2I \rightarrow (R_1R_2)O + I^-$$

where R_1 and R_2 represent alkyl radicals.

Experiment 220 Preparation of diethyl ether: continuous process

Formula: $(C_2H_5)_2O$.

Physical properties: volatile, inflammable liquid, boiling point 35 °C; density 0.7 g cm^{-3}.

Apparatus: distilling flask (500 cm^3); condenser; adapter; Buchner flask; tap funnel; thermometer (360 °C); sand-tray; **ICE**. (Standard-joint apparatus BDEGGHJ.)

Material: ethanol; sodium chloride; anhydrous calcium chloride.

Arrange the apparatus as shown in Figure 56. Note that the bulb of the thermometer dips into the liquid and also that a rubber tube leads any ether vapour to the waste pipe away from the region of the flame.

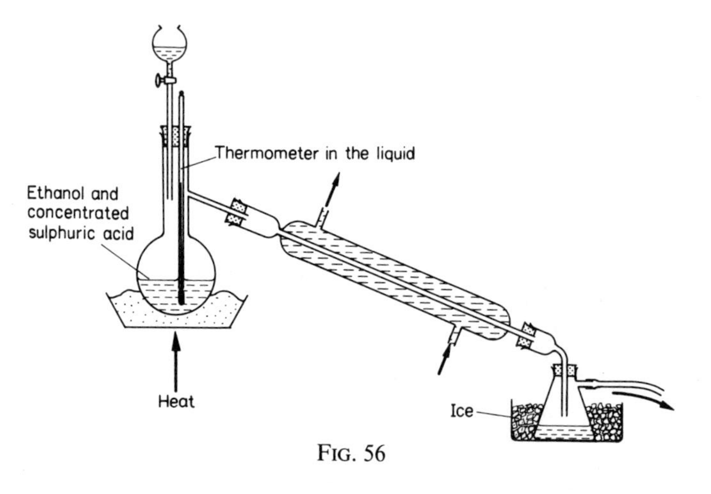

FIG. 56

Pour 50 cm^3 of ethanol into the distilling flask and gradually add, with shaking, 40 cm^3 of concentrated sulphuric acid. Attach the flask to the apparatus and pour 100 cm^3 of ethanol into the funnel.

Heat on the sand-tray, and when the thermometer records 140 °C allow ethanol to drip into the flask at about the speed at which the liquid is distilling. Adjust the Bunsen to maintain the temperature within two or three degrees of 140 °C and continue distilling until all the ethanol has been added from the funnel.

Transfer the distillate to the separating funnel and shake cautiously with 50 cm^3 of sodium hydroxide solution to absorb sulphurous acid which has been formed by some reduction of sulphuric acid. In view of the volatile nature of ether, care must be taken when shaking the mixture to avoid a high pressure in the funnel by frequently opening the stopper. At all times the high inflammability of ether requires precautions against fire.

Run off the lower layer of aqueous solution and retain the ether layer. Add a saturated solution of sodium chloride to remove traces of ethanol, and, after shaking, run off the lower layer and discard.

Run the ether layer into a flask and add enough granulated calcium chloride to cover the bottom of the flask. Leave, loosely corked, in a cool place overnight to dry.

Arrange a distilling flask attached to a sloping condenser which is fitted with an adapter dipping into a dry Buchner flask as before. Transfer the dry ether to the flask and close the neck with a bored cork and thermometer. Heat the flask on a water-bath containing water which has been brought to boiling point elsewhere, so avoiding a flame near ether during distillation. Collect the fraction when the temperature is steady at 35 °C. Yield 30 g.

Experiment 221 Preparation of diethyl ether from iodoethane

Apparatus: flask (150 cm^3); condenser; distilling flask (100 cm^3); Buchner flask; adapter; measuring cylinder; water-bath; weighing bottle.

Material: ethanol; sodium; iodoethane.

Take 50 cm^3 of alcohol in the 150 cm^3 flask. Weigh 3 g of sodium in a weighing bottle and add it in small pieces to the ethanol, allowing half-minute intervals between additions.

Fit the flask with a reflux condenser and then place it on a water-bath. Measure 10 cm^3 of iodoethane and pour it down the condenser tube (see Figure 48).

Heat the flask until crystals of sodium iodide are formed, then transfer the liquid to a distilling flask and distil over a water-bath, collecting the fraction distilling at 35 °C.

The yield is small, but the action is important because it enabled Williamson to establish the formula of ether.

$$C_2H_5O^- + C_2H_5I \rightarrow (C_2H_5)_2O + I^-$$

Experiment 222 Reactions of diethyl ether

Apparatus: bellows.

Material: diethyl ether (ethoxyethane); iodine.

Generally speaking, ethers are chemically very stable and inert. It requires the drastic action of hydrogen iodide, prepared *in situ* by moist white phosphorus and iodine, to decompose an ether, e.g.

$$(C_2H_5)_2O + 2HI \rightarrow 2C_2H_5I + H_2O$$

(*a*) *Volatility of diethyl ether.* Take about 20 cm^3 of ether in a beaker, and stand the beaker in a pool of water on a piece of wood. Blow a rapid stream of air from bellows into the ether. Note the ice, which on occasions is sufficient to bind the beaker to the wood. The vapour pressure of ether is high because it is near to its boiling point (35 °C).

(*b*) *Solvent action.* Add a crystal of iodine to about 5 cm^3 of diethyl ether. Note how easily the iodine dissolves. Pour the solution into a test-tube half full of water. Note that the ether is immiscible in water, and that the iodine remains to a large extent in the ether.

(*c*) *Inflammability.* Take about 1 cm^3 of ether in a test-tube. Dip a glass rod into the liquid and hold the rod near a flame. Care must always be exercised when ether is being used.

26
Aldehydes and Ketones

The oxidation of alcohols has already been considered briefly, see p. 207. It is now necessary to distinguish between two types of alcohols.

Primary alcohols consist of an alkyl radical linked to a hydroxyl group, and the structural formula shows the —OH group attached to a terminal carbon atom, e.g.

$$\begin{array}{ccccccccc} & & H & & H & & & & \\ & & | & & | & & & & \\ H & — & C & — & C & — & O & — & H \\ & & | & & | & & & & \\ & & H & & H & & & & \end{array} \quad \text{ethanol}$$

$$\begin{array}{ccccccccccc} & & H & & H & & H & & & & \\ & & | & & | & & | & & & & \\ H & — & C & — & C & — & C & — & O & — & H \\ & & | & & | & & | & & & & \\ & & H & & H & & H & & & & \end{array} \quad \text{propanol}$$

Secondary alcohols contain a $>$CHOH group, linked to two alkyl radicals, i.e. the —OH group is not attached to a terminal carbon atom. Since ethanol contains only two carbon atoms, it is immaterial to which carbon atom the —OH group is attached. In the case of propanol, however, the —OH group may be attached either to a terminal carbon atom or to the middle one, and it is necessary to distinguish the two compounds; the former is called 1-propanol (or just propanol) and the latter, 2-propanol (or isopropyl alcohol or propan-2-ol).

$$\begin{array}{ccccccccccc} & & H & & H & & H & & & & \\ & & | & & | & & | & & & & \\ H & — & C & — & C & — & C & — & O & — & H \\ & & | & & | & & | & & & & \\ & & H & & H & & H & & & & \end{array} \quad \text{1-propanol, a primary alcohol}$$

$$\begin{array}{ccccccccc} & & H & & H & & H & & \\ & & | & & | & & | & & \\ H & — & C & — & C & — & C & — & H \\ & & | & & | & & | & & \\ & & H & & O & & H & & \\ & & & & | & & & & \\ & & & & H & & & & \end{array} \quad \text{2-propanol, a secondary alcohol}$$

Oxidation. Whether an alcohol is primary or secondary, the first product of oxidation is formed by the removal of two hydrogen atoms, one of them being the hydrogen of the —OH group and the other a hydrogen attached to the same carbon as the —OH group.

```
   H  H  H              H  H     H
   |  |  |              |  |    /
H—C—C—C—O—H → H—C—C—C           propanal (propionalde-
   |  |  |              |  |    \\       hyde), an aldehyde
   H  H  H              H  H     O

   H  H  H              H     H
   |  |  |              |     |
H—C—C—C—H      → H—C—C—C—H   propanone (acetone), a
   |  |  |              |  ||  |       ketone
   H  O  H              H  O  H
      |
      H
```

An aldehyde is thus the first product of oxidation of a primary alcohol.

A ketone is thus the first product of oxidation of a secondary alcohol.

An aldehyde, therefore, contains the characteristic —CHO group attached to an alkyl radical.

A ketone contains the group >CO attached to *two* alkyl radicals, i.e.

```
R1
  \
   C=O
  /
R2
```

Preparation of aldehydes and ketones. It is usual to prepare aldehydes by simple oxidation of primary alcohols as described below. Ketones may be prepared by a similar method using secondary alcohols.

Experiment 223 Preparation of formaldehyde (methanal)

Formula: HCHO.

Physical properties: gas with an irritating smell, boiling point $-21\ ^\circ C$, soluble in water to give the solution formalin.

Apparatus: as in Figure 57.

Material: methanol; platinized asbestos.

Arrange the apparatus as shown. Fill the siphon tube and warm the methanol in flask A gently so that a steady stream of alcohol vapour and air is carried through the combustion tube containing the catalyst (platinized asbestos). At the same time heat the combustion tube, gently at first, and then strongly. Note the glow of the platinum, indicating that oxidation of alcohol is exothermic. Continue to heat for about ten minutes. The aqueous solution of formaldehyde (methanal) which collects in flask B may be tested as described in Experiment 226, p. 222.

$$2CH_3OH + O_2 \rightarrow 2HCHO + 2H_2O$$

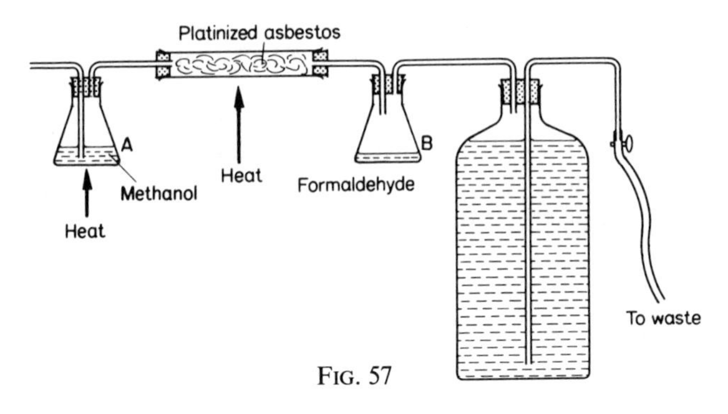

FIG. 57

Experiment 224 Preparation of acetaldehyde (ethanal)

Formula: CH_3CHO.

Physical properties: colourless liquid with pungent smell, boiling point 21 °C, density 0.8 g cm^{-3}.

Apparatus: large flask (1 dm^3); cork with tap funnel and bent tube; condenser; sand-tray; flask (250 cm^3); measuring cylinder. (Standard-joint apparatus FGGHMQ).

Material: ethanol; sodium dichromate. **ICE.**

Arrange the apparatus as shown in Figure 58. Connections must be well-made to avoid losses.

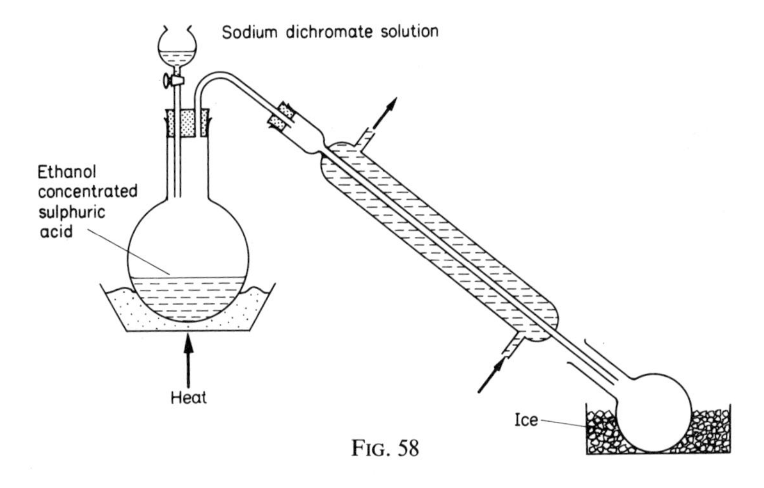

FIG. 58

Weigh 75 g of sodium[1] dichromate and place it in the large flask together with 100 cm^3 of water. Fit the flask to the rest of the apparatus.

Mix 30 cm^3 of concentrated sulphuric acid and 60 cm^3 of ethanol, adding the acid to the alcohol slowly and with shaking, in the 250 cm^3 flask, and cool the mixture. Transfer the mixture to the funnel.

Heat the dichromate solution to a moderate temperature and then allow the mixture in the funnel to run in slowly, shaking occasionally. When action begins, remove the flame until the first vigorous action has moderated, then continue heating until about 60 cm^3 of distillate has been collected. By that time practically all the aldehyde will have distilled. The product will contain acetaldehyde (ethanal), alcohol and water.

$$3C_2H_5OH + Cr_2O_7^{2-} + 8H^+ \rightarrow 3CH_3CHO + 7H_2O + 2Cr^{3+}$$

Experiment 225 Purification of acetaldehyde (ethanal)

Apparatus: flask (250 cm^3); two absorption bottles; condenser; delivery tube; apparatus being arranged as in Figure 59; water-bath; measuring cylinder; apparatus for preparing dry ammonia; Buchner funnel.

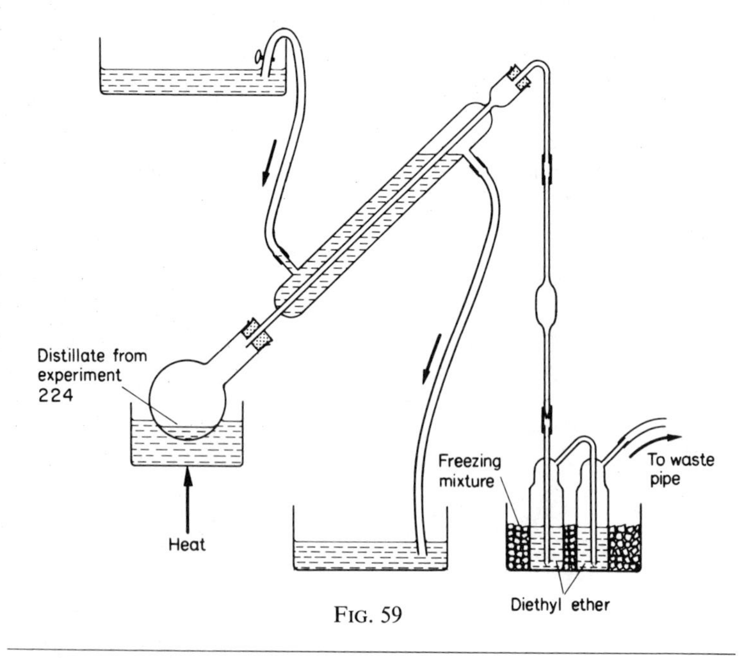

FIG. 59

[1] Sodium dichromate is much more soluble than the potassium salt.

Material: distillate from Experiment 224, diethyl ether (ethoxyethane); **ICE**; sodium chloride; anhydrous calcium chloride; calcium oxide.

Acetaldehyde forms an addition compound with ammonia, and this is the basis of the separation of the aldehyde from its mixture with alcohol and water.

Take the distillate in the 250 cm^3 flask and attach it to the condenser. Pass water at 35 °C through the condenser and maintain it at this temperature by siphoning water slightly above this temperature from a trough elevated above the apparatus. The waste-flow from the condenser can be warmed with a little boiling water and returned to the trough.

Heat the flask on the water-bath. The temperature of the condenser allows the volatile aldehyde to pass into the cold ether, where it dissolves, but condenses the ethanol and water which return into the flask. After half an hour, detach the tube connecting the condenser to the absorption bottles.

Prepare ammonia and dry it by passing it through a tower filled with calcium oxide. Pass the dry ammonia into the aldehyde-ether solution for half an hour to ensure that the solution is saturated. The crystals are acetaldehyde-ammonia (1-aminoethanol). Yield 15 g. Filter the crystals in a Buchner funnel and keep them in a dry, corked test-tube. The regeneration of pure aldehyde from the crystals is tedious, but if required the procedure is as follows.

Transfer the crystals to a distilling flask attached to a sloping condenser. Fit an adapter to the lower end of the condenser leading to a flask cooled in ice. To the crystals add about four times the bulk of a mixture of equal parts of water and concentrated sulphuric acid. Close the neck of the flask with a cork and heat it on a water-bath, starting with cold water and stopping distillation when the water boils. Dry the distillate with anhydrous calcium chloride and redistil on a water-bath maintained at about 30 °C.

$$CH_3{-}C(H){=}O + NH_3 \rightarrow CH_3{-}C(H)(NH_2){-}OH$$

PROPERTIES OF ALDEHYDES

The formula of acetaldehyde (ethanal, a typical aldehyde) is

$$H{-}C(H)(H){-}C(H){=}O$$

The hydrogen atom of the —CHO group is readily oxidized to —OH to give the group —COOH, the characteristic group of acids. For this reason aldehydes are *reducing agents*.

The double bond of the C=O group readily breaks in the presence of certain hydrogen compounds, the hydrogen atom attaching itself to the oxygen to form a —OH group and the rest of the compound attaching itself to the carbon.

$$RCHO + HX \rightarrow R{-}\overset{\displaystyle H}{\underset{\displaystyle X}{\overset{|}{\underset{|}{C}}}}{-}OH$$

The resulting compound, having been formed by 'adding on' to the aldehyde, is called an *addition* compound.

The oxygen atom of —CHO may be removed by the hydrogen of compounds containing the amino group (—NH_2) to form water, and the resulting compound is called a *'condensation'* compound.

$$RCHO + H_2NZ \rightarrow RCH{=}NZ + H_2O$$

PROPERTIES OF KETONES

The formula of a typical ketone is R_1COR_2, where R_1 and R_2 are alkyl groups, e.g. CH_3COCH_3, acetone (propanone).[1] The molecule does not contain a hydrogen atom favourably placed for oxidation, and consequently ketones are *not* reducers.

The molecule does, however, contain the C=O group and the formation of *addition* compounds and *condensation* compounds would be expected: this in fact is the case.

Experiment 226 Aldehydes as reducing agents

Material: Fehling's solution; formaldehyde (methanal); acetaldehyde (ethanal); acetone (propanone).

(*a*) *Reduction of silver oxide.* Take a test-tube containing 2 cm^3 of silver nitrate solution. Add some sodium hydroxide solution to precipitate silver oxide. Decant the liquid and add ammonia solution drop by drop and with continuous shaking until the precipitate has almost dissolved. Divide the solution into two portions and add respectively two drops of formaldehyde solution (formalin) and one drop of acetaldehyde. Leave the test-tube in a beaker of hot water and observe the mirror of silver in both cases.

$$2Ag(NH_3)_2^+ + RCHO + H_2O \rightarrow 2Ag\downarrow + RCOO^- + 3NH_4^+ + NH_3$$

(*b*) *Reduction of Fehling's solution.* Fehling's solution is prepared

---

[1] Acetone (propanone, CH_3COCH_3) is a colourless liquid with a characteristic odour; it has a boiling point of 56 °C and a density of 0.8 g cm^{-3}.

as follows. Dissolve 17 g of copper(II) sulphate crystals in water and make up to 250 cm^3: this is referred to as 'Solution A'. Dissolve 35 g of sodium hydroxide and 87 g of potassium sodium tartrate (2,3-dihydroxybutanedioate, Rochelle salt) in water and make up to 250 cm^3: this is 'Solution B'. Mix equal volumes of solution A and solution B as required for an experiment; the resulting Fehling's solution is a clear, deep-blue solution containing copper in the form of a tartrate anion.

Take a test-tube about a quarter full of the deep-blue solution. Add a few drops of formalin and boil the mixture. Note the appearance of a yellow, orange, and finally red precipitate of copper(I) oxide. The copper in the salt has been reduced from the divalent to the monovalent condition. Repeat the experiment, substituting acetaldehyde for formaldehyde, and note the similarity of the reactions. In both cases the aldehyde has been oxidized to the corresponding acid.

Repeat both the above reactions using a ketone, acetone, in place of an aldehyde and show that neither the diamminesilver ion nor Fehling's solution is reduced.

Experiment 227 Preparation of an addition compound; the hydrogensulphite compound of acetone

Material: acetone (propanone); sodium carbonate-10-water; siphon of sulphur dioxide; **ICE**.

Crush some sodium carbonate crystals and take enough to form a layer about 5 mm deep in a conical flask. Add just enough water to cover the crystals. Pass a steady stream of sulphur dioxide into the mass, using the delivery tube as a stirrer. Carbon dioxide is displaced, and the solution effervesces and becomes a light-green colour; this is a saturated solution of sodium hydrogensulphite. Add about 2 cm^3 of acetone, shake well, and leave to stand, preferably in ice. The crystals are the hydrogensulphite compound of acetone[1].

$$CH_3COCH_3 + HSO_3^- \rightarrow CH_3{-}\overset{\displaystyle OH}{\underset{\displaystyle OSO_2^-}{\overset{|}{\underset{|}{C}}}}{-}CH_3$$

Note (*i*) All aldehydes and ketones form hydrogensulphite compounds.

(*ii*) All aldehydes and ketones form addition compounds with hydrogen cyanide.

---

[1] Difficulty may be experienced in obtaining these crystals, which are very soluble in water. Benzaldehyde may be substituted for acetone for the purpose of illustration (Experiment 233).

(*iii*) Aldehydes (except formaldehyde) form addition compounds with ammonia.

(*iv*) Ketones do *not* form addition compounds with ammonia.

Experiment 228 Action of ammonia on formaldehyde

Apparatus: water-bath.

Material: formalin (40% *m*/*V* formaldehyde, methanal).

Mix a test-tube one-quarter full of formalin with an equal volume of concentrated solution of ammonia. Pour the mixture into a watch-glass and leave to evaporate to dryness on a water-bath. The white solid is not an addition compound, but the complex condensation compound hexamethylenetetraamine, $(CH_2)_6N_4$, used in medicine as a diuretic under the name urotropine or hexamine.

Experiment 229 Preparation of condensation compounds of acetaldehyde and acetone

Apparatus: small corked flask; measuring cylinder.

Material: hydroxyammonium chloride; sodium hydroxide; acetone; sodium acetate-3-water; 2,4-dinitrophenylhydrazine; semicarbazide hydrochloride (semicarbazinium chloride).

(*a*) *Condensation with hydroxylamine*

Make a solution containing 3 g of sodium hydroxide in 10 cm^3 of water, in a corked flask. Dissolve 5 g of hydroxyammonium chloride in 10 cm^3 of water and add this to the sodium hydroxide solution. Add 8 cm^3 of acetone, gradually and with shaking, cork the flask and leave it overnight. The colourless needles are crystals of the oxime of acetone, acetoxime (propanone oxime).

$$(CH_3)_2C{=}O + NH_3OH^+ \rightarrow (CH_3)_2C{=}NOH + H_2O + H^+$$

Note: Any aldehyde or ketone may be used.

(*b*) *Condensation with phenylhydrazine*

Phenylhydrazine itself is not suitable, as the compounds formed with either acetone or acetaldehyde do not crystallize readily, but the substituted nitro-compound of phenylhydrazine gives excellent results.

Take enough of the dinitrophenylhydrazine to cover the bottom of a test-tube. Add dilute hydrochloric acid, gradually and with shaking, until the solid is almost dissolved. Filter to obtain a clear solution. Add one drop of either acetaldehyde or acetone to a few drops of the prepared solution. The crystals formed are the hydrazones (actually, the dinitrophenylhydrazones).

$$CH_3CHO + H_2NNHC_6H_3(NO_2)_2 \rightarrow CH_3CH{=}NNHC_6H_3(NO_2)_2 + H_2O$$

$$(CH_3)_2CO + H_2NNHC_6H_3(NO_2)_2 \rightarrow (CH_3)_2C{=}NNHC_6H_3(NO_2)_2 + H_2O$$

Note: Any aldehyde or ketone may be used.

(*c*) *Condensation with semicarbazide*

Semicarbazide has the formula $NH_2NHCONH_2$ and is used in the form of its hydrochloride. Aldehydes and ketones give crystalline derivatives which may serve to identify the parent compounds by their melting points.

Mix four spatula-loads of the semicarbazide with five of crystalline sodium acetate into 5 cm^3 of water, and warm the mixture gently in a small beaker to obtain a clear solution. To this warm solution add four spatula-loads of the aldehyde or ketone (or 1 cm^3 of a liquid) in 5 cm^3 of ethanol. Continue warming for fifteen minutes and then cool and filter off the crude semicarbazone; it may be recrystallized from ethanol.

Aldehydes readily polymerize, i.e. two or more molecules unite to form a complex molecule having the same percentage composition but two or more times the relative molecular mass of the aldehyde. Aldehydes polymerize in presence of acid or alkali; formaldehyde polymerizes when its solution is evaporated or even when left to stand for some time. Polymers of aldehydes will, in general, dissociate into the simpler molecules when distilled, but this is not possible when the polymer has been made by the action of sodium hydroxide (see below).

Experiment 230 Preparation of polymers of formaldehyde and acetaldehyde

Material: formalin; acetaldehyde (ethanal); sodium hydroxide.

(*a*) *Preparation of paraformaldehyde.* Fill a watch-glass with formalin and leave to evaporate on a water-bath. The white solid is paraformaldehyde (methanal trimer).

$$3HCHO \rightarrow (CH_2O)_3$$

(*b*) *Preparation of paraldehyde.* Take ten drops of acetaldehyde in a dry test-tube. Add one drop of concentrated sulphuric acid and cool to room temperature. Add enough water to treble the volume. The suspension is the immiscible liquid, paraldehyde (ethanal trimer).

$$3CH_3CHO \rightarrow (C_2H_4O)_3$$

Heat the contents of the tube. The oily liquid dissolves as it changes back to acetaldehyde.

(*c*) *Preparation of aldehyde resin.* To a test-tube a quarter full of acetaldehyde, add half that volume of water, and then a piece of sodium hydroxide about the size of a pea. Warm gently. The yellow resinous suspension is a polymer of unknown composition; it cannot be changed back to acetaldehyde.

Note: Formaldehyde differs from other aldehydes in its action with sodium hydroxide. Instead of polymerizing, it forms a mixture of methanol and sodium formate. As an aldehyde on reduction gives an alcohol, and on oxidation gives an acid, it appears that half of the formaldehyde is oxidized at the expense of the other half, which is reduced (disproportionation).

$$2HCHO + OH^- \rightarrow CH_3OH + HCOO^-$$

A similar action occurs with aldehydes of the aromatic series (see p. 229) and is known as Cannizzaro's Reaction.

Experiment 231 Further reactions of aldehydes and ketones

Material: Schiff's reagent; acetaldehyde (ethanal); acetone (propanone); aqueous iodine-potassium iodide solution.

(*a*) *Reaction with Schiff's reagent.* Schiff's reagent is made by passing sulphur dioxide into a solution of magenta dye until the solution is colourless. Add a drop of acetaldehyde; the colour is restored. *Any aldehyde restores the colour. Ketones do not.*

(*b*) *Iodoform reaction.* Repeat Experiment 213, but using either acetaldehyde or acetone in place of ethanol. Show that both these compounds give iodoform. Other aldehydes and ketones *do not* react in this way.

BENZALDEHYDE

Introduction of —CHO group into the benzene nucleus. The usual method of preparing aldehydes in the aliphatic series is to oxidize the corresponding alcohol (see p. 207). Groups in the side-chain behave generally as they would in aliphatic compounds, and it would be expected that the group $—CH_2OH$ (as in phenylmethanol) would oxidize to —CHO and this in fact is the case; benzyl alcohol oxidizes to benzaldehyde (benzenecarbaldehyde). Likewise it would be expected that as chloroethane hydrolyses to ethanol, so the group $—CH_2Cl$ would hydrolyse to $—CH_2OH$ and this also is true. If then, benzyl chloride ((chloromethyl)benzene) were hydrolysed and oxidized, benzaldehyde would be the product, and a single agent to effect both of these actions is copper(II) nitrate solution.

Experiment 232 Preparation of benzaldehyde

Formula: C_6H_5CHO.

Physical properties: pale yellow liquid with a pleasant smell of almonds, boiling point 180 °C.

Apparatus: flask (500 cm^3); condenser; separating funnel; Buchner funnel and flask; distilling flask; sand-tray; thermometer (360 °C). (Standard-joint apparatus AGHP.)

Material: benzyl chloride; copper(II) nitrate; supply of carbon dioxide; diethyl ether; siphon of sulphur dioxide; anhydrous calcium chloride.

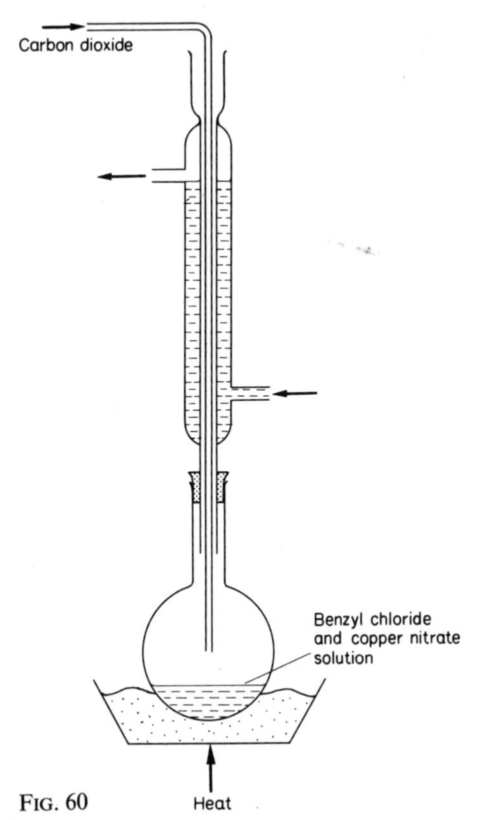

FIG. 60

In the large flask mix 20 cm^3 of benzyl chloride, 17 g of copper(II) nitrate and 200 cm^3 of water. Attach a reflux condenser[1] and down the inner tube insert a long piece of glass tubing. Pass carbon dioxide into the flask to displace air and then, while continuing to pass in

[1] Avoid using a condenser with a tapered end.

carbon dioxide, heat the flask on a sand-tray (see Figure 60). The action takes about eight hours to complete, and if the preparation is interrupted the condenser tube should be corked until the action can be resumed. The oil changes slowly to a yellow colour and the copper(II) nitrate solution becomes paler. Transfer the yellow oil to a separating funnel and extract with 90 cm^3 of ether taken in three portions. Distil the ether-benzaldehyde solution on a water-bath to remove the ether.

Prepare a saturated solution of sodium hydrogensulphite in a conical flask by just covering 45 g of sodium carbonate-10-water with water and then passing in sulphur dioxide until an apple-green solution is formed. Shake the oil with this solution and leave it to stand for a quarter of an hour, then filter.

Return the crystals to the large flask and add dilute sulphuric acid in excess (400 cm^3) to decompose the hydrogensulphite compound and steam distil (see Figure 69).

Extract the aqueous distillate with ether as before and dry the ether solution with pieces of calcium chloride. Distil, first over hot water (to remove the ether) and then on a gauze. Yield 6 g.

$$2C_6H_5CH_2Cl + 2NO_3^- \rightarrow 2C_6H_5CHO + 2Cl^- + H_2O + NO + NO_2$$

Experiment 233 Reactions of benzaldehyde

Apparatus: separating funnel; watch-glass; distilling flask and condenser; stoppered bottle; water-bath.

Material: siphon of sulphur dioxide; Fehling's solution; potassium hydroxide; diethyl ether; 2,4-dinitrophenylhydrazine hydrochloride.

(*a*) Take a drop of benzaldehyde on a watch-glass and spread the liquid. Leave it to stand in the air. Note crystals of benzoic acid showing the rapid oxidation of benzaldehyde.

(*b*) Make a solution containing diamminesilver ions in a clean test-tube (see Experiment 226(*a*)). Add a drop of benzaldehyde and note the silver mirror.

(*c*) Show that benzaldehyde does not reduce Fehling's solution.

(*d*) Prepare a saturated solution of sodium hydrogensulphite (see Experiment 227). Add two drops of benzaldehyde to about 5 cm^3 of the solution. The crystals are the hydrogensulphite compound of the aldehyde.

(*e*) Take a small specimen tube with a well-fitted cork. Mix five drops of concentrated ammonia solution with one drop of benzaldehyde and set aside. Note the crystals of the *condensation* compound hydrobenzamide $(C_6H_5CH)_3N_2$. Contrast with the action of acetaldehyde (*addition* compound, acetaldehyde-ammonia).

(*f*) Take enough 2,4-dinitrophenylhydrazine hydrochloride to cover the bottom of a test-tube. Add one and a half times that bulk of sodium acetate and about 1 cm^3 of water to make a solution. Add two drops of benzaldehyde and obtain crystals of the hydrazone,

$$C_6H_5CH{=}NNHC_6H_5$$

This reaction is characteristic of all aldehydes and ketones.

(*g*) Cannizzaro's Reaction. Take a bottle with a well-fitted stopper[1] (about 125 cm^3 is suitable). Mix 20 cm^3 of benzaldehyde and a solution of 20 g of potassium hydroxide in 10 cm^3 of water. Shake to emulsify and leave overnight.

$$2C_6H_5CHO + OH^- \rightarrow C_6H_5COO^- + \quad C_6H_5CH_2OH$$

Dissolve the solid product in the least amount of water (about 80 cm^3) and extract with 20 cm^3 of ether. Run off the lower layer (potassium benzoate) and acidify with concentrated hydrochloric acid to obtain a precipitate of benzoic acid.

Distil the ether-benzyl alcohol solution over hot water to remove the ether, and then heat on a gauze. Collect a few drops of the distillate. Take two drops, add ten drops of water and four drops of concentrated nitric acid. Note the smell of benzaldehyde due to oxidation of the alcohol. Boil and obtain benzoic acid on cooling.

ACETOPHENONE

Experiment 234 Preparation of acetophenone

Formula: $C_6H_5COCH_3$.

Physical properties: white solid, melting point 20.5 °C.

Apparatus: flask (500 cm^3); reflux condenser; tap-funnel; glass tube (see Figure 57); thermometer (360 °C); distilling flask; measuring cylinder.

Material: benzene; acetyl chloride; fresh anhydrous aluminium chloride; anhydrous calcium chloride; **ICE**.

The preparation is an example of the Friedel and Crafts reaction. Acetophenone (phenylethanone) is a ketone with the general properties of an aliphatic ketone. *Care! The experiment should be performed in a fume chamber and the apparatus dried thoroughly.*

The preparation depends for its success on the quality of the anhydrous aluminium chloride, and although it can be prepared in the laboratory by heating aluminium strongly in a stream of chlorine, the preparation in the quantity required is tedious. It is best to buy the

[1] If a glass stopper is used it should be smeared with vaseline.

compound as required, preferably in small quantities. Transfer 50 g to the 500 cm^3 flask and cover it immediately with 35 cm^3 of benzene. Attach the flask to a reflux condenser, and arrange the apparatus as shown in Figure 52, but with the flask standing in ice. Take a measured 35 cm^3 of acetyl chloride in the tap-funnel and allow it to drip slowly into the flask. When all the acetyl chloride has been added, leave the apparatus in position for an hour and then pour the contents into about 150 cm^3 of ice-cold water. Decant the clear liquid from the solid residue into a separating funnel. Add 20 cm^3 of benzene, shake well and allow to settle. Run off the lower layer and discard it. Wash the retained solution (of acetophenone in benzene) with water, and again run off the lower layer, then run the acetophenone-benzene solution into a flask, add anhydrous calcium chloride and leave to dry. Decant the clear liquid into a distilling flask and distil. When the temperature reaches 130 °C, empty the condenser and continue with the distillation. Collect the fraction between 195 °C and 200 °C and leave it to solidify. The yield is small.

(If ethyl iodide were used in place of acetyl chloride, the ethyl group would be introduced into the benzene molecule to give ethylbenzene.)

$$C_6H_6 + CH_3COCl \rightarrow C_6H_5COCH_3 + HCl$$

BENZOPHENONE

Experiment 235 Benzophenone (diphenylmethanone) and its oxime

Material: benzophenone (diphenylmethanone).

Benzophenone, like acetophenone, may be prepared by the Friedel and Crafts reaction.

Formula: $C_6H_5COC_6H_5$.

Physical properties: melting point 48 °C.

Apparatus: as in Figures 47 and 48.

Material: benzene; benzoyl chloride; fresh anhydrous aluminium chloride; hydroxyammonium chloride; anhydrous calcium chloride.

Into a small, round-bottomed flask put 6 g of fresh anhydrous aluminium chloride and add 15 cm^3 of benzene; set up the apparatus *for refluxing in the fume chamber.* Add 4 cm^3 of benzoyl chloride down the condenser and then fit an absorption tube containing calcium chloride at the top of the condenser. Heat the flask in a beaker of water to 50 °C for thirty minutes. Allow the flask and its contents to cool before pouring the liquid into 40 cm^3 of water in a small separating funnel: the benzophenone remains in the excess benzene, which is the uppermost layer. After shaking, discard the water layer and then mix the benzene layer with an equal volume of sodium hydroxide solution. Separate the mixture and dry the benzene layer with anhydrous calcium chloride for fifteen minutes.

The solution of benzophenone in benzene is distilled in a small distillation apparatus until crystals of benzophenone have formed. To the crystals add 3 g of hydroxyammonium chloride, 10 cm^3 of ethanol and 2 cm^3 of water. Next add 5.5 g of sodium hydroxide (in flake or pellet form) gradually so that the contents do not boil too vigorously.

To complete the formation of benzophenone oxime, reflux the mixture for ten minutes, then cool it and pour into 15 cm^3 of concentrated hydrochloric acid and 100 cm^3 of water.

$$C_6H_5COCl + C_6H_6 \rightarrow C_6H_5COC_6H_5 + HCl$$
$$C_6H_5COC_6H_5 + NH_2OH \rightarrow (C_6H_5)_2C{=}NOH + H_2O$$

The precipitate of the oxime may be filtered off at the pump, washed with water and recrystallized from methanol (melting point 142 °C).

Benzophenone oxime on treatment with thionyl chloride (sulphur dichloride oxide) in ethereal solution undergoes the Beckmann rearrangement, yielding benzanilide (N-phenylbenzamide): carefully distil off the diethyl ether and then add water; recrystallization can be from ethanol.

$$(C_6H_5)_2C{=}NOH \rightarrow C_6H_5CONHC_6H_5$$

27
Carboxylic Acids

The name 'fatty acids' is given to the series of which formic (methanoic) and acetic (ethanoic) acids are the first members, although it is only among higher members that the term 'fatty' is appropriate; nowadays they could be called alkanoic acids. Natural fat contains the glyceryl esters of palmitic (hexadecanoic) and stearic (octadecanoic) acids, from which the acids themselves may be obtained by hydrolysis (see Experiment 357, p. 440).

Acetic acid may be obtained in the laboratory by hydrolysing its ester (p. 242), acyl chloride (p. 245), nitrile (p. 250) or amide (p. 249).

Formic acid is usually prepared by removing the elements of carbon dioxide from oxalic (ethanedioic) acid:

$$\begin{matrix} \text{COOH} \\ | \\ \text{COOH} \end{matrix} \rightarrow \text{HCOOH} + \text{CO}_2$$

N.B. Oxalic (ethanedioic) acid is considered in Experiments 349 and 350, p. 435.

Experiment 236 Preparation of formic acid[1]

Formula: HCOOH.

Physical properties: colourless liquid, boiling point 101 °C, density 1.2 g cm^{-3}.

Apparatus: distilling flask (250 cm^3); condenser; sand-tray; thermometer (360 °C).

Material: oxalic acid; glycerol.

Weigh 30 g of glycerol (1,2,3-propanetriol) in an evaporating basin, and dehydrate by heating on a sand-tray until a thermometer, which is used as a stirrer, indicates a temperature of 175–180 °C.

Weigh three portions (30 g each) of oxalic acid.

Pour the glycerol into the distilling flask and connect it to a sloping

[1] The relation between the preparation and the purification of formic acid is conveniently shown by groups of students performing Experiments 236, 237 and 238 simultaneously.

condenser. Add one of the portions of oxalic acid and close the neck of the flask with a cork and thermometer, adjusting the latter so that it dips *into* the liquid.

Heat the mixture gradually on the sand-tray to 110 °C and maintain it at that temperature until the evolution of gas (carbon dioxide) has moderated, and collect the distillate of formic acid. Allow the temperature to fall to 80 °C and add a second portion of oxalic acid. Heat as before, and when the action slackens, allow to cool and add the third portion.

The distillate is an aqueous solution of formic acid (about 50%) and may be used as formic acid for most purposes.

Use the product for the preparation of lead(II) formate and for reactions of the acid.

The glycerol may be considered to act as a catalyst in this reaction, an intermediate compound of oxalic acid and glycerol being formed, followed by loss of carbon dioxide. This gives a compound of formic acid and glycerol which hydrolyses to eliminate glycerol. The reaction is shown by the simplified equation on p. 232.

Experiment 237 Preparation of lead(II) formate

Formula: $Pb(HCOO)_2$.

Physical properties: white needle crystals.

Apparatus: water-bath; filter-funnel (in steam jacket for preference).

Material: Formic (methanoic) acid; lead(II) carbonate.

Take an evaporating basin half full of formic acid (either pure formic acid or the product from Experiment 236) and heat on a water-bath. Add lead carbonate in small portions, stirring continuously, until some of the solid remains undissolved. Filter the hot solution into another basin and leave it to crystallize. Filter and dry.

$$2HCOOH + PbCO_3 \rightarrow Pb(HCOO)_2 + CO_2 + H_2O$$

(i) Heat a little of the product in an ignition tube. Ignite the gas (carbon monoxide) and note the residue of lead(II) oxide.

(ii) Warm a little of the formate with concentrated sulphuric acid. Test for carbon monoxide. The solid left is lead(II) sulphate.

Write equations for both these reactions.

Experiment 238 Preparation of pure formic acid from lead(II) formate

Apparatus: as shown in Figure 61; glass-wool.

Material: lead(II) formate.

Care! The experiment should be done in a fume chamber.

Place the powdered lead formate in the tube, holding the compound loosely in position by a plug of glass-wool at each end. Clamp the tube in a sloping position and connect the upper end to a supply of hydrogen sulphide. Place a receiver at the lower end. Pass a stream of hydrogen sulphide through the tube and warm the tube with a luminous flame. (The temperature should be maintained at about 100 °C.) Formic acid is liberated and collects in the receiver. The yield is small.

$$Pb(HCOO)_2 + H_2S \rightarrow 2HCOOH + PbS$$

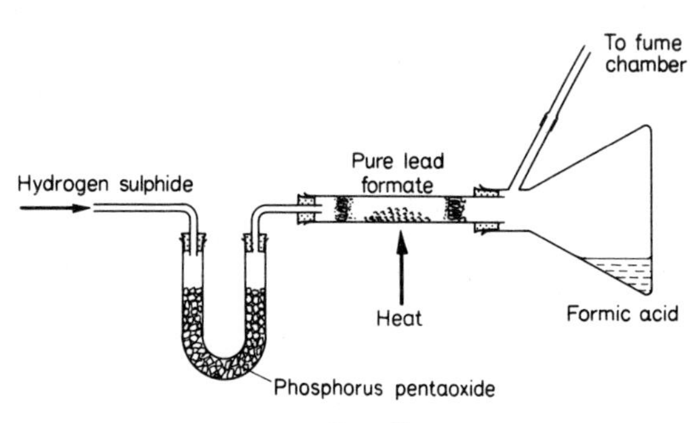

FIG. 61

REACTIONS OF FORMIC ACID AND ACETIC ACID

Formic acid, HCOOH, and acetic acid, CH_3COOH, contain the characteristic carboxyl group, —COOH. Purely acid properties depend on the terminal hydrogen atom of this group. Both of these acids in aqueous solution turn litmus red and neutralize bases to form salts, and although weaker (i.e. less ionized) than mineral acids, they are stronger than carbonic acid and displace carbon dioxide from carbonates. **Formic acid differs from acetic acid in being a reducing agent.** This property is accounted for by the structural formula which shows that formic acid is both aldehydic and acidic.

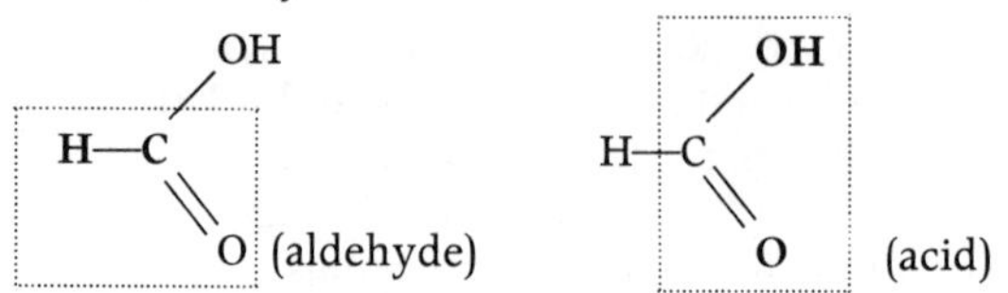

When acting as a reducing agent, it is oxidized to carbon dioxide and water.

Experiment 239 Reactions of formic acid

Material: formic (methanoic) acid; sodium formate (methanoate); ethanol.

(*a*) Mix about five drops each of formic acid and ethanol, add three drops of concentrated sulphuric acid, and warm gently. The pleasant smelling volatile liquid is ethyl formate.

$$HCOOH + C_2H_5OH \rightarrow HCOOC_2H_5 + H_2O$$

(*b*) Add a few drops of formic acid to a test-tube a quarter full of an acidified solution of potassium permanganate (manganate(VII)), and warm gently. The permanganate is *reduced.*

$$2MnO_4^- + 11H^+ + 5HCOO^- \rightarrow 2Mn^{2+} + 8H_2O + 5CO_2\uparrow$$

(*c*) Add five drops of formic acid to five drops of mercury(II) chloride solution and warm. Mercury(II) chloride is *reduced* to mercury(I) chloride. Add ammonia solution to test for mercury(I) chloride (black precipitate obtained).

$$2HgCl_2 + HCOOH \rightarrow Hg_2Cl_2 + 2HCl + CO_2$$

(*d*) Prepare a test-tube about a quarter full of ammoniacal silver oxide (Experiment 113(*e*), p. 102). Add two or three drops of formic acid and warm gently. Silver is formed as a black powder by *reduction* of the diamminesilver ions.

$$2Ag(NH_3)_2^+ + HCOO^- \rightarrow 2Ag\downarrow + NH_4^+ + 3NH_3 + CO_2$$

(*e*) To a test-tube about one-eighth full of formic acid add about half that volume of concentrated sulphuric acid. Heat gently, and light the gas at the mouth of the tube. The formic acid has been *dehydrated* to give carbon monoxide.

$$HCOOH + (H_2SO_4) \rightarrow CO + (H_2SO_4 \cdot H_2O)$$

Formic acid is the only fatty acid in which the OH group of the carboxyl group and a hydrogen atom are linked to the same carbon atom. It is this proximity which allows dehydration to occur so readily.

$$\begin{matrix} H \\ \quad\diagdown \\ \quad\quad C{=}O \rightarrow CO + H_2O \\ \quad\diagup \\ HO \end{matrix}$$

(*f*) Repeat (*e*), using sodium formate in place of formic acid. Note the formation of carbon monoxide as before.

(*g*) Iron(III) chloride test. Add sodium hydroxide solution, drop by drop, to a test-tube one-quarter full of iron(III) chloride solution until a permanent precipitate of iron(III) hydroxide is obtained. Filter off the precipitate. The filtrate is called 'neutral' iron(III) chloride. Add ammonia solution to about 10 cm^3 of formic acid until it is just alkaline to litmus then boil the solution for a minute or two to remove excess ammonia, and cool it. Add 'neutral' iron(III) chloride. A deep red coloration is obtained, which, on boiling, gives a red-brown precipitate of basic formate.

Experiment 240 The preparation of acetic acid

Apparatus: as in Figure 48, then as in Figure 49.

Material: concentrated sulphuric acid; sodium dichromate; ethanol.

Put 50 cm^3 of water in the flask and carefully add 30 cm^3 of concentrated sulphuric acid to it with swirling and cooling. Next add 50 g of sodium dichromate, being careful in handling this material because it is very corrosive. Set up the apparatus for boiling under reflux (Figure 48) and slowly pour a mixture of 15 cm^3 of ethanol in 50 cm^3 of water down the condenser; shake the flask, and cool it if a vigorous reaction ensues. Then heat the flask on a water-bath for about twenty minutes. Allow the flask and contents to cool slightly before rearranging the apparatus for a distillation (Figure 49). Heat the flask over a gauze and collect about 100 cm^3 of distillate in the receiver.

$$C_2H_5OH + H_2O \rightarrow CH_3COOH + 4H^+ + 4e^-$$

Experiment 241 Reactions of acetic acid

Material: acetic (ethanoic) acid; sodium acetate (ethanoate).

(*a*) Mix five drops each of acetic acid and ethanol, add three drops of concentrated sulphuric acid and warm gently. Pour the product into a test-tube about half-full of water. The pleasant smelling ester is ethyl acetate.

$$CH_3COOH + C_2H_5OH \rightarrow CH_3COOC_2H_5 + H_2O$$

(*b*) Repeat Experiment 239(*b*), using acetic acid in place of formic acid, and show that potassium permanganate is not reduced.

(*c*) Show that acetic acid does not reduce mercury(II) chloride.

(*d*) Show that acetic acid does not reduce ammoniacal silver oxide.

(*e*) Show that acetic acid is not dehydrated by concentrated sulphuric acid.

(*f*) Show that the product of the action of concentrated sulphuric acid on sodium acetate is acetic acid.

(*g*) Show that a neutral solution of an acetate with neutral iron(III) chloride gives similar results to those obtained in Experiment 239 (*g*).

Note: Acetic acid is typical of a series of acids. Formic acid is the only fatty acid which shows reducing properties.

Experiment 242 The reactions of tartrates

Material: tartaric (2,3-dihydroxybutanoic) acid or potassium sodium tartrate.

(*a*) Put a spatula-load of tartaric acid into 1 cm^3 of concentrated sulphuric acid and then warm the mixture: much charring occurs and carbon monoxide and sulphur dioxide are released.

(*b*) Add a drop of sodium hydroxide solution to 5 cm^3 of silver nitrate solution and then dilute ammonia solution until the initial precipitate has just redissolved. Now add a few drops of a neutral tartrate solution and warm the mixture: a silver mirror should be produced by reduction of the silver ions.

(*c*) *Fenton's test.* To a solution of the tartrate add a few drops of saturated iron(II) sulphate solution and then a few drops of hydrogen peroxide solution followed by excess sodium hydroxide solution. A deep violet coloration is produced which is intensified by the addition of one drop of iron(III) chloride.

(*d*) To a solution of the tartrate add calcium chloride solution: calcium tartrate is precipitated.

(*e*) Add the tartrate solution to copper(II) sulphate solution and then add sodium hydroxide: observe the absence of copper hydroxide precipitate because of the formation of a deep blue complex copper tartrate ion.

AROMATIC ACIDS

Experiment 243 Preparation of benzoic acid by oxidation

Formula: C_6H_5COOH (benzenecarboxylic acid).

Physical properties: white crystals, sparingly soluble in cold water, moderately soluble in hot water, melting point 121.4 °C, sublimes if rapidly heated.

Apparatus: 500 cm^3 round-bottomed flask; condenser.

Materials: benzyl chloride ((chloromethyl) benzene); potassium permanganate (manganate(VII)); anhydrous sodium carbonate; hydrated sodium sulphite (sulphate(IV)); dilute hydrochloric acid.

Pour 5 cm^3 of benzyl chloride into the flask and add 5 g of anhydrous sodium carbonate and 10 g of potassium permanganate. Then add 200 cm^3 of water. Reflux the mixture until oily drops of benzyl chloride are no longer visible in the condenser, i.e. 1–1½ hours.

$$C_6H_5CH_2Cl + H_2O \rightarrow C_6H_5CH_2OH + HCl$$

$$3C_6H_5CH_2OH + 4MnO_4^- \rightarrow 3C_6H_5COO^- + 4MnO_2\downarrow + OH^- + 4H_2O$$

To the hot solution, add 20 g of hydrated sodium sulphite, followed by 150 cm^3 of dilute hydrochloric acid, added in small portions with shaking. Alternatively, sulphur dioxide may be bubbled through the solution from a siphon.

$$MnO_2 + SO_2 \rightarrow Mn^{2+} + SO_4^{2-}$$
$$C_6H_5COO^- + H_2O + SO_2 \rightarrow C_6H_5COOH + HSO_3^-$$

On cooling, benzoic acid separates. A purer product may be obtained by recrystallizing from hot water.

Experiment 244 Preparation of benzoic acid from an ester

Apparatus: as in Figure 48.

Material: methyl benzoate (benzenecarboxylate); concentrated sodium hydroxide solution.

Put about 20 cm^3 methyl benzoate into the round-bottomed flask and add about 100 cm^3 concentrated sodium hydroxide solution; a few anti-bumping granules may ensure smooth refluxing. Heat the flask under a reflux condenser for about thirty minutes (as in Figure 48). Allow the flask and its contents to cool back to room temperature. Then slowly add about 25 cm^3 of concentrated sulphuric acid; if this does not complete precipitation of benzoic acid add some dilute sulphuric acid.

Filter off the precipitate using a Buchner funnel and flask, and wash it thoroughly with water. A portion of the precipitate may be purified by recrystallization from boiling water: dissolve it in the minimum quantity of boiling water, filter the solution whilst it is still hot and then allow the solution to cool.

$$C_6H_5COOCH_3 + OH^- \rightarrow C_6H_5COO^- + CH_3OH$$
$$C_6H_5COO^- + H^+ \rightarrow C_6H_5COOH$$

Experiment 245 Reactions of benzoic acid

Material: ethanol; calcium oxide; phosphorus tri- (or penta-) chloride; zinc powder; benzoic acid.

(*a*) Take about a spatula-load of benzoic acid in a test-tube and fill with water to a depth of 5 cm. Note the apparent insolubility. Boil to obtain a solution. Cool under the tap to reprecipitate crystals of the acid. Show that the crystals dissolve readily in sodium hydroxide solution to form sodium benzoate, $C_6H_5COO^-Na^+$, and are reprecipitated on addition of dilute acid.

(*b*) In a dry test-tube mix about a spatula-load of benzoic acid with four times its bulk of calcium oxide. Heat strongly and cautiously note the smell of benzene. Light the vapour at the mouth of the test-tube and observe the smoky flame. The alkali has removed the elements of carbon dioxide to form calcium carbonate and benzene.

$$C_6H_5COOH + CaO \rightarrow C_6H_6 + CaCO_3$$

(*c*) Take enough of the acid to cover the bottom of a test-tube. Add five drops of ethanol and two drops of concentrated sulphuric acid. Warm gently and recognize the smell of the ester, ethyl benzoate, $C_6H_5COOC_2H_5$.

(*d*) In a dry test-tube take enough benzoic acid to cover the bottom of the tube with a thin layer. Add an equal portion of phosphorus pentachloride or two drops of the trichloride. The liquid is benzoyl chloride (benzenecarbonyl chloride) which has an irritant vapour.

$$3C_6H_5COOH + PCl_3 \rightarrow 3C_6H_5COCl + H_3PO_3$$

(*e*) Add five drops of ethanol to the benzoyl chloride prepared previously. Warm gently and recognize ethyl benzoate.

(*f*) Heat a spatula-load of benzoic acid very gently with about three times the bulk of zinc powder. Note a slight smell of benzaldehyde (benzenecarbaldehyde) due to partial reduction of the acid.

(*g*) *Preparation of benzamide.* Grind together 10 g of ammonium carbonate and 5 cm^3 of benzoyl chloride (*Care!*) in a mortar in a fume chamber. If, after ten minutes, the smell of benzoyl chloride still remains, add a few drops of concentrated ammonia solution and regrind the mixture. Stir the product into a beaker containing 100 cm^3 of water to precipitate the benzamide (benzenecarboxamide). A pure sample may be obtained by recrystallizing from hot water.

Experiment 246 Reactions of salicylic acid

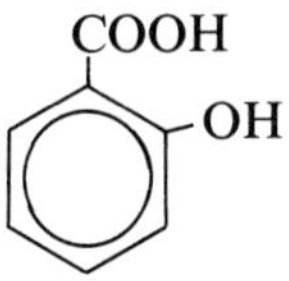

COOH
OH

Material: calcium oxide; salicylic acid (2-hydroxybenzenecarboxylic acid); aspirin; methanol.

(*a*) Heat about a spatula-load of the acid with four times its bulk of calcium oxide. Note the smell of phenol. Compare the action with the similar result when benzoic acid was used.

(*b*) Shake two or three crystals with 10 cm^3 of water in a test-tube. Add a drop of iron(III) chloride solution. Compare the violet coloration with that obtained with phenol. Salicylic acid is both an acid and a phenol and each group retains its individual properties.

(*c*) Take enough salicylic acid to cover the bottom of a test-tube. Add five drops of methanol and two drops of concentrated sulphuric acid. Warm gently and recognize the 'oil of wintergreen' smell of the ester, methyl salicylate.

$$C_6H_4(OH)COOCH_3$$

(*d*) Aspirin is acetyl salicylic acid, in which the acetyl group $-COCH_3$, has displaced the hydrogen atom of the $-OH$ group.

$$C_6H_4(OCOCH_3)COOH$$

Place an aspirin tablet in a test-tube and add water to a depth of about 4 cm. Add a drop of iron(III) chloride solution. Note the slight violet coloration. Add two drops of concentrated sulphuric acid and boil. Allow to cool and observe the intense violet coloration due to hydrolysis of aspirin to salicylic acid and the subsequent action of the acid on iron(III) chloride.

(*e*) Take a test-tube one-quarter full of equal parts of water and concentrated sulphuric acid. Add an aspirin tablet and boil. Note the smell of acetic acid due to hydrolysis.

28
Esters, Acyl Chlorides, Amides and Nitriles

1. The hydrogen atom of the carboxyl (—COOH) group may be replaced by an alkyl radical to give *an ester.* Thus acetic (ethanoic) acid, CH_3COOH forms $CH_3COOC_2H_5$ ethyl acetate (ethanoate).

2. The —OH group of the —COOH group may be replaced by:

(*a*) —Cl to form an *acyl chloride,* CH_3COCl

or (*b*) —NH_2 to form *an amide,* CH_3CONH_2

ESTERS

Experiment 247 Preparation of ethyl acetate

Formula: $CH_3COOC_2H_5$.

Physical properties: pleasant smelling immiscible liquid, boiling point 77 °C, density 0.9 g cm^{-3}.

Apparatus: distilling flask (500 cm^3); tap funnel; adapter; oil-bath; flask (250 cm^3); measuring cylinder; thermometer (200 °C). (Standard-joint apparatus FGHMQ.)

Material: glacial acetic acid; ethanol; anhydrous calcium chloride.

The apparatus is arranged as in Figure 62. Prepare two solutions, as follows.

(*a*) To 25 cm^3 of ethanol add gradually 25 cm^3 of concentrated sulphuric acid with shaking. Pour this mixture into the distilling flask.

(*b*) Mix 50 cm^3 of ethanol with 50 cm^3 of glacial acetic acid. Pour this mixture into the funnel.

Raise the temperature of the oil-bath to 140 °C and maintain it at that temperature throughout. Allow the mixture in the funnel to drip into the flask at the same rate at which the distillate collects in the receiver. Transfer the distillate to a separating funnel and shake with a solution of sodium carbonate, taking care to avoid a high pressure of carbon dioxide by inverting the funnel and leaving the stop-tap open. Test the upper layer with litmus paper to make sure that all traces of acetic acid and sulphurous acid (present as impurities) have been removed. Run off the lower layer and discard it.

To the ester in the funnel, add a solution of 50 g of calcium chloride in 50 cm^3 of water. Shake well; this removes the ethanol remaining as an impurity.

Discard the lower layer and run the ester into a flask. Add enough anhydrous calcium chloride to cover the bottom of the flask, and leave it corked overnight.

Distil and collect the fraction between 74 °C and 79 °C. Yield 45 g.

$$CH_3COOH + C_2H_5OH \rightarrow CH_3COOC_2H_5 + H_2O$$

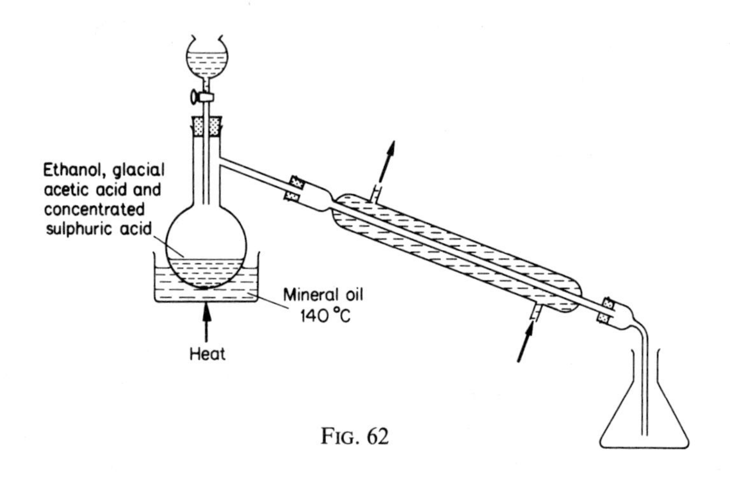

FIG. 62

Experiment 248 Hydrolysis of an ester: saponification

Apparatus: 250 cm^3 flask; condenser; porous pot; water-bath; measuring cylinder; distilling flask.

Material: ethyl acetate; solid sodium hydroxide.

An ester may be hydrolysed either by:

(i) the action of an acid, in which case the protons of the acid catalyses the hydrolysis,

$$CH_3COOC_2H_5 + H_2O \rightarrow CH_3COOH + C_2H_5OH$$

or (ii) the action of an alkali, in which case the products are the alcohol and the sodium salt of the acid,

$$CH_3COOC_2H_5 + OH^- \rightarrow CH_3COO^- + C_2H_5OH$$

This method is known as saponification because of its application to soap manufacture when fats are the esters hydrolysed. (See p. 440.)

Dissolve about 20 g of sodium hydroxide in 50 cm^3 of water in the flask and add some fragments of porous pot. Attach a reflux condenser. Pour 25 cm^3 of ethyl acetate down the inner tube of the condenser. Boil gently on a gauze until the ester has reacted and thus appeared to go into solution (about forty-five minutes).

Transfer to a distilling flask connected to a sloping condenser and distil off half of the liquid: this consists of ethanol and water. Confirm by the iodoform reaction (Experiment 202, p. 198).

Transfer the remaining liquid from the flask to an evaporating basin and heat to dryness on the water-bath. Powder the solid and distil with concentrated sulphuric acid to obtain acetic acid.

$$CH_3COOC_2H_5 + OH^- \rightarrow CH_3COO^- + C_2H_5OH$$

$$CH_3COO^- + H^+ \rightarrow CH_3COOH$$

The experiment can be extended to ethyl benzoate and phenyl benzoate: in the former case the residue in the flask after distillation can be acidified directly to precipitate benzoic acid; in the latter case the residue in the flask after refluxing should be shaken with diethyl ether to extract the phenol and then acidified to precipitate the benzoic acid.

Experiment 249 Preparation of ethyl benzoate (Fischer-Speier method)

The Fischer-Speier method avoids the use of concentrated sulphuric acid as a dehydrating agent and is of particular value where the acid or the alcohol used in esterification contains a benzene or similar nucleus which may sulphonate. By using dry hydrogen chloride, the possibility of nuclear sulphonation is eliminated; a further advantage is that loss of alcohol by ether formation is averted.

Formula: $C_6H_5COOC_2H_5$.

Physical properties: colourless, sweet smelling liquid, immiscible with water, boiling point 212 °C, density 1.05 g cm^{-3}.

Apparatus: flask (100 cm^3); condenser; apparatus for generating and drying hydrogen chloride; drying tube; separating funnel; distilling flask (50 cm^3); thermometer (360 °C); measuring cylinder.

Material: benzoic acid; ethanol; carbon tetrachloride (tetrachloromethane); anhydrous calcium chloride.

Arrange the apparatus as in Figure 49. Weigh 25 g of benzoic acid and transfer it to the flask. Add 40 cm^3 of ethanol. Heat the mixture on a gauze while hydrogen chloride is passed into it at a fairly rapid rate. When the liquid boils, the rate of entry of hydrogen chloride may be decreased to a steady flow since, by that time, the liquid should be saturated with the gas. Continue boiling in this way for two hours. Allow to cool. Pour a few drops of the liquid into water; only an oily ester should form. If white solid (benzoic acid) appears, the reaction is incomplete and should be continued.

Pour the liquid into a separating funnel. Use a measured 200 cm^3 of water to wash the flask and pour the washings into the funnel. Add about 15 cm^3 of carbon tetrachloride and shake to dissolve the ethyl benzoate in this (the lower) layer. After the two layers are clearly defined, run off the lower layer temporarily into a beaker, discard the upper (aqueous layer) and return the carbon tetrachloride layer to the funnel. Add about 50 cm^3 of sodium carbonate solution and shake, taking care that the pressure of carbon dioxide is kept low by frequent opening of the stop-cock. When evolution of carbon dioxide ceases, the liquid is free from hydrochloric acid. Allow to stand until the two layers are clearly defined, then run off the lower layer into a dry conical flask. Add about a spatula-load of anhydrous calcium chloride, stopper the flask and leave overnight to dry.

Distil the liquid after adding two or three pieces of porous pot to minimize bumping, using an air-cooled condenser, and keeping the temperature rising gradually to about 76 °C when carbon tetrachloride begins to distil. Allow this distillation to complete itself at a moderate rate and, when the temperature rises to 205 °C, change the receiver. Collect the distillate between 205 and 213 °C. Yield about 20 g.

$$C_6H_5COOH + C_2H_5OH \rightarrow C_6H_5COOC_2H_5 + H_2O$$

Experiment 250 Preparation of phenyl benzoate

Formula: $C_6H_5COOC_6H_5$.

Physical properties: colourless crystals, melting point 69 °C.

Apparatus: 250 cm^3 wide-mouthed conical flask; cork.

Materials: phenol; sodium hydroxide; benzoyl chloride (*Care!*).

Make a solution in the conical flask of 7 g of phenol, 10 g of sodium hydroxide and 100 cm^3 of water. Add 12 cm^3 of benzoyl chloride, cork and shake vigorously until the smell of benzoyl chloride has almost disappeared, i.e. for about fifteen minutes. The phenyl benzoate may be filtered off at the pump and recrystallized from ethanol. Yield: about 10 g.

$$C_6H_5OH + OH^- \rightarrow C_6H_5O^- + H_2O$$

$$C_6H_5COCl + C_6H_5O^- \rightarrow C_6H_5COOC_6H_5 + Cl^-$$

ACYL CHLORIDES

Experiment 251 Preparation of acetyl chloride

Formula: CH_3COCl.

Physical properties: colourless fuming liquid, boiling point 55 °C, density 1.1 g cm^{-3}.

Apparatus: two 250 cm^3 distilling flasks; condenser; tap funnel; calcium chloride tube; water-bath; thermometer (100 °C). (Standard-joint apparatus EFGGHMNL.)

Material: phosphorus trichloride; anhydrous calcium chloride.

Arrange the apparatus as shown in Figure 63.

Take 50 cm^3 of glacial acetic acid in the distilling flask, and 25 cm^3 of phosphorus trichloride in the tap funnel, which should be stoppered immediately.

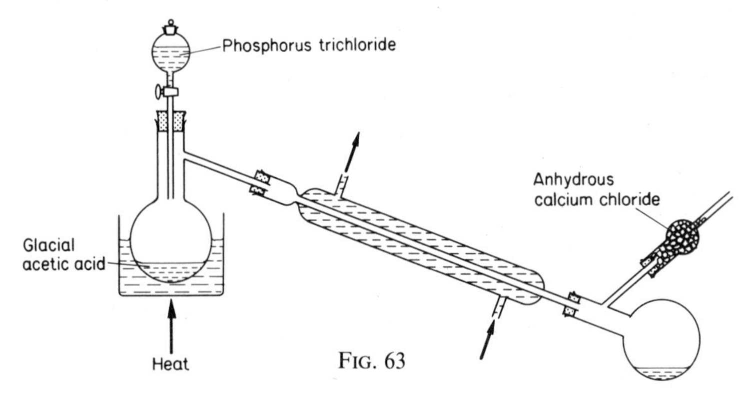

FIG. 63

Immerse the bulb of the flask in the water in the bath and allow the phosphorus trichloride to drip slowly into the acid.

When all has been added, raise the temperature of the bath to 40 °C until the evolution of hydrogen chloride has subsided, and then heat to boiling.

When nothing further distils over, attach the receiver-distilling flask and its contents to the condenser in place of the first distilling flask. Close the neck with a thermometer and cork and heat as before. Collect the portion distilling between 53 °C and 56 °C, and store the acetyl (ethanoyl) chloride in a stoppered bottle. Yield 40 g.

$$6CH_3COOH + 4PCl_3 \rightarrow 6CH_3COCl + P_4O_6 + 6HCl$$

Experiment 252 Reactions of acetyl chloride

Material: acetyl (ethanoyl) chloride; ethanol; sodium chloride;

aniline (phenylamine). Use three or four drops of acetyl chloride for each reaction.

(*a*) To the acetyl chloride add five drops of ethanol. Allow to cool and add about a quarter of a test-tube full of a concentrated solution of sodium chloride. The immiscible liquid which rises to the surface is ethyl acetate, recognized by its fragrant smell. This is a useful method of making esters, no dehydrating agent being needed.

$$CH_3COCl + C_2H_5OH \rightarrow CH_3COOC_2H_5 + HCl$$

(*b*) Add about five drops of water to the second portion and place the test-tube in a stand. Note the accelerating action as hydrolysis causes rise of temperature. Test the gas with silver nitrate solution on a glass rod.

$$CH_3COCl + H_2O \rightarrow CH_3COOH + HCl$$

(*c*) Reaction with ammonia. (*Care!*) This vigorous reaction should be carried out exactly as stated. Take the third test-tube containing acetyl chloride and place the tube in a stand. Take three or four drops of concentrated ammonia solution in another test-tube and empty this small quantity into the acetyl chloride. The thick white fumes are ammonium chloride and the solid residue acetamide (ethanamide).

$$CH_3COCl + 2NH_3 \rightarrow CH_3CONH_2 + NH_4Cl$$

(*d*) To the fourth portion add three drops of aniline. The solid formed is acetanilide (N-phenylacetamide). When cool, add enough water to fill the tube one-third full, boil for a few minutes and allow to cool. Acetanilide crystallizes as colourless flakes.

$$CH_3COCl + C_6H_5NH_2 \rightarrow C_6H_5NHCOCH_3 + HCl$$

Experiment 253 Preparation of benzoyl chloride

Apparatus: as in Figure 48.

Material: benzoic (benzenecarboxylic) acid; thionyl chloride (sulphur dichloride oxide).

$$C_6H_5COOH + SOCl_2 \rightarrow C_6H_5COCl + SO_2 + HCl$$

Set up the apparatus in the fume chamber as in Figure 48. Fix an absorption tube containing calcium chloride at the top of the reflux condenser. Put 20 g of benzoic acid, 15 cm^3 of thionyl chloride and a few anti-bumping granules in the flask. Heat the flask for one hour and then let it cool.

Re-arrange the apparatus for distillation, this time fixing the absorption tube into the receiving flask so that moisture cannot get into the apparatus. Have a thermometer, not a tap-funnel, in the distillation flask. Distil the contents of the flask, collecting the benzoyl chloride (benzenecarbonyl chloride), which boils at 196 °C, in a clean dry vessel: the water can be allowed to drain out of the condenser whilst the benzoyl chloride is distilled.

N.B. The product has a very irritating odour and is lachrymatory.

AMIDES

Experiment 254 Preparation of acetamide

Formula: CH_3CONH_2.

Physical properties: white solid with an odour of mice, melting point 82 °C.

Apparatus: flask (250 cm^3); condenser; thermometer (360 °C); measuring cylinder. (Standard-joint apparatus GGHILM.)

Material: ammonium acetate (ethanoate).

The decomposition of ammonium acetate by heat is shown by the equation

$$CH_3COONH_4 \rightarrow CH_3CONH_2 + H_2O$$

The water formed, however, hydrolyses ammonium acetate (salt of a weak acid and a weak base) and this would become the main action unless a method were designed to control it.

$$CH_3COO^-NH_4^+ + H_2O \rightleftharpoons CH_3COOH + NH_3$$

There are two such methods.

(*a*) by heating ammonium acetate in a sealed tube. Here the pressure of ammonia and acetic acid vapour favours their recombination to form ammonium acetate,

(*b*) by adding excess acetic acid and heating the mixture for some time under ordinary pressure. Here the increased concentration of acetic acid has the same effect as in (*a*).

In either case the ammonium acetate is prevented from hydrolysing appreciably, and the main action becomes the formation of acetamide as shown in the first equation. Both (*a*) and (*b*) are good examples of reversible reactions (see Chapter 2).

Add 70 cm^3 of glacial acetic acid to 50 g of ammonium acetate in the 250 cm^3 flask. Attach a reflux condenser and boil gently on a gauze for four hours (see Figure 64).

After cooling, transfer the mixture to a distilling flask attached to a sloping condenser and close the neck of the flask with a cork and thermometer. Raise the temperature to 130 °C and discard the distillate (chiefly water and acetic acid).

Distillation must be controlled at a very slow rate to make as complete as possible the removal of excess acetic acid and water.

Replace the condenser by the inner tube of a condenser, and continue distillation, collecting the distillate in beakers (Figure 65). Take separate fractions first up to 180 °C and then in 10 cm^3 portions.

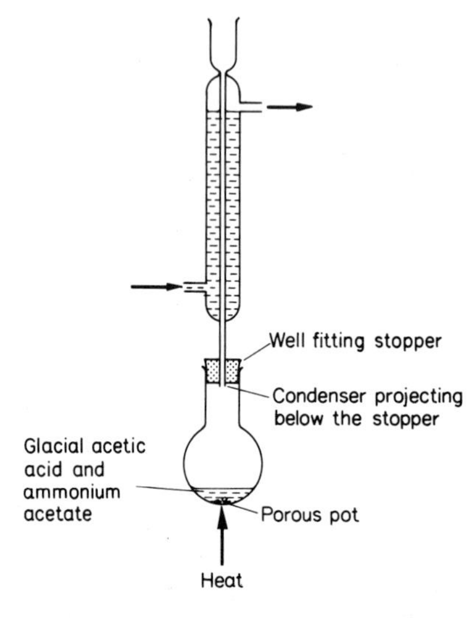

FIG. 64

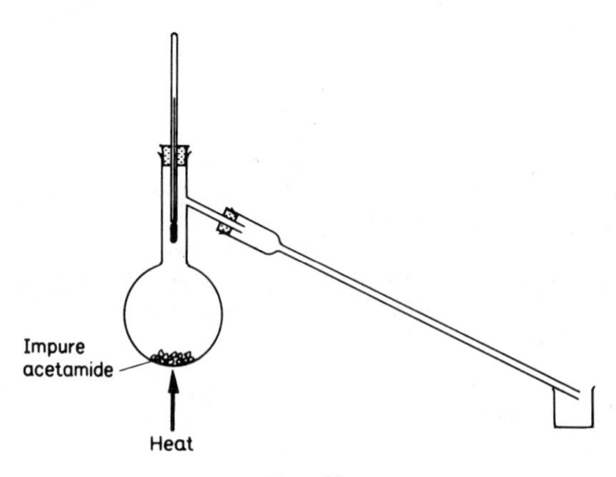

FIG. 65

When crystallization has occurred in one portion, use a small crystal to seed the rest. Dry the crystals between filter papers and keep in a dry stoppered flask. Note the peculiar smell. Yield 25 g.

Experiment 255 Reactions of acetamide

Material: acetamide (ethanamide); sodium nitrite (nitrate(III)).

(*a*) Take about half a spatula-load of acetamide and add about 5 cm^3 of sodium hydroxide solution. Warm gently and recognize ammonia.

$$CH_3CONH_2 + OH^- \rightarrow CH_3COO^- + NH_3$$

Evaporate the solution to dryness and warm the crystals (sodium acetate) with a drop or two of concentrated sulphuric acid. Confirm acetic acid by its smell.

$$CH_3COO^-Na^+ + H_2SO_4 \rightarrow CH_3COOH + Na^+HSO_4^-$$

(*b*) Dissolve about a spatula-load of sodium nitrite in the minimum of water. Keep the test-tube cold under the tap and add enough dilute hydrochloric acid to double the volume. Immediately add about half a spatula-load of acetamide. The effervescence is due to nitrogen.

$$CH_3CONH_2 + HNO_2 \rightarrow CH_3COOH + H_2O + N_2$$

Nitrous acid has thus replaced the $-NH_2$ group by a $-OH$ group, an action which occurs generally to compounds containing the $-NH_2$ group.

(*c*) Boil about a spatula-load of acetamide with about 5 cm^3 of dilute hydrochloric acid, and recognize the smell of acetic acid in the vapour. Ammonium chloride is also formed; to show this add an excess of sodium hydroxide solution and boil. Note the ammonia in the vapour. Compare this action with test (*a*). Hydrolysis has occurred in both cases.

$$CH_3CONH_2 + H^+ + H_2O \rightarrow CH_3COOH + NH_4^+$$

(*d*) When acetamide is dehydrated, acetonitrile (methyl cyanide) is formed (see Experiment 256).

Experiment 256 Preparation of acetonitrile (methyl cyanide)

Formula: CH_3CN.

Physical properties: pleasant smelling liquid, boiling point 82 °C, density 0.8 g cm^{-3}.

Apparatus: retort (150 cm^3).

Material: acetamide (ethanamide); phosphorus pentaoxide (phosphorus(V) oxide).

Into a small retort introduce 10 g of acetamide. Make a paper funnel and use it to introduce about twice the mass of phosphorus pentaoxide. Mix as well as possible and heat moderately, using a large luminous flame. Collect the acetonitrile as a slightly discoloured liquid (see Figure 66). Yield very small.

$$2CH_3CONH_2 + P_4O_{10} \rightarrow 2CH_3CN + 4HPO_3$$

Divide the liquid into two portions.

(*a*) To one portion add sodium hydroxide solution and heat. Test for ammonia with litmus paper.

$$CH_3CN + OH^- + H_2O \rightarrow CH_3COO^- + NH_3$$

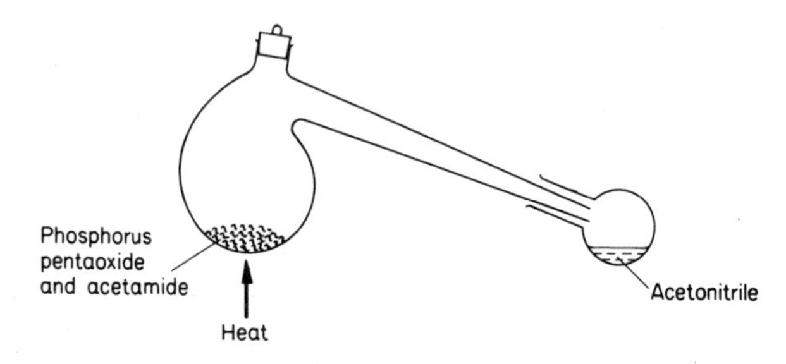

FIG. 66

It is because the compound hydrolyses to acetic acid (which is neutralized to sodium acetate) that the name acetonitrile is given as an alternative to methyl cyanide.

(*b*) Add to the other portion a piece of granulated zinc and about 1 cm^3 of concentrated hydrochloric acid. Leave until effervescence has almost ceased. Add sodium hydroxide solution until the solution is alkaline and then heat. The fishy ammoniacal smell is due to ethylamine. This method of reduction of a nitrile is known as Mendius' reaction and is used as a preparation of amines.

$$CH_3CN + 4H^+ + 2Zn \rightarrow CH_3CH_2NH_2 + 2Zn^{2+}$$

N.B. Urea is considered in Experiment 361, p. 444.

29
Nitro Compounds and Sulphonic Acids

NITRO COMPOUNDS

Experiment 257 Preparation of nitrobenzene

Formula: $C_6H_5NO_2$.

Physical properties: yellow oil, boiling point 206 °C, density 1.2 g cm^{-3}.

Apparatus: flask (500 cm^3); thermometer (360 °C); separating funnel; water-bath; distilling flask (150 cm^3); condenser; flask (250 cm^3); measuring cylinder; trough.

Material: benzene; anhydrous calcium chloride. *Care! This experiment should be carried out in a fume chamber.*

(*a*) *Small-scale preparation.* Take about 10 cm^3 of concentrated nitric acid in a boiling-tube and add that volume of concentrated sulphuric acid. Cool the mixture under the tap. Add about 5 cm^3 benzene in small portions, shaking gently at each addition. Leave the boiling-tube standing in water at a temperature of 50 °C for half an hour. Decant as much as possible of the upper layer (nitrobenzene) into a test-tube and shake gently with sodium carbonate solution to neutralize traces of acids. Decant and then discard the upper (aqueous) layer, and treat again with sodium carbonate solution until there is no further effervescence. Decant as before. Add calcium chloride to the yellow oil and leave until clear, then decant into a dry test-tube. The nitrobenzene is pure enough for later conversion into dinitrobenzene or into aniline (phenylamine).

(*b*) *Larger-scale preparation.* Measure 55 cm^3 of concentrated sulphuric acid and add slowly to 50 cm^3 of concentrated nitric acid in a 250 cm^3 flask. Mix well and cool under the tap. When cool, transfer to a separating funnel. Take 40 cm^3 of benzene in a 500 cm^3 flask, stand the flask in a trough of water, and place the stem of the separating funnel in the neck of the flask. Allow the mixture of acids to drip into the benzene at such a rate that the temperature does not exceed 50 °C; a thermometer may be left standing in the flask throughout the experiment for this purpose. (The 'nitration' of benzene is exothermic, and a high rise in temperature would cause further nitration to give dinitrobenzene at the expense of the mono-nitro compound.) Warm on a water-bath for a quarter of an hour, shake well, then heat for a further quarter of an hour. Allow to cool. Transfer to a separating funnel and run off the lower (acid) layer, and discard it. Shake the product in the separating funnel with a dilute solution of sodium carbonate (to neutralize any free acid), then separate, this time retaining the lower layer. (Note that the density of nitrobenzene is less than that of the acid mixture but greater than that of the sodium carbonate solution.) Return the product to the funnel and wash with water. Run off the cloudy liquid into a flask and add a few pieces of calcium chloride. Leave until the liquid is clear, then transfer it to a distilling flask. Distil, and collect between 205 °C and 208 °C. Yield 35 g.

$$C_6H_6 + HNO_3 \rightarrow C_6H_5NO_2 + H_2O$$

The nitrating mixture furnishes nitronium (nitryl) ions (NO_2^+) which are the operative nitrating agents.

$$HNO_3 + 2H_2SO_4 \rightleftharpoons NO_2^+ + 2HSO_4^- + H_3O^+$$

$$C_6H_6 + NO_2^+ \rightarrow C_6H_5NO_2 + H^+$$

INTRODUCTION OF A SECOND GROUP INTO THE BENZENE NUCLEUS

When one hydrogen atom of the benzene nucleus has been displaced by a group, a second hydrogen atom may be displaced by a similar or by a different group. There are three possible positions in the benzene nucleus for this second group to occupy, namely, on the next carbon atom, or on the next but one, or on the carbon opposite. If the carbon atoms are numbered for reference, starting with 1 for the position occupied by the group originally present and progressing in a clockwise direction, the second group may occupy position 2, 3 or 4. Positions 5 and 6 are positionally identical with 3 and 2 respectively. The prefix ortho-, meta-, para- is given to those substitution compounds which have their groups respectively in positions 1, 2; 1, 3; or 1, 4.

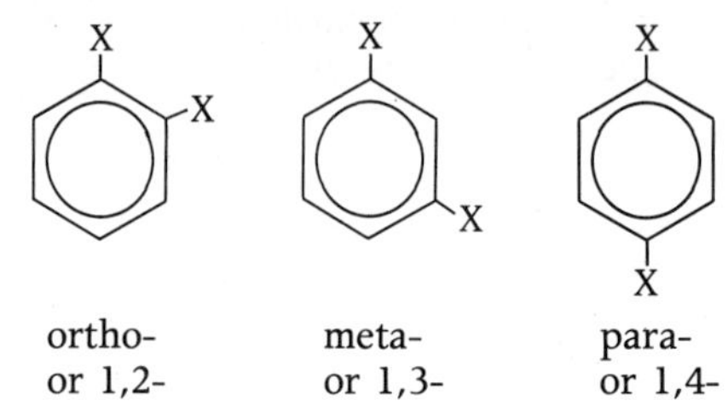

POSITION OCCUPIED BY THE SECOND GROUP

The group which is first into the benzene nucleus determines the position which the second group will occupy. If the first group is either $—NO_2$, $—SO_3H$, —CHO, or —COOH, the next incoming group, whatever the group might be, will displace the hydrogen atom which is attached to the carbon atom in the meta position (alternate or 3-position). If the first group is either $—NH_2$, —OH, —Cl, —Br, —I, or an alkyl radical, the next incoming group will displace the hydrogen atom attached to the carbon in the ortho position (adjacent or 2-position) in some of the molecules and the hydrogen atom attached to the carbon in the para position (opposite or 4-position) in the remaining molecules. Thus in the latter case the product will not be a single compound but a mixture of two compounds, and a separation based on their physical differences will be necessary.

Experiment 258 Preparation of 1,3-dinitrobenzene

Formula: $C_6H_4(NO_2)_2$.

Physical properties: pale yellow solid, melting point 90 °C.

Apparatus: flask (250 cm^3); large basin; water-bath; filter funnel (preferably a Buchner); measuring cylinder.

Material: nitrobenzene; concentrated nitric acid; ethanol.

(*a*) *Small-scale preparation.* Mix five drops each of concentrated nitric acid and concentrated sulphuric acid in a boiling-tube. Add five drops of nitrobenzene, shake the mixture, and heat almost to boiling for about a minute. Pour into a test-tube half filled with water. The solid is 1,3-dinitrobenzene. Filter off the solid, and add it to about 10 cm^3 of ethanol. Heat the test-tube and contents in a beaker of boiling water until the solid has dissolved, then pour the solution into a crystallizing dish to cool. Filter to obtain pale yellow crystals of 1,3-dinitrobenzene.

(*b*) *Larger-scale preparation.* Measure 24 cm^3 of concentrated nitric acid into a 250 cm^3 flask. Add 20 cm^3 of concentrated sulphuric acid and allow the mixture to cool. Add 25 cm^3 of nitrobenzene in portions of about 5 cm^3, shaking the mixture well between additions. Place the flask on the water-bath and heat for about a quarter of an hour. Test for the completion of the reaction by pouring a few drops of the liquid into a beaker of water to find whether it will solidify; if it does not, continue to heat. Pour the whole of the liquid into about 250 cm^3 of water in a basin, and leave

$$C_6H_5NO_2 + NO_2^+ \rightarrow C_6H_4(NO_2)_2 + H^+$$

until cold. Filter, and recrystallize the yellow solid from ethanol. The product may be used for the preparation of 3-nitroaniline (3-nitrophenylamine; see Experiment 259). Yield 25 g.

Experiment 259 Preparation of 3-nitroaniline

Partial reduction. When 1,3-dinitrobenzene is reduced with tin and hydrochloric acid, both nitro-groups are reduced to amino- ($—NH_2$) groups. The product has, therefore, a benzene nucleus with two $—NH_2$ groups in the meta position to each other; the compound is 1,3-diaminobenzene.

It is, however, possible to reduce one of the $—NO_2$ groups and leave the other. This resulting compound has the formula

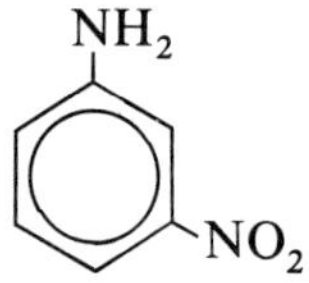

NH_2

NO_2

and is called 3-nitroaniline (3-nitrophenylamine).

The reducing agent to effect this partial reduction is hydrogen sulphide dissolved in a concentrated solution of ammonia.

Physical properties: light brown needle crystals, melting point 114 °C.

Apparatus: flask (500 cm^3); wash-bottle; water-bath; measuring cylinder; Buchner funnel and flask; crystallizing dish.

Material: 1,3-dinitrobenzene; ethanol.

Care! This experiment must be done in a fume chamber.

In the flask mix 20 g of 1,3-dinitrobenzene, 75 cm^3 of ethanol, and 10 cm^3 of concentrated ammonia solution.

Pass hydrogen sulphide first through water in the wash-bottle and then into the mixture, shaking the flask well from time to time. After half an hour turn off the supply of hydrogen sulphide and heat the flask on the water-bath for about a quarter of an hour.

Pass hydrogen sulphide in again for a quarter of an hour, then heat again and continue until the gas has been passed in for a full hour (total time including periods of heating is about two hours). Add 200 cm^3 of water to precipitate the 3-nitroaniline from alcoholic solution, then filter through the Buchner funnel. Discard the filtrate, and return the solid to the flask. Take 200 cm^3 of moderately concentrated hydrochloric acid, made by adding 150 cm^3 of water to 50 cm^3 of the concentrated acid, and heat almost to boiling. Use the acid in portions of 50 cm^3 at a time to dissolve the 3-nitroaniline from the solid and decant each time through the Buchner; the un-

39
Complexometric Titrations

Edta or ethylenediaminetetraacetic acid is usually employed as its disodium salt; it is also called diaminoethanetetraacetic acid, ethanediaminetetraacetic acid and bis[di(carboxymethyl)amino] ethane. The formula of the acid is

$$\begin{array}{l} CH_2N(CH_2COOH)_2 \\ | \\ CH_2N(CH_2COOH)_2 \end{array}$$

and its anhydrous sodium salt, of relative molecular mass 336, is sometimes designated as Na_2H_2Y. It forms very stable complexes with metal ions in solution usually in a 1:1 molar ratio. The indicator for its titrations is Eriochrome Black T, a complex organic substance that is itself a weak complexing agent. The principles of the titration are as for acid-alkali titrations and if a graph is plotted of the negative logarithm (numerical value of the metal ion concentration in mol dm^{-3}) against the volume of edta, a graph is obtained of the same shape as in Figure 77.

Experiment 336 Determination of the total hardness of water

Material: 0.01 M edta; a buffer solution made from 57 cm^3 of concentrated ammonia solution and 7 g of ammonium chloride made up with distilled water to 100 cm^3.

$$H_2Y^{2-} + M^{2+} \rightarrow YM^{2-} + 2H^+$$

Titrate 50 cm^3 of tap water to which 1 cm^3 of buffer solution and 0.5 cm^3 of indicator have been added with 0.01 M edta solution: the colour change is from wine red to pure blue at the end-point. Calculate the hardness of water in parts per million (mg kg^{-1}) of calcium (as carbonate).

dissolved portion consists chiefly of sulphur and unchanged 1,3-dinitrobenzene.

The solution, which is 3-nitrophenylammonium chloride, is then made alkaline to set free the base. Add a concentrated solution of ammonia until there is no further precipitation, and filter. Test the filtrate with ammonia solution to ensure that the action is complete, then discard the filtrate. Add 200 cm^3 of water to the solid and heat to boiling. Decant the solution into a large dish; repeat with any residual solid. When cold, filter off the crystals. Yield 10 g.

SULPHONIC ACIDS

Experiment 260 Preparation of the sodium salt of benzenesulphonic acid

Apparatus: flask (250 cm^3) fitted with a reflux condenser; sand-tray; large flask (1.5 dm^3); Buchner funnel and flask; large evaporating basin; measuring cylinder; water-bath; muslin bag.

Material: benzene; calcium hydroxide.

Take 50 cm^3 of benzene in the 250 cm^3 flask and add 50 cm^3 of concentrated sulphuric acid. Fit the flask with a reflux condenser[1] and heat on a sand-tray until the two liquid layers merge into one—about three hours. Allow to cool, then pour the contents of the flask (chiefly benzenesulphonic acid and excess sulphuric acid) into about 1 dm^3 of water in the large flask. Heat the solution and add calcium hydroxide as a thick paste until a test drop of the liquid is alkaline to litmus. Filter while hot, first through a muslin bag, and then through a Buchner funnel. (The solution is chiefly the calcium salt of benzenesulphonic acid, and the solid chiefly calcium sulphate.) To obtain the sodium salt, first evaporate the solution to about half its volume and then add a concentrated solution of sodium carbonate until a test portion of the filtrate no longer gives a precipitate with sodium carbonate solution. Filter the mixture through a Buchner funnel, several times, until the filtrate is clear. (The solution is sodium benzenesulphonate and the solid is calcium carbonate.) Evaporate the solution on a gauze until crystals begin to deposit, and then continue on a water-bath until general crystallization occurs. Set the basin aside, and, when cold, filter off the crystals and dry them on a pad of filter papers or on a porous plate. Sodium benzenesulphonate may be used to prepare phenol (see Experiment 218). Yield 60 g.

$$C_6H_6 + H_2SO_4 \rightarrow C_6H_5SO_3H + H_2O$$

---

[1] A wide inner tube without taper at the lower end must be used.

Experiment 261 To make sodium alkylbenzenesulphonate

Apparatus: wide-necked or three-necked flask (100 cm^3).

Material: an alkylbenzene hydrocarbon (this may be obtained from Unilever Ltd), e.g. dodecylbenzene; oleum (20% free sulphur trioxide; concentrated sodium hydroxide solution; **ICE**.

Care! Carry out this preparation in a fume chamber, wear rubber gloves and safety spectacles.

Put 20 cm^3 of the hydrocarbon into the flask and cool it in a bath of iced water. Fit up a stirrer and thermometer in the flask: keep the contents below 10 °C. From a tap-funnel add, with vigorous stirring, 25 cm^3 of oleum, but do not let the temperature rise above 55 °C—add more crushed ice if necessary to the bath. When the mixing is complete, replace the cold water by warm water and keep the temperature between 50 and 55 °C for thirty minutes. After this substitute cold water for the hot water and cool the flask below 20 °C.

$$R{-}C_6H_4 + SO_3 \rightarrow R{-}C_6H_4{-}SO_3H$$

The next step is to neutralize the sulphonic acid. Weigh out 10 g of finely crushed ice into a 100 cm^3 beaker and put this beaker in the cooling bath. Stir the ice and add 7.5 cm^3 of the sodium hydroxide solution. Add the sulphonic acid until the pH falls to 7, but do not let the temperature rise above 50 °C. The soapless detergent produced may be used for experiments with hard water.

30
Amines and Diazo Compounds

The structure of a primary amine shows an alkyl radical united with an amino (—NH_2) group, e.g. CH_3NH_2, methylamine or $C_2H_5NH_2$, ethylamine.

A primary amine may be considered as ammonia in which one atom of hydrogen in the molecule has been displaced by an alkyl radical, e.g.

```
H     H        H     CH3
 \   /          \   /
  N      →        N
  |               |
  H               H
```

In Experiment 256, ethylamine was obtained by reducing methyl cyanide.

The usual method of preparing amines is to treat the appropriate amide with bromine and potassium hydroxide (Hofmann's Reaction).

The sum total of the reaction is

$$RCONH_2 + Br_2 + 4OH^- \rightarrow RNH_2 + 2H_2O + 2Br^- + CO_3^{2-}$$

where R is an alkyl radical. The stages of the reaction may be shown by the equations

$$CH_3CONH_2 + Br_2 + OH^- \rightarrow \underset{\text{Nibromoacetamide}}{CH_3CONHBr} + Br^- + H_2O$$

$$CH_3CONHBr + OH^- \rightarrow \underset{\text{methyl isocyanate}}{CH_3NCO} + Br^- + H_2O$$

$$CH_3NCO + 2OH^- \rightarrow \underset{\text{methylamine}}{CH_3NH_2} + CO_3^{2-}$$

N.B. Urea is considered in Experiment 361, p. 444.

Experiment 262 Preparation of methylamine and its chloride

Formula: CH_3NH_2.

Physical properties: alkaline gas with a smell resembling ammonia and rotting fish, soluble in water.

Apparatus: 250 cm^3 flask; large distilling flask (500 cm^3); tap-funnel; adapter; condenser; measuring cylinder; pieces of porous pot; water-bath; thermometer (100 °C). (Standard-joint apparatus FGHMQ).

Material: acetamide (ethanamide); bromine; potassium hydroxide; ethanol.

Make a solution of 15 g of potassium hydroxide in 150 cm^3 of water.

Take 15 g of acetamide in the 250 cm^3 flask and add 13.5 cm^3 of bromine, taking care to avoid bromine vapour reaching the eyes: *the experiment should be performed in a fume chamber.* Add the potassium hydroxide solution in portions of about 20 cm^3, keeping the flask cool under the tap and shaking the mixture at intervals. The reddish-brown liquid changes gradually to yellow, and contains N-bromacetamide.

Transfer the liquid to a tap-funnel fitted to a 500 cm^3 distilling flask. Prepare a solution of 40 g of potassium hydroxide in 75 cm^3 of water and pour this into the flask. Add a few fragments of porous pot. Attach the flask to a sloping condenser fitted with an adapter which dips into a conical flask. Make a solution of 25 cm^3 of concentrated hydrochloric acid and 75 cm^3 of water, and pour this into the conical flask. Adjust the apparatus so that the end of the adapter is just under the surface of the acid, and throughout the preparation make periodical adjustments to keep it so.

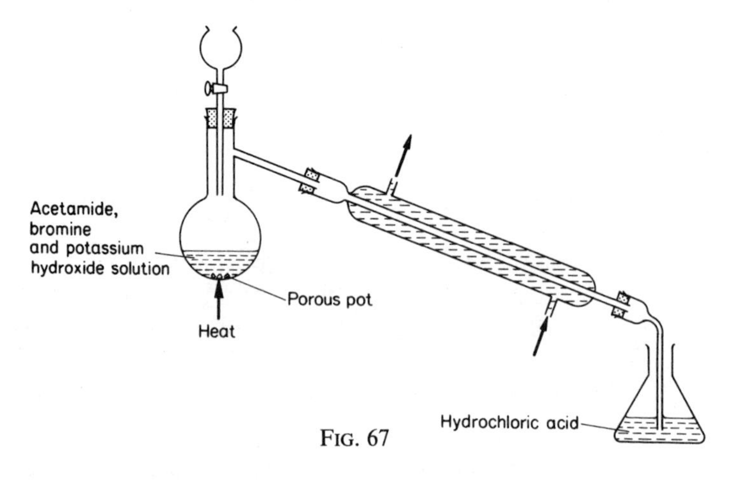

FIG. 67

Heat the distilling flask on a water-bath maintained at 60 °C and allow the liquid in the tap-funnel to drip in slowly. After all the liquid has been added continue to heat the flask, keeping it at this temperature, until the yellow colour disappears (about thirty minutes).

Replace the water-bath by a gauze and boil the liquid (Figure 67). Water and methylamine distil into the hydrochloric acid. When about two-thirds of the liquid has distilled, detach the apparatus, transfer the liquid to a dish and evaporate the total distillate (methylammonium chloride) first on a gauze and, when crystals appear, on a water-bath. The crystals contain a little ammonium chloride. Transfer the crystals to a dry flask and dissolve the methylammonium chloride by adding five successive portions of 20 cm^3 of ethanol and warming each time on a water-bath. Pour each extraction into an evaporating basin and heat the total solution on a water-bath until crystals form, then set aside to crystallize. Yield 7 g.

Experiment 263 Reactions of primary amines

Material: methylammonium chloride; ethylammonium chloride; sodium nitrite (nitrate(III)).

(*a*) Warm a few crystals of methylammonium chloride with sodium hydroxide solution. Note the fishy smell of the free amine.

$$RNH_3^+ + OH^- \rightarrow RNH_2\uparrow + H_2O$$

Compare

$$NH_4^+ + OH^- \rightarrow NH_3\uparrow + H_2O$$

Repeat with ethylammonium chloride.

(*b*) In one test-tube make about 5 cm^3 of a dilute (light blue) solution of copper(II) sulphate. In another test-tube take about half a spatula-load of methylammonium chloride and add about 10 cm^3 of sodium hydroxide solution. Add the second solution to the first until the precipitate which forms redissolves to form a deep blue solution. This action is similar to the action of ammonia on copper salts, and is a reminder of the close structural relationship of ammonia and methylamine.

(*c*) Dissolve about one spatula-load of sodium nitrite in the minimum of water. Keep the test-tube cool under the tap and add enough dilute hydrochloric acid to double the volume. Pour this solution into a test-tube containing enough ethylammonium chloride to cover the bottom. The effervescence is nitrogen and the test recalls a similar one with acetamide. In both cases the $-NH_2$ group is replaced by the $-OH$ group.

$$C_2H_5NH_3^+ + HNO_2 \rightarrow C_2H_5OH + N_2 + H^+ + H_2O$$

(*d*) All primary amines give the isocyanide test with chloroform and potassium hydroxide dissolved in ethanol (see Experiment 201(*c*)). The test is not recommended with aliphatic amines because of the poisonous nature of the products.

Experiment 264 Preparation of aminoacetic acid (glycine)

Formula: $CH_2(NH_2)COOH$ or H_2NCH_2COOH.

Physical properties: colourless, crystalline solid, soluble in water, almost insoluble in alcohol, melting point 233 °C with decomposition.

Apparatus: Mechanical stirrer; tap-funnel; 500 cm^3 beaker or flask; water-bath; boiler for supply of steam; measuring cylinder.

Material: chloroacetic (chloroethanoic) acid; concentrated ammonia solution; ethanol.

Arrange the apparatus as in Figure 68, using a water-turbine or other means to drive the stirrer. Measure 300 cm^3 of the ammonia solution and pour it into the beaker or flask.

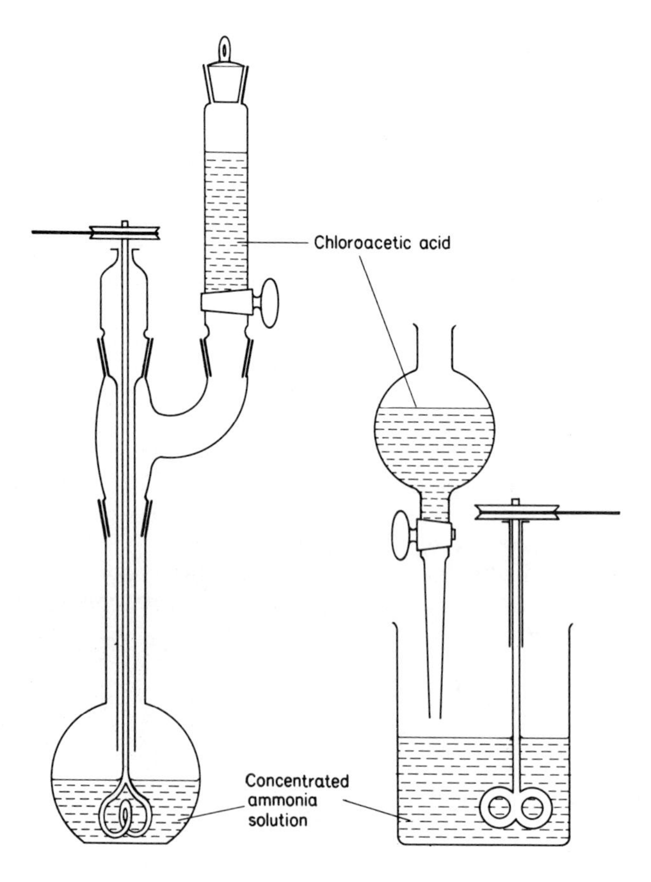

FIG. 68

Weigh 30 g of chloroacetic acid, taking care not to handle it directly as it has a blistering action on the skin. Dissolve in 30 cm^3 of water and transfer the solution to a tap-funnel. While the ammonia solution is being stirred, allow the chloroacetic acid solution to drip slowly from the tap-funnel until all has been added (about ten minutes) and continue stirring for a further ten minutes. Leave standing in a fume chamber overnight. Replace the stirrer by a steam inlet tube, place the flask and contents on a water-bath and heat, in a fume chamber, while steam is passed into the solution. Heat until ammonia ceases to be expelled. The solution contains mainly aminoacetic (aminoethanoic) acid and ammonium chloride.

$$CH_2ClCOOH + 2NH_3 \rightarrow CH_2(NH_2)COOH + NH_4Cl$$

Remove the steam inlet tube and, while the flask is still being heated on the water-bath, add (in portions of about a spatula-load at a time with thorough mixing) copper(II) carbonate until there is no further effervescence; then add a slight excess and leave on the water-bath for about half an hour. Filter the deep blue solution into a large evaporating basin and evaporate on a water-bath until crystals form; continue until a test portion on a glass rod shows ready crystallization. The blue crystals are the copper salt of aminoacetic acid.

Filter the crystals and wash them with small quantities of ethanol until the filtrate is no longer blue. Dissolve the crystals in a minimum of hot water, transfer the solution to a flask, and, while it is still hot, pass in hydrogen sulphide. Filter off the copper(II) sulphide and, if the filtrate is still blue, treat it further with hydrogen sulphide. The solution contains aminoacetic acid.

$$[CH_2(NH_2)COO]_2Cu + H_2S \rightarrow 2CH_2(NH_2)COOH + CuS$$

Evaporate the colourless solution on the water-bath until crystals form and allow to crystallize. Remove the crystals and dry on porous paper. Yield 7 g.

Experiment 265 Reactions of aminoacetic acid (glycine)

Material: sodium carbonate; sodium nitrite (nitrate(III)); dilute acetic (ethanoic) acid; copper(II) sulphate; 10% solution of aminoacetic acid.

Aminoacetic acid is one of a series of amino-acids, an important property of which is the ability to link together the NH_2 group of one molecule with the COOH of a second molecule, with the condensation of a molecule of water to give first a dipeptide and eventually a polypeptide.

$$H_2NCH_2COOH + H_2NCH_2COOH \rightarrow H_2NCH_2CONHCH_2COOH + H_2O$$

Proteins consist of long chains of linked amino-acids. (See Chapter 49.)

Use about 5 cm^3 of the aminoacetic acid solution for each of the following tests.

(*a*) Test the solution with litmus paper (red and blue). The almost neutral condition is the result of the opposing effects of the $—NH_2$ and —COOH groups; the amino-acid is an *ampholyte* (amphoteric electrolyte).

(*b*) Using the solution tested in (*a*), add a few drops of a concentrated solution of sodium carbonate. The evolution of carbon dioxide indicates the presence of the —COOH group

$$2CH_2NH_2COOH + CO_3^{2-} \rightarrow 2CH_2NH_2COO^- + CO_2 + H_2O$$

(*c*) Dissolve half a spatula-load of sodium nitrite in the minimum of water, add a few drops of dilute acetic acid and cool the solution. Add this to a second portion of aminoacetic acid. Observe the effervescence (of nitrogen); this indicates the presence of the $—NH_2$ group.

$$CH_2NH_2COOH + HNO_2 \rightarrow CH_2OHCOOH + H_2O + N_2\uparrow$$

Glycollic (*hydroxyacetic or hydroxyethanoic*) *acid* is formed.

(*d*) To a third portion of aminoacetic acid, add a few drops of copper(II) sulphate solution. Observe the deep blue colour of the copper salt of the acid.

Experiment 266 Preparation of nylon

Material: 1,6-diaminohexane (5% *m*/*V* in water); adipyl (hexanedioyl) chloride (5% *m*/*V* in carbon tetrachloride).

Put a small volume of the chloride solution at the bottom of a small beaker and very carefully pour on top of it the solution of the amine: nylon forms at the boundary layer. With a glass rod, scoop out the nylon at the centre and pull it vertically upwards, winding the thread on to the rod.

$$\underset{\text{The monomers}}{H_2N(CH_2)_6NH_2 + ClCO(CH_2)_4COCl}$$

$$\rightarrow \underset{\text{A long chain polymer}}{\ldots NH(CH_2)_6NHCO(CH_2)_4CO \ldots} + 2HCl$$

INDIRECT SUBSTITUTION

Of the groups which are introduced into the benzene ring by an indirect method, $—NH_2$ is the most important. Not only is aniline (phenylamine, $C_6H_5NH_2$) of interest, but from aniline other compounds of importance can be produced.

Experiment 267 Preparation of aniline

Formula: $C_6H_5NH_2$.
Physical properties: colourless oil, boiling point 184 °C, density 1 g cm^{-3}, usually discoloured by oxidation products.
Apparatus: large flask (1.5 dm^3); air condenser; water-bath; separating funnel; measuring cylinder; distilling flask (250 cm^3); steam distillation apparatus as in Figure 69. (Standard-joint apparatus BGHIPQ.)

Material: nitrobenzene; tin; sodium hydroxide; diethyl ether.

(*a*) *Small-scale preparation.* Take five drops of nitrobenzene in a test-tube and add five times that volume of concentrated hydrochloric acid and about 2 g of tin. Stand the test-tube in boiling water. Keep the test-tube at this temperature, shaking the tube from time to time, until there are no longer oily drops of nitrobenzene. The action is vigorous at first, and, if necessary, the test-tube should be removed until the action moderates. Prepare a solution of sodium hydroxide by dissolving about 2.5 g in 10 cm^3 of water, and add this solution until the precipitate which first forms is almost dissolved. Cool the test-tube under the tap, then add about 5 cm^3 of ether. Shake the mixture gently, allow to stand, and decant the ether layer into a dry test-tube. Stand the test-tube in hot water (no flame) to drive off the ether. The aniline so obtained will be impure, but may be used for the tests described later (Experiment 268).

(*b*) *Larger-scale preparation.* Take a large flask and mix in it 40 cm^3 of nitrobenzene and 85 g of granulated tin. Fit the flask with a reflux air condenser and heat on a water-bath for a quarter of an hour. From a measured 160 cm^3 of concentrated hydrochloric acid add 10 cm^3 and shake the mixture. Every five minutes add another 10 cm^3 of acid and shake. When all the acid has been added, leave the flask warming on the water-bath for half an hour. During this period prepare a solution of 140 g of sodium hydroxide in 200 cm^3 of water. (*Care should be taken in handling this concentrated solution of caustic alkali.*) Add this solution, in portions of about 50 cm^3 at a time, until the precipitated tin(IV) hydroxide has almost dissolved and the solution is alkaline. The aniline will then be free and appear as an oil. Distil the mixture in steam, by passing steam from the boiler into the mixture while the latter is being heated on a sand-tray or gauze. Collect the condensed vapours in a receiver (see Figure 69) until no more oily drops pass down the condenser.

Transfer the distillate to a separating funnel and extract the aniline with ether. Use 90 cm^3 of ether in three portions of about 30 cm^3 each. Shake the first portion of ether[1] with the aqueous aniline, lift-

[1] See Chapter 2 for notes on extraction by ether and Experiment 20 for another experiment using steam distillation.

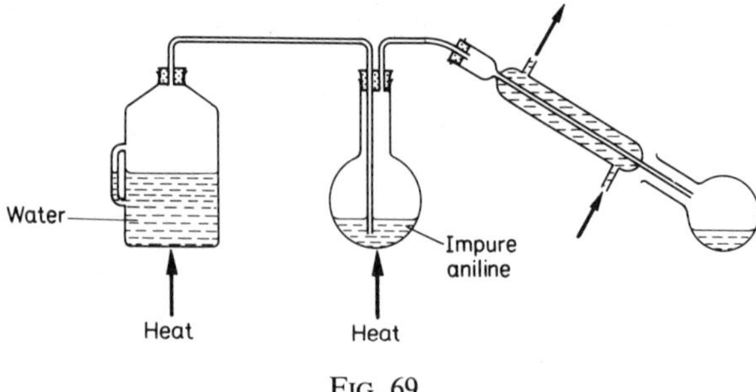

FIG. 69

ing the stopper from time to time to avoid a high pressure in the funnel, then run off the lower (aqueous) layer into a beaker and the upper (ether) layer into a flask. Return the aqueous layer to the funnel, add the second portion of ether and proceed as before, and so on with the third. To dry the ether-aniline solution, add about 10 g of sodium hydroxide and leave the flask, corked, until the liquid is clear. Decant the liquid into a distilling flask fitted to a sloping condenser, and proceed as in Experiment 180, p. 170. Yield 25 g.

$$2C_6H_5NO_2 + 12H^+ + 3Sn \rightarrow 2C_6H_5NH_2 + 4H_2O + 3Sn^{4+}$$

STEAM DISTILLATION

Aniline (phenylamine) boils at 184 °C. When steam at 100 °C is passed into aniline a mixture of aniline and water distils over at a temperature below 100 °C. The theoretical considerations are as follows.

(i) Aniline and water are immiscible.

(ii) Mixtures of immiscible liquids have a vapour pressure equal to the sum of the separate pressures of the constituents.

(iii) The boiling point of a liquid (or a mixture of liquids) is the temperature at which the vapour pressure (or total vapour pressure) is equal to the atmospheric pressure.

(iv) During distillation in these circumstances, the vapour contains both constituents, and contains them in the ratio by volume of their separate vapour pressures at the temperature at which the mixture is boiling. Thus if the atmospheric pressure is 760 mmHg, and if a mixture of aniline and water boils at 98 °C, the vapour pressure due to water vapour will be 710 mmHg (see the table on p. 455). By difference, the pressure due to aniline vapour will be 50 mmHg. The ratio of water vapour to aniline vapour will be given by the expression

$$\frac{\text{volume of water vapour}}{\text{volume of aniline vapour}} = \frac{710/760}{50/760} = \frac{14.2}{1}$$

(v) The mass of each constituent in the distillate will be the product of its volume and its vapour density. Thus,

$$\frac{\text{mass of water}}{\text{mass of aniline}} = \frac{(710/760) \times 9}{(50/760) \times 46.5} = \frac{2.75}{1}$$

Experiment 268 Reactions of aniline

Material: aniline (phenylamine); bleaching powder or sodium hypochlorite (chlorate(I)) solution; chloroform (trichloromethane); **ICE** or solid ammonium chloride; phenol; acetyl (ethanoyl) chloride.

Two drops of aniline are sufficient for each reaction.

(*a*) Add three drops of water. Note the immiscibility. To the mixture add five drops of concentrated hydrochloric acid. The clear solution contains anilinium chloride (phenylammonium chloride).

$$C_6H_5NH_2 + HCl \rightarrow C_6H_5NH_3^+ + Cl^-$$

Now add sodium hydroxide solution until the mixture is alkaline. The emulsion is due to the reappearance of free aniline.

(*b*) Add to a test-tube full of water and shake thoroughly. Add a drop of a prepared solution of bleaching powder in water or sodium hypochlorite solution. The violet coloration serves as a test for aniline. Repeat the test, using a much smaller quantity of aniline, and show that the test is a very delicate one.

(*c*) Repetition of the isocyanide test as given in Experiment 201 is not recommended.

(*d*) Add two drops of water and five drops of concentrated hydrochloric acid, and, to the solution of anilinium chloride so made, add bromine water until the solution is just yellow. Shake the solution and leave to stand. The yellow crystals are tribromoaniline (tribromophenylamine).

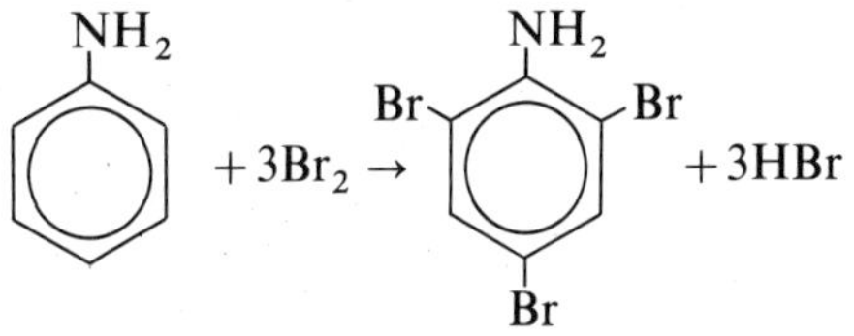

$$C_6H_5NH_2 + 3Br_2 \rightarrow C_6H_2Br_3NH_2 \text{ (2,4,6-tribromo; } NH_2 \text{ with Br, Br ortho and Br para)} + 3HBr$$

(*e*) Prepare anilinium chloride solution as above and cool it at least to 8 °C in ice or in a beaker containing water and solid ammonium chloride. Prepare a test-tube one-quarter full of a saturated solution of sodium nitrite. Take a third test-tube, add phenol to cover the bottom of the tube and add sodium hydroxide solution to dissolve it. Leave the test-tubes to cool for a few minutes, then mix gradually the anilinium chloride and sodium nitrite. Finally add the phenol to obtain a red dye. (See p. 269).

(*f*) Add two drops of acetyl chloride. Warm gently, allow to cool and then add water. The solid is acetanilide. (See below.)

Experiment 269 Preparation of acetanilide (N-phenylacetamide or N-phenylethanamide)

Formula: $C_6H_5NHCOCH_3$.

Physical properties: white crystalline solid, melting point 112 °C.

Apparatus: flask (250 cm^3); inner tube of condenser; thermometer and tubing as shown in the diagram (Figure 70); Buchner funnel and flask; large basin or trough; large (1 dm^3) flask. (Standard-joint apparatus GILM.)

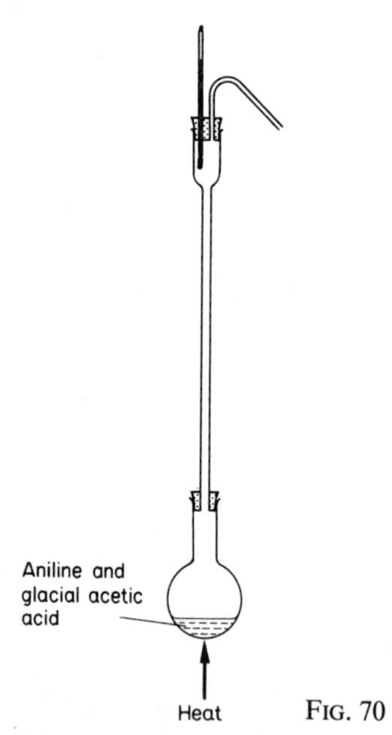

FIG. 70

Material: freshly distilled aniline; ethanol.

In the flask mix 25 cm^3 of aniline and 30 cm^3 of glacial acetic acid. Attach the air condenser as shown. Heat on a gauze and adjust the flame so that the thermometer in the condenser reads 105 °C. Heat for two hours.

Pour the hot liquid into about 500 cm^3 of water in a large basin. Filter the crystals of acetanilide at the pump and wash with water. Transfer the crystals to a large flask (1 dm^3) and fill the flask three-quarters full of water. Add 10 cm^3 of ethanol and boil until the solid has dissolved (if some remains undissolved it may be treated later). Pour the solution into a large basin and leave to crystallize. Yield 25 g.

$$C_6H_5NH_2 + CH_3COOH \rightarrow \underset{\text{(acetanilide)}}{C_6H_5NHCOCH_3} + H_2O$$

Experiment 270 Preparation of benzanilide

Material: benzoyl chloride; (benzenecarbonyl chloride); aniline (phenylamine); concentrated sodium hydroxide solution.

This preparation illustrates the process of benzoylation, but as benzoyl chloride is lachrymatory *the first stage must be confined to the fume chamber*. To 10 cm^3 of aniline in a corked flask add 75 cm^3 of concentrated sodium hydroxide and then 15 cm^3 of benzoyl chloride. Shake the flask vigorously for fifteen minutes and observe the separation of a coarse white powder. Using a Buchner funnel and flask, filter off the ester at the pump and rinse it thoroughly with water.

The crude product can be recrystallized from ethanol by putting it in a flask fitted with a condenser and *just* covering the ester with ethanol. On heating, most of the ester dissolves but more ethanol may be added cautiously to ensure complete dissolution. The boiling solution must then be filtered and allowed to cool: the filtrate should yield colourless crystals.

Introduction of groups by diazotization of aniline. Sandmeyer reaction. In the preparation of phenol from aniline, the aniline was treated with nitrous acid at 0 °C and gave a diazonium salt.

Diazonium salts are important intermediary compounds for the introduction of —Cl, —Br, —I, —CN. Thus from aniline, chlorobenzene, bromobenzene, iodobenzene and benzonitrile can readily be prepared. Since the —CN group can be reduced to $—CH_2NH_2$, or hydrolysed to —COOH, benzylamine ($C_6H_5CH_2NH_2$) and benzoic acid may be added to the list. Note that —Cl and —Br can also be directly introduced (Experiment 205, p. 200) but iodine must be introduced via the diazo compound. The —CN group may be introduced either by a method similar to that used for the preparation of phenol (Experiment 218, p. 210), namely, by fusing sodium benzenesulphonate with potassium cyanide, or via the diazo compound.

In general the method consists of diazotizing the aniline and then mixing the diazonium salt with the copper(I) compound containing the group to be introduced. The product is then isolated by steam distillation.

Experiment 271 Preparation of 4-chlorotoluene (Sandmeyer)

Apparatus: large flask (1.5 dm^3); flask (250 cm^3); tap-funnel;

steam distillation apparatus; distilling flask; freezing mixture (**ICE**-salt) in a large basin or trough; measuring cylinder; thermometer (360 °C).

Material: 4-methylaniline (4-methylphenylamine); sodium nitrite (nitrate(III)).

Dissolve 25 g of 4-methylaniline in a solution of 60 cm^3 of concentrated hydrochloric acid and 40 cm^3 of water in a 250 cm^3 flask, and stand in a freezing mixture.

Take 250 cm^3 of concentrated hydrochloric acid in the 1.5 dm^3 flask, add 10 g of copper(II) oxide and heat to dissolve. Add 15 g of copper foil or turnings to the copper(II) chloride solution and boil in a fume chamber until the dark colour has disappeared. Cool this solution of copper(I) chloride and stand the flask loosely corked in a freezing mixture.

Dissolve 18.5 g of sodium nitrite in 40 cm^3 of water in the conical flask and transfer it to the tap funnel.

When the 4-methylanilinium chloride solution has cooled to 0 °C, allow the sodium nitrite solution to drip slowly into it at such a rate that the temperature does not rise above 8 °C. The 4-methylaniline has then been diazotized.

$$CH_3C_6H_4NH_3^+ + HNO_2 \rightarrow CH_3C_6H_4N{=}N^+ + 2H_2O$$

Add the diazonium compound in portions of about 10 cm^3 at a time to the copper(I) chloride solution, shaking well between additions, to obtain a golden yellow solid—a loose compound of the diazonium salt and the copper(I) chloride.

$$CH_3C_6H_4N{=}N^+ + Cl^- \rightarrow CH_3C_6H_4Cl + N_2\uparrow$$

Experiment 272 Small-scale preparation and reactions of benzenediazonium chloride

Apparatus: thermometer (100 °C); measuring cylinder.

Material: aniline (phenylamine); copper(I) chloride in concentrated hydrochloric acid; ethanaol; **ICE**; phenol; 1.2 M potassium iodide solution.

Measure 13 cm^3 of concentrated hydrochloric acid in a measuring cylinder and add an equal volume of water. Add this to 5 cm^3 of aniline in a boiling-tube and stand the test-tube in a beaker containing ice. Make a solution of sodium nitrite containing 4.5 g of the solid in about 10 cm^3 of water, and add this, a little at a time, stirring with a thermometer and keeping the temperature between 5 and 10 °C. (This temperature is low enough to prevent much decomposition without being so low as to retard the action appreciably.)

$$C_6H_5NH_2 + HNO_2 + H^+ \rightarrow \underset{\text{benzenediazonium ion}}{C_6H_5N_2^+} + 2H_2O$$

Use about 5 cm^3 of this solution for each of the following experiments.

(*a*) Boil. Nitrogen is evolved and the smell of phenol observed.

$$C_6H_5N_2^+ + H_2O \rightarrow C_6H_5OH + H^+ + N_2\uparrow$$

(*b*) Add about 2 cm^3 of potassium iodide solution. On warming, there is an effervescence and a brown liquid, iodobenzene, is formed.

$$C_6H_5N_2^+ + I^- \rightarrow C_6H_5I + N_2\uparrow$$

(*c*) Add about 2 cm^3 of a copper(I) chloride solution in hydrochloric acid and warm. Nitrogen is evolved and an oily liquid, chlorobenzene, is formed.

$$C_6H_5N_2Cl \rightarrow C_6H_5Cl + N_2\uparrow$$

(*d*) *Gattermann's reaction.* Prepare a small quantity of finely divided copper by sprinkling zinc dust into copper(II) sulphate solution. Decant the liquid and wash the copper with dilute sulphuric acid, then with water. Add the copper to the diazonium chloride. Note the effervescence (nitrogen). The main product (chlorobenzene) could be separated, if the experiment were done on a larger scale, by steam distillation.

$$C_6H_5N_2^+ + Cl^- \rightarrow C_6H_5Cl + N_2\uparrow$$

(*e*) Add a solution of phenol in sodium hydroxide. A light red precipitate of sodium hydroxyazobenzene ((4-hydroxyphenyl) azobenzene) is formed.

$$C_6H_5N_2^+ + C_6H_5O^- + Na^+ \rightarrow C_6H_5{-}N{=}N{-}C_6H_4{-}O^-Na^+ + H^+$$

This formation of a dye by linking a diazonium compound to a phenol is called *coupling*. A similar dye is formed when an alkaline solution of resorcinol (1,3-benzenediol) is added to a solution of benzenediazonium chloride: the dye is 'Resorcin red'.

(*f*) Add a few drops of aniline (phenylamine) and shake the mixture. The yellow precipitate is diazo-aminobenzene (N-(phenylazo) phenylamine).

$$C_6H_5N_2^+ + C_6H_5NH_2 \rightarrow C_6H_5N_2NHC_6H_5 + H^+$$

PART IV

Volumetric Analysis

NOTE ON UNITS

$$1\ \text{cm}^3 \equiv 1\ \text{ml} = 1\ \text{cc}$$
$$1000\ \text{cm}^3 \equiv 1\ \text{dm}^3 = 1\ \text{litre (l)}$$

The symbol M is used to denote a concentration in mol dm^{-3}.

31
Introduction

As its name implies, volumetric analysis relies on methods involving accurate measurement of volumes of liquids, though one or more weighings may also be needed. Gravimetric analysis involves only weighings. Of the two methods of analysis, gravimetric analysis is the more accurate but volumetric analysis is much more rapidly carried out. Volumetric analysis is, however, by no means inaccurate and the error involved in an analysis carried out by an experienced worker should not exceed 0.2 %, but in a typical school experiment is about 1 %.

STANDARD SOLUTIONS

In general, a volumetric analysis is carried out by preparing a standard solution of the given material (or using a solution supplied) and determining the volume of it needed to react exactly with a known volume of another solution of accurately known concentration, in a chemical reaction for which the equation is known. The completion of the reaction is traced by some means, usually by an indicator showing change of colour. To take a simple case, the concentration of a solution of potassium hydroxide could be determined by adding to it the indicator methyl orange, and then allowing the alkali to react with a standard solution of hydrochloric acid which is slowly added until the solution just becomes orange, the acid being then just in excess. This process of adding one standard solution to another to determine equivalent volumes is called *titration*.

A *standard solution* is one of which the concentration is known. Any kind of unit of mass or volume may be used, but for scientific purposes grams per litre (g dm^{-3}) is the most convenient.

For volumetric analysis, the system of working in 1 mol dm^{-3} (or some multiple or submultiple of this concentration) is almost universal. A 1 M solution is a particular kind of standard solution and is defined in the following way.

A solution is said to be 1 M if it contains one mole of the substance in 1 dm^3 (litre) of solution.

For convenience such an expression as 'a 1 M solution of sodium hydroxide' is usually written concisely as 1 M NaOH. Solutions which are 0.1 M, 0.05 M and 0.02 M are also commonly used.

A *litre* is the special name for a cubic decimetre of material and is 1000 cm^3; it should not be used to express results of high precision.

A *mole* is the amount of substance of a system which contains as many elementary entities as there are atoms in 0.012 kilogram of carbon-12. The elementary unit must be specified: it may be an empirical formula for an ionic substance, an electron, a molecule, an atom or an ion.

The *Avogadro Constant* is the number of particles in 0.012 kg of carbon-12, or one mole of any other material, and is 6.02×10^{23} per mole.

1. *Sulphuric acid* H_2SO_4; relative molecular mass (M_r) is 98, a 1 M solution contains 98 g dm^{-3} (on the old system of normality there are 2 g of replaceable hydrogen in 98 g of acid and so 1 N H_2SO_4 contains 49 g dm^{-3} of the acid).
2. *Sodium hydroxide* NaOH; relative molecular mass is 40, a 1 M solution contains 40 g dm^{-3}.
3. *Potassium permanganate* $KMnO_4$; this compound, when reacting in acid solution (p. 319), gains electrons thus:

$$MnO_4^- + 8H^+ + 5e^- \rightarrow Mn^{2+} + 4H_2O$$

   A solution would contain 158 g dm^{-3} but such a solution cannot be used in practice because of limitations imposed by the low solubility of the compound; 0.02 M potassium permanganate (manganate(VII)) solutions containing 3.16 g dm^{-3} are usually employed.
4. *Silver nitrate* $AgNO_3$; relative molecular mass is 170, a 1 M solution would contain 170 g dm^{-3} but usually 0.1 M $AgNO_3$ (17 g dm^{-3}) is used.

The justification for the concentration of other solutions will be found at appropriate places in the text. It should be noted particularly that the *concentration of a 1 M solution must be calculated from the formula of the compound allowing for any water of crystallization.* This is most important when a compound is capable of existing in two or more different forms. Consider the following varieties of sodium carbonate.

1. Sodium carbonate-10-water (decahydrate). The relative molecular mass corresponding to $Na_2CO_3 \cdot 10H_2O$ is $(2 \times 23) + 12 + (3 \times 16) + 10(2 + 16) = 286$. It is usually employed in 0.05 M solution, which will contain $0.05 \times 286 = 14.3$ g dm^{-3}.

2. Sodium carbonate (anhydrous). The relative molecular mass corresponding to Na_2CO_3 is 106. It is usually employed in 0.05 M solution which will contain 5.3 g dm^{-3}.

Sodium carbonate possesses two relative molecular masses each appropriate to the particular variety of crystal; consequently, it is important to know the formula of the crystals in order to make up a standard solution.

It follows from the definition of 1 M solutions that they do not contain chemically equivalent masses of the compounds per dm^3 of solution and, consequently, that equal volumes of different 1 M solutions are not chemically equivalent to one another, unless the equation involves one mole of each of those substances.

In the equation

$$NaOH + HCl \rightarrow NaCl + H_2O$$

there is one mole of each of the reagents but in the reaction

$$2NaOH + H_2SO_4 \rightarrow Na_2SO_4 + 2H_2O$$

the ratio of the number of moles is 2:1. In 1 dm^3 (1000 cm^3) of a 1 M solution of a material there is one mole of the material; thus in v cm^3 of an m M solution there are $vm/1000$ moles of material. The indicator shows when the two substances have completely reacted and at this point,

$$\frac{v_A m_A}{v_B m_B} = \frac{a}{b}$$

if the equation is

$$aA + bB \rightarrow cC + dD$$

because the experimentally determined ratio of the number of moles of material is equal to the ratio of the numbers of entities of the materials.

STOICHIOMETRIC AND IONIC EQUATIONS

In volumetric analysis, there is a clear distinction between a stoicheiometric equation (i.e. one which states mass relationships) and an ionic equation for the same reaction. The former is necessary to find out the masses of reacting substances and the relationship between volumes of reactants, whereas an ionic equation gives an indication of the actual reaction which is taking place in the solution. We can use as an example the reaction between sodium hydroxide solution and hydrochloric acid. The substance we weigh, sodium hydroxide, contains sodium, hydrogen and oxygen in the proportions indicated by the formula, NaOH, and contains sodium ions and hydroxide ions in equal numbers.

	NaOH	is	$Na^+ + OH^-$
1 mole weighs	40 g		23 g

Similarly, hydrogen chloride consists of molecules containing hydrogen and chlorine atoms in equal numbers and in solutions furnishes hydrogen ions and chloride ions. In solution

$$\underset{\substack{36.5\text{ g}\\ 1\text{ dm}^3\text{ 1 M}\\ \text{solution}}}{HCl} \quad \text{is} \quad \underset{1\text{ g}}{H^+} + \underset{35.5\text{ g}}{Cl^-}$$

Stoichiometrically,

$$\underset{\substack{40\text{ g}\\ 1\text{ dm}^3\text{ 1 M}}}{NaOH} + \underset{\substack{36.5\text{ g}\\ 1\text{ dm}^3\text{ 1 M}}}{HCl} \rightarrow NaCl + H_2O$$

$$OH^- + H^+ \rightarrow H_2O$$

The ionic equation is not very useful for calculating molar masses.

TYPES OF IONIC REACTIONS IN VOLUMETRIC ANALYSIS

In the following chapters in volumetric analysis, we shall meet four types of ionic reactions.

1. PROTON DONATION AND PROTON ACCEPTANCE (NEUTRALIZATION)

Here, the hydrogen ion (proton) is accepted by the base (proton acceptor) to form a neutral molecule, e.g.

$$\underset{\text{(from acid)}}{H^+} + \underset{\text{(from base)}}{OH^-} \rightarrow \underset{\text{undissociated}}{H_2O}$$

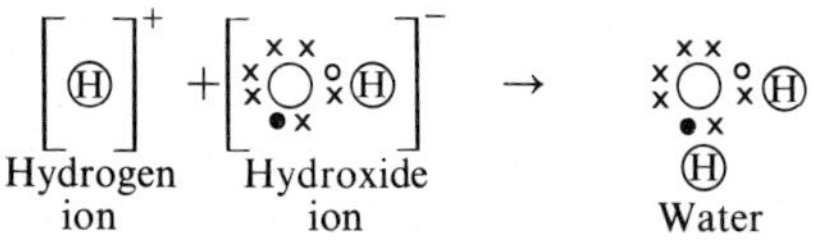

$$\left[\text{H}\right]^+ + \left[\overset{\times\times}{\underset{\bullet\times}{{}^{\times}_{\times}\text{O}{}^{\circ}_{\times}}}\text{H}\right]^- \rightarrow \overset{\times\times}{\underset{\bullet\times}{{}^{\times}_{\times}\text{O}{}^{\circ}_{\times}}}\text{H}$$

Hydrogen ion | Hydroxide ion | Water (with second H bonded via • ×)

• = electron given up by (say) sodium atom when sodium atom became an ion.

× = electron originally in oxygen atom.

o = electron originally in the hydrogen atom associated with the oxygen atom.

Electrons are identical no matter what their source and this scheme is of historical interest in that it explains the charges on the ions and why water is neutral. See Chapter 33, p. 299.

2. ELECTRON LOSS AND ELECTRON GAIN (OXIDATION AND REDUCTION)

The process of electron loss is termed *oxidation* and of electron gain, *reduction*. For example, one iron(II) ion with two positive charges loses one electron to the oxidizing agent and becomes an iron(III) ion with three positive charges.

$$Fe^{2+} \rightarrow Fe^{3+} + e^-$$

When the permanganate oxidizes in acidic solution, five electrons are transferred.

$$MnO_4^- + 8H^+ + 5e^- \rightarrow Mn^{2+} + 4H_2O$$

Combining these two equations, we have

$$5Fe^{2+} + MnO_4^- + 8H^+ \rightarrow 5Fe^{3+} + Mn^{2+} + 4H_2O$$

Stoicheiometrically, five moles of iron(II) sulphate would be oxidized to iron(III) sulphate by one mole of potassium permanganate in acidic solution, five moles of electrons being lost by the iron(II) sulphate to the permanganate. Although very much smaller in mass, one electron is electrically equivalent to one hydrogen ion.

From these considerations,

$$5FeSO_4 \equiv KMnO_4 \equiv 5e^-$$

See Chapters 34–37, p. 319 et seq.

3. COMBINATION OF POSITIVE AND NEGATIVE IONS (PRECIPITATION)

There are two main reasons why neutral entities formed by the combination of ions can exist. They may, on the one hand, be covalent substances such as water, or, on the other hand, the charged particles may exist in a space lattice or crystal, where one ion is firmly held, being surrounded by ions of a differing charge. In neutralization, we considered the first of these cases and, in precipitation, we are mainly concerned with the second. When, for example, a silver ion comes into contact with a chloride ion in solution, union takes place by virtue of the ions being held in a space lattice (Figure 71). This ionic association removes virtually all the silver and chloride ions from the solution.

$$Ag^+ + Cl^- \rightarrow Ag^+Cl^-(s) \quad \text{(often written as } AgCl\downarrow)$$

Thus

$$Ag \equiv AgNO_3 \equiv e^-.$$

See Chapter 38, p. 346.

4. COMPLEX ION FORMATION

A simple ion (a copper ion, Cu^{2+}, is taken to be simple for this purpose because the hydration is usually neglected) may combine with molecules, e.g. ammonia, to yield complex ions. Thus

$$Cu^{2+} + 6NH_3 \rightarrow [Cu(NH_3)_6]^{2+}$$

The extent of this reaction is usually measured by partition coefficient methods (see Experiment 19, p. 23) and with dilute solutions of ammonia may only involve four molecules of ammonia for each copper ion. Alternatively, two ions may unite to give a complex ion.

Bonding between the species involved may take place in more than one direction, e.g. ethylenediamine (1,2-ethanediamine), $CH_2(NH_2)CH_2NH_2$, may bond by both of its nitrogen atoms to a metallic ion. Such a compound is called a chelate. A derivative of ethylenediamine in which the four hydrogen atoms attached to the nitrogen atoms are replaced by $-CH_2COOH$ radicals is employed in titrations.

See Chapter 39, p. 365.

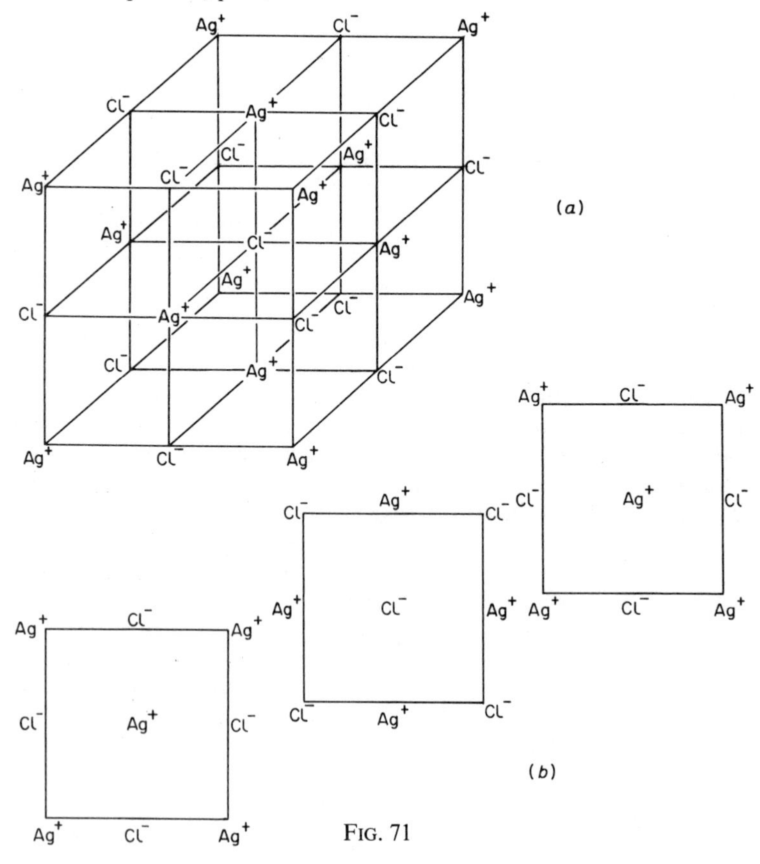

FIG. 71

INSTRUMENTS OF VOLUMETRIC ANALYSIS

The course of a volumetric analysis is usually as follows. A standard solution of the material to be analysed is made up by weighing in a *weighing bottle* an appropriate quantity of the material, making it up to 250 cm^3 of aqueous solution in a *measuring flask*, transferring 25 cm^3 of this solution to a conical flask by means of a *pipette*, and titrating it with a standard (often 0.1 M) solution of some reagent from a *burette*, using a suitable indicator. Each of these measuring instruments and the manner of its use will now be considered. A measuring cylinder may occasionally be used but only for very approximate work.

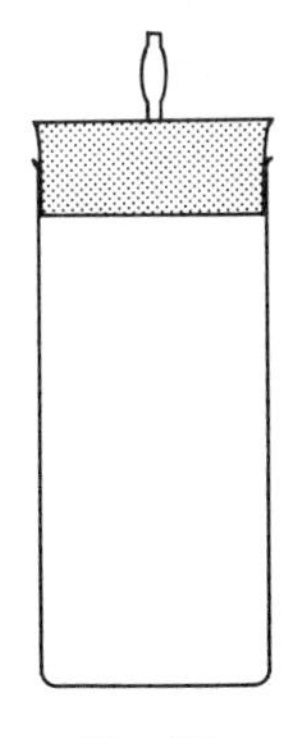

FIG. 72

WEIGHING BOTTLE

This consists of a cylindrical glass vessel with an accurately ground stopper (Figure 72), in which the materials can be weighed out of contact with the open atmosphere. The weighing bottle is usually heated in a steam oven before use to ensure dryness and is then allowed to cool in a desiccator. It should be handled in a dry cloth to avoid contamination with grease from the fingers. It may be used in one of two ways. The first method is to weigh the bottle empty, powder the material given in a clean, dry mortar and then weigh out an amount of material exactly. This has the disadvantage of being tedious and exposing the material to the atmosphere while adjustments of amount are being made. A spatula of plastic material or stainless metal is used to transfer the substance. The second method is to weigh the bottle containing an amount of the material known to be roughly suitable for the purpose in hand, to transfer the material to a measuring flask as described in the next section and weigh the bottle containing a trace of residual material, after which the actual mass transferred to the measuring flask can be obtained by difference. The second method is much quicker and is preferable except where solutions of an exact concentration are being prepared directly: the first method is then essential. Plastic weighing bottles can frequently be substituted for the more fragile glass ones but must be avoided with organic solvents.

The degree of accuracy necessary in the weighings will be considered later (p. 284).

MEASURING FLASK

Measuring flasks of 250 cm^3 capacity are usually employed (Figure 73) because the amount of solution used in a single titration is usually 25 cm^3 and several such titrations may be carried out by drawing from the 250 cm^3 of solution prepared. Measuring flasks of 100 cm^3, 500 cm^3, 1000 cm^3 and 2000 cm^3 are also in frequent use.

It should be noted that a measuring flask is made to *contain* a volume of liquid and will not *deliver* that volume, because some liquid is inevitably retained as a film on the sides of the flask. Measuring flasks are usually graduated at 20 °C and should only be used at temperatures close to this.

When a solution is to be made up the measuring flask should first be well rinsed with several small quantities of the solvent (usually distilled water) that is to be used. This removes any traces of impurities. A small beaker should be similarly rinsed and the material carefully transferred to it from the weighing bottle. The solvent is then added from a wash-bottle down the sides of the beaker so that there is no splashing. Gentle stirring with a glass rod will hasten the process of solution, but the rod should not be removed from the beaker unless all solution is first washed from its surface into the beaker. (If the solvent is used hot, the solution must finally be cooled to room temperature. The solution must also, of course, remain unchanged by heat.) When the solute is completely dissolved, the glass rod should be placed in a funnel which rests in the neck of the measuring flask and the solution poured down it from the beaker into the flask. The entire interior surface of the beaker should then be washed several times with the solvent and the washings transferred to the flask down the rod which will also be washed by them. (If the material is being weighed by difference, the weighing bottle should now be weighed again. If not, the weighing bottle should be washed out with solvent in the same way as the beaker and the washings added to the contents of the flask.) The measuring flask is then filled with the solvent from a wash-bottle until the bottom level of the meniscus is at the mark (Figure 73). A pipette should be used to add the last drop or two. (When deciding whether the level is correctly adjusted, lower the eye until the mark is at your eye-level and so avoid error due to parallax.) The flask should then be stoppered and *shaken vigorously* for some few minutes to make the solution uniform. After shaking, the solution will be *below* the mark: some of the solution is retained as a film on the stopper and neck of the flask. The total volume is still, however, 250 cm^3, and 25 cm^3 taken from it will be accurately one-tenth. Do not be tempted to make the solution up to the mark again. This proceeding makes the volume 250 cm^3 plus an unknown added volume.

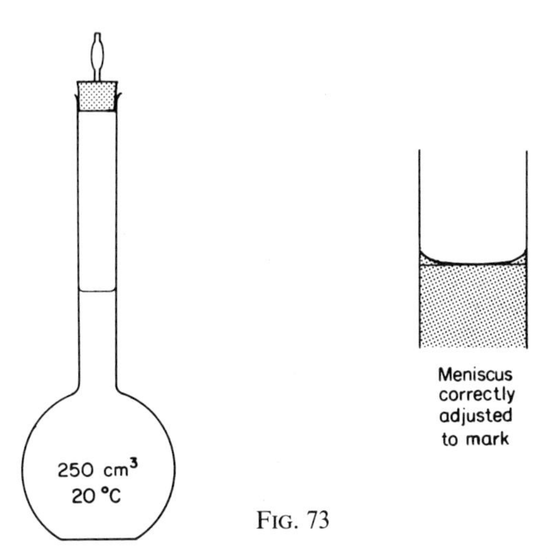

FIG. 73

If time is an important factor the solution can be made uniform more quickly by pouring it into a large dry flask in which it can easily be swirled round for a short time. *It is essential that the solution should be of uniform concentration throughout.* Whenever the words 'shake well' appear in the text they mean this essential process of making the solution homogeneous.

PIPETTE

The pipette (Figure 74) is designed to *deliver* a certain volume of liquid. When filled to the mark it contains more than this volume, a little of the liquid being retained after delivery as a film on the sides of the pipette and in the tip. The actual volume *delivered* from the pipette should be constant, and it is therefore important to observe certain conditions when using the pipette so that the small volume of liquid retained in it is constant.

The pipette is filled above the mark by sucking solution into it and this liquid is allowed to drain away. A pipette with suction device or a burette should always be used for poisonous or corrosive solutions.

If a liquid is taken into the mouth it should be spat out at once and the mouth washed out. This process of filling and allowing the liquid to run away should be repeated to ensure that the pipette contains nothing but the solution which is to be measured. The pipette is then filled above the mark and the liquid is retained by pressing the forefinger on the open end of the stem. The pipette is then raised so that the mark is at eye-level and, by controlled release of the finger

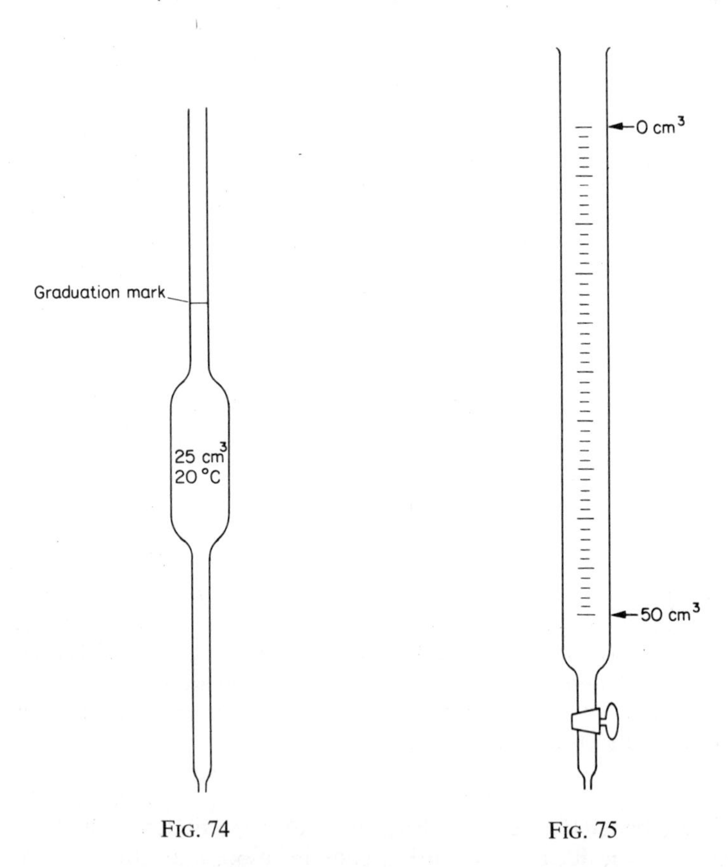

FIG. 74

FIG. 75

from the stem, the liquid is allowed to fall slowly until the bottom level of the meniscus is at the mark. The tip of the pipette is then placed inside a conical flask which should be dry or should contain only distilled water. By removing the finger from the stem, the liquid may be delivered from the pipette into the flask. A 25 cm^3 pipette should deliver its contents in about 20 seconds and the point of the pipette should be touching the side of the vessel. If delivery is more rapid than this, the volume delivered is not constant. A pipette which delivers too rapidly should have its extreme tip heated very gently in a flame, when the hole will close slightly. A little liquid will be retained in the tip; no attempt should be made to expel it as the pipette will already have *delivered* 25 cm^3 of liquid which is now available for titration with a standard solution from a burette, using a suitable indicator. The greatest difference between deliveries from a 25 cm^3 pipette should not exceed 0.025 cm^3.

BURETTE

The burette is illustrated in Figure 75. Burettes employed in volumetric analysis usually have a capacity of 50 cm^3 and are graduated in cm^3 and 0.1 cm^3. A glass-stoppered burette is to be preferred and the stopper should turn smoothly. If it sticks, the socket and stopper should be dried and a *very thin* smear of Vaseline placed on the stopper. If too much is used, the vaseline may block the hole in the stopper. A rubber band may be used to prevent the stopper from sliding out of the socket. A Mohr burette is quite satisfactory for most purposes but it should not be used with iodine solutions because the rubber connection is attacked by iodine. The burette is first washed out with the solution it is to contain, the washings being allowed to run away *through the jet* to wash this part also. A second washing is desirable to ensure complete elimination of impurities. It is then filled up to the region of the zero mark with the solution and *the jet is filled* by opening the tap for a second or two. Time is then allowed for drainage down the sides of the burette, after which, with the surface of the liquid at eye-level, the reading of the *bottom level* of the meniscus is taken. A white sheet of paper held behind the liquid at an angle of 45 degrees will help to define the meniscus and a reading within 0.05 cm^3 may be obtained. The titration is then completed and the new reading taken, after which the volume of liquid delivered is found by difference.

As a titration must be accurate to one drop of reagent and the volume of it needed is at first known only very approximately, it is almost always a saving of time to carry out a rough titration first in the following way, after which accurate titration can be quickly performed. To the solution and indicator in the conical flask, add the solution from the burette 1 cm^3 *at a time*, until excess is present as shown by the change in the indicator. Suppose that the indicator changes between the 23rd and 24th cm^3 added. Then in subsequent titrations up to 22 cm^3 of solution can be safely added from the burette, after which adjustment to the accurate end-point must be made drop by drop: the smallest possible addition at this slow rate will be about 0.05 cm^3. It is usually quicker to carry out a deliberately approximate titration first, although with some indicators a fairly accurate first titration is possible because the indicator shows signs of the arrival of the end-point. In the second and subsequent titrations, the flask should be shaken at intervals and the reagent should be added in quantities of not more than about 5 cm^3 because the large local excess of reagent which may result is apt to induce undesirable variants of the main reaction.

A burette (including the tap) should always be well washed out after use. If an alkaline solution has been in the burette, about 10 cm^3 of dilute acid should be run into the burette after running out the alkali, and the burette then well washed out with water.

SOURCES OF ERROR

The following are the chief sources of error in volumetric analysis.

SOLUTION NOT HOMOGENEOUS

This is a frequent source of error. After the solution has been made up to the mark, it is essential either to shake very well or to pour the solution into a large flask and swirl.

INACCURACY OF INSTRUMENTS USED

Measuring flasks, burettes and pipettes of reasonable price are necessarily manufactured by mass-production methods and inaccuracies are certain to arise. An experienced analyst can calibrate his apparatus and so practically eliminate errors from this source. Actually the least accurate of the instruments is the burette for the following reasons.

1. An error of a drop may arise in the titration because this is the least amount that can be added. The volume of a drop from an ordinary school burette may be about 0.05 cm^3, but the error may be reduced by averaging three close titrations.
2. The burette may drain irregularly. For this reason, burettes should be treated at intervals with a soapless detergent to remove grease.
3. The second decimal place in the readings can only be obtained approximately.

The probable error in using the burette is about 1 part in 500, and this is usually the greatest error an experienced analyst will encounter. The other instruments are generally more accurate, provided that they are consistently used as previously described in the text.

ERRORS IN WEIGHING

As the error in using the burette is about 1 part in 500, there is no point in weighing out material extremely accurately. If one gram or more is being weighed, a mistake of one unit in the third decimal place will introduce an error of one part in 1000, which is much less than the burette error.

IMPURITY OF MATERIALS

It is obvious that if analyses are based on solutions made up from impure chemicals, the results will be unreliable. Chemicals of analytical reagent quality are so pure that the errors they introduce, compared with those from other sources, are almost always negligible. Moisture is nearly always present to the extent of 0.2 to 0.5% in any powder not specially dried. This impurity can usually be removed by storing the substance before use for a few hours in a desiccator.

Errors may arise from the action of light, atmospheric carbon dioxide, dust particles or oxygen on standard solutions. These may be minimized by the use of coloured glass bottles, tightly fitting stoppers, soda-lime tubes to absorb carbon dioxide, and in other ways. If a solution is kept for a long time, it should be standardized at intervals.

INACCURACY IN THE END-POINT RECORDED

If too much indicator is added to the solution to be titrated, a certain amount of the reagent added will be needed to cause the colour change to take place. It is often helpful to perform a 'blank' experiment to ascertain the volume of reagent necessary to affect the indicator which has been added to a volume of water approximately equal to the final volume of solution likely to be obtained in the actual titration. Furthermore, where a change in colour indicates the end-point of a reaction, practice is often necessary before the change can be clearly recognized. It is sometimes useful to have on the

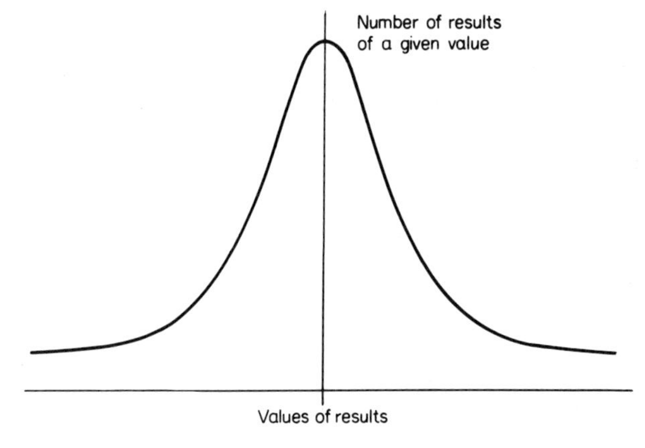

FIG. 76

bench for comparison a flask containing a few drops of the unchanged indicator added to water, so that any alteration in colour can be easily observed.

ACCURACY

The results of one person performing the same experiment many times or of many people doing the same experiment once will fall in a pattern known as a normal distribution curve. The sharper the peak, the more accurately the experiment has been performed (see Fig. 76).

The results obtained by volumetric analysis are usually accurate to 3 significant figures, but if the sequence of digits is above 500 then it is more realistic to quote only 2 significant figures because of the unreliability of the third one.

32
Indicators

IONIZATION

The aqueous solutions of some substances will readily conduct an electric current and decomposition occurs as a result. These substances are termed electrolytes, and acids, alkalis, and the majority of salts belong to this class. Thus a solution of hydrogen chloride in water conducts an electric current and is decomposed into hydrogen (which is evolved at the cathode or negative pole) and chlorine (at the anode or positive pole). This is explained by assuming that ions are present in the liquid before any current is passed through it. These ions are electrically charged atoms or groups of atoms, ions of metals or metallic groups being positively charged and ions of non-metals and non-metallic groups being negatively charged. The amount of charge is directly proportional to the valency of the atom or group. Since the majority of the volumetric processes depend upon the interaction of ions, we shall often represent a substance, not by its usual formula, but by its ionic formula. This indicates the ions furnished by that substance when dissolved in water. Thus:

HCl		$H^+ + Cl^-$
H_2SO_4		$2H^+ + SO_4^{2-}$
$NaOH$	may be represented	$Na^+ + OH^-$
$Ca(OH)_2$	as indicated opposite	$Ca^{2+} + 2OH^-$
KNO_3		$K^+ + NO_3^-$
$BaCl_2$		$Ba^{2+} + 2Cl^-$

The plus sign between them merely indicates that the ions are present in the same solution in those proportions. It does not imply any chemical bond between the ions. They are free to move anywhere in the solution. When an electric current is passed, however, the negative ions or anions move to the positive pole and the positive ions or cations to the negative pole, and the phenomenon of electrolysis is observed.

DEFINITIONS

An acid is a solution of a substance which contains hydrogen, and when the acid is dissolved in water it furnishes hydrogen ions as the only cations, e.g.

$$HCl \text{ is } H^+ + Cl^-$$

A base is a substance which will react with hydrogen ions to give a salt and water only. The alkalis are substances which, when dissolved in water, furnish hydroxide ions as the only anions, e.g.

$$NaOH \text{ is } Na^+ + OH^-$$

Neutralization is a reaction between an acid and an alkali producing a salt and water only, e.g.

$$Na^+ + OH^- + H^+ + Cl^- \rightarrow Na^+ + Cl^- + H_2O$$

It will be clearly seen that, because the sodium chloride produced is in the form of ions of sodium and chlorine, neutralization consists essentially of the action between hydrogen ions and hydroxide ions to form molecules of water which are undissociated.

$$H^+ + OH^- \rightarrow H_2O$$

WEAK AND STRONG ELECTROLYTES

According to Arrhenius the ions are furnished by a reversible reaction in which an undissociated molecule splits up into ions to an extent which varies considerably from substance to substance and according to the dilution of the solution. The greater the dilution, the greater the dissociation.

Strong electrolytes are dissociated to a considerable extent even in fairly concentrated solution, whereas weak electrolytes are only slightly dissociated. Strong electrolytes are believed to be completely dissociated into ions even in concentrated solution, and it is because the mobility (i.e. the free movement of the ions) is restricted in concentrated solution that they appear to be to some extent undissociated. The following table shows how the dissociation varies for two weak electrolytes and also how the dilution affects the degree of dissociation.

Degree of Dissociation (at 18 °C)

	0.1 M	0.01 M	0.001 M
Ammonia in aqueous solution	0.0133	0.0415	0.125
Acetic (ethanoic) acid	0.0133	0.0415	0.125

Thus in 0.1 M acetic acid 1.33% of the acetic acid consists of hydrogen ions and acetate ions.

DISSOCIATION CONSTANT

Consider an electrolyte AB which ionizes when in solution.

$$AB \rightleftharpoons A^+ + B^-$$

The equilibrium constant, *K*, is given by

$$K = \frac{[A^+][B^-]}{[AB]}$$

where the symbol in square brackets is used to designate the concentration of a species in mol dm^{-3}.

The equilibrium constant in this case is termed the dissociation constant of the electrolyte. This can be calculated from the above table thus:

Dissociation constant for acetic acid:

$$CH_3COOH \rightleftharpoons H^+ \qquad + CH_3COO^-$$

Concentration in 0.1 M solution in mol dm^{-3}.

0.09867 0.00133 0.00133 respectively

$$\therefore K = \frac{(0.00133)^2}{0.09867} = 1.8 \times 10^{-5} \text{ mol dm}^{-3}$$

IONIC PRODUCT OF WATER

Neutralization consists of the formation of molecules of undissociated water by the union of hydrogen ions and hydroxide ions.

$$H^+ + OH^- = H_2O$$

Like other ionic reactions this is to some extent reversible.

It is true that pure water is practically a non-conductor of electricity, but from conductivity experiments it can be shown that pure water does contain minute quantities of both hydrogen ions and hydroxide ions. Then

$$H_2O \rightleftharpoons H^+ + OH^-$$

represents the equilibrium between the ions and undissociated water, and the dissociation constant *K* would be given by

$$K = \frac{[H^+][OH^-]}{[H_2O]}$$

The proportions of the ions are so small, however, that $[H_2O]$ can be considered to be constant without serious error and the expression $[H^+][OH^-]$ is therefore also constant and is termed the ionic product of water. The value of this product at 25 °C is very nearly 10^{-14} mol^2 dm^{-6}, and because the ions are present in equal amounts (one molecule of water gives one hydrogen ion and one hydroxide ion) it follows that pure water contains a concentration of hydrogen ion of 10^{-7} mol dm^{-3}.

$$K_w = [H^+][OH^-] = 10^{-14} \text{ mol}^2 \text{ dm}^{-6} \text{ (at 25 °C)}$$

The concentrations are expressed thus:

$[H^+] = 10^{-7}$ mol dm^{-3} or 10^{-7} g of hydrogen ions per 1000 cm^3.

$[OH^-] = 10^{-7}$ mol dm^{-3} or 17×10^{-7} g of hydroxyl ions per 1000 cm^3.

Although this ionic product is exceptionally small, it is of great importance because it is this constant which is used to trace the alteration of the hydrogen ion concentration which takes place during neutralization.

It is important to note that however strongly acidic an aqueous solution may be, it is never completely free from hydroxide ions because there must be enough hydroxide ions to maintain the equation

$$[H^+][OH^-] = 10^{-14} \text{ mol}^2 \text{ dm}^{-6}$$

Similarly, an alkaline solution is never completely free from H^+. When H^+ and OH^- are present in equal proportions, the concentration of each can only be 10^{-7} mol dm^{-3}, and this constitutes a neutral solution.

HYDROLYSIS

The presence in water of hydroxyl ions and hydrogen ions accounts for the phenomenon of hydrolysis. This action can be considered as the reverse of neutralization and occurs when the ions of water react with the ions derived from a substance in solution. Thus, sodium carbonate in solution furnishes carbonate ions and sodium ions.

$$Na_2CO_3 \text{ is } 2Na^+ + CO_3^{2-}$$

Thus in sodium carbonate solution the following ions are present:

$$Na^+, \quad CO_3^{2-}, \quad H^+, \quad OH^-$$

although the last two are only present to a very small extent. Now carbonic acid is a very weak acid and dissociates only very slightly; hence the following reaction occurs until there is left only that small quantity of hydrogen ions necessary to maintain the very small equilibrium constant for carbonic acid:

$$2H^+ + CO_3^{2-} \rightleftharpoons H_2CO_3$$

As hydrogen ions are removed from the solution by the formation of this undissociated acid, more hydrogen ions are liberated by the dissociation of water to maintain the value of the ionic product

$$[H^+][OH^-] = 10^{-14} \text{ mol}^2 \text{ dm}^{-6}$$

Finally an equilibrium is set up, leaving in solution an excess of hydroxide ions over hydrogen ions which causes the solution to be alkaline.

By a similar type of reasoning it can be shown that a sodium hydrogencarbonate solution (which is a solution of an *acid* salt) is alkaline. Thus

$$H_2O \rightleftharpoons H^+ + OH^-$$
$$HCO_3^- + H^+ \rightleftharpoons \underset{\text{only slightly dissociated}}{H_2CO_3}$$

The removal of the hydrogen ions to form undissociated carbonic acid leaves an excess of hydroxide ions; hence the solution is alkaline.

Similarly ammonium chloride furnishes ammonium and chloride ions in solution.

$$NH_4Cl \quad \text{is} \quad NH_4^+ + Cl^-$$

Ammonia in solution exists in equilibrium with only a few ions:

$$NH_4^+ + OH^- \rightleftharpoons NH_3 + H_2O$$

Thus hydroxide ions are removed from the solution, leaving an excess of hydrogen ions which cause the solution to be acidic.

It follows that the solution of a normal salt in water, i.e. a salt formed by replacing all the replaceable hydrogen by a metal, is by no means certain to be neutral. Whether such a salt will form an acidic or an alkaline solution can usually be predicted by an examination of the possible reactions of its ions with hydrogen ions or hydroxide ions.

pH VALUE

It was suggested by Sϕrensen that a useful method of denoting the hydrogen ion concentration of a solution would be by using the logarithm of the concentration with the sign reversed.

By this means the concentration of an acidic or an alkaline solution can be expressed in the same terms, since if the pH value is known, the concentration of the hydroxide ion can be at once determined from the equation

$$[H^+][OH^-] = 10^{-14}\ mol^2\ dm^{-6}$$

By definition

$$pH = -\log(\text{numerical value of } [H^+] \text{ measured in mol dm}^{-3})$$

e.g. let $[H^+] = 10^{-2}$ mol dm^{-3}, i.e. 0.01 g dm^{-3} as would be the case in a solution of a strong acid, such as hydrochloric, of concentration 0.01 M. It is assumed that the strong acid is completely dissociated. Then

$$\log[H^+] = -2$$
$$\therefore pH = 2$$

pH VALUE OF 0.1 M ACETIC ACID

In 0.1 M acetic acid, degree of dissociation = 0.0133. Hence

$$[H^+] = \tfrac{1}{10} \times 0.0133\ mol\ dm^{-3}$$
$$= 0.00133\ mol\ dm^{-3}$$
$$pH = -\log 0.00133$$
$$= -[\bar{3}.1239]$$
$$= 2.9$$

(Compare 0.001 M HCl; pH = 3.0, assuming complete dissociation.)

pH VALUE OF 0.01 M SODIUM HYDROXIDE

$$[OH^-] = \tfrac{1}{100} \text{ or } 10^{-2}\ mol\ dm^{-3}$$

but

$$[OH^-][H^+] = 10^{-14}\ mol\ dm^{-3}$$
$$\therefore [10^{-2}][H^+] = 10^{-14}$$
$$[H^+] = 10^{-12}$$
$$pH = 12$$

It follows that an exactly neutral solution is one of pH value 7, because this solution will contain equal concentrations (expressed as mol dm^{-3}) of hydrogen and hydroxide ion. The smaller the pH value, the more acidic the solution and the larger the pH value, the more alkaline the solution (see p. 42).

Another advantage of using the pH system is that changes in concentration too large to be represented graphically can easily be traced by using pH value. As an example, consider the addition of 25 cm^3 1 M HCl to 25 cm^3 1 M NaOH. The hydrogen ion concentration of the alkaline solution if completely ionized is 10^{-14} mol dm^{-3}, whereas that of the 1 M acid is 1 mol dm^{-3}, and at exact neutrality the hydrogen ion concentration is 10^{-7} mol dm^{-3}.

Using pH value this change can be shown graphically as in Figure 77. It can be seen there is a rapid change of pH value as the solution changes from being slightly alkaline to being slightly acidic.

TO CALCULATE THE CHANGE OF pH VALUE AS ONE DROP OF EXCESS ACID FALLS INTO THE NEUTRAL SOLUTION

If 25 cm^3 of 1 M HCl have been added to 25 cm^3 of 1 M NaOH the solution will be exactly neutral and its pH value will be 7. If 1 drop = 0.05 cm^3, it will contain 0.05×10^{-3} g of hydrogen ions. This is added to 50 cm^3 of neutral solution, and hence

$$[H^+] = \frac{1000}{50} \times 0.05 \times 10^{-3} \text{ g dm}^{-3}$$

$$= 10^{-3} \text{ mol dm}^{-3}$$

$$\therefore \text{pH} = 3$$

Neglecting the readjustment due to the existing small amount, which is merely 10^{-7} g dm^{-3}, it is seen that one drop of acid *at this stage* produces an increase in $[H^+]$ of 10000 times its value. By a similar argument it can be shown that a further 9 drops (0.5 cm^3) would be required to change the pH value to 2.

THEORY OF INDICATORS

Most of the indicators used in acidimetry and alkalimetry are weak acids and their degree of dissociation is greatly affected by alteration of hydrogen ion concentration of the solution, thus producing a change in colour. Consider phenolphthalein, a very weak organic acid; its dissociation can be regarded as the splitting up of the molecule (let it be HA) into a hydrogen ion and a negatively charged ion.

$$\underset{\text{(colourless)}}{HA} \rightleftharpoons H^+ + \underset{\text{(pink)}}{A^-}$$

The colour change of an indicator may not be due solely to changes in extent of ionization, but this is a good approximation. If dissociation occurs to any marked extent it is clear that a pink colour will be observed, whereas if the undissociated acid is mainly present, no colour will be observed. Consider the effect of adding an acidic solution, the hydrogen ion concentration of which is high (compared with the concentration of H^+ furnished by the indicator) so that the dissociation of the indicator is suppressed. Hence the indicator will be practically unionized and no colour will be observed. On adding an alkaline solution, however, hydrogen ions from the indicator will react with some of the hydroxide ions present to form undissociated water, and hence the dissociation of the weak acid will be increased with a corresponding increase in the coloured ions A^- which will make the solution pink. Colour changes such as this take place at different concentrations with different indicators, and below is a list of some of the indicators in common use together with the range of pH value over which the colour change takes place.

pH range	*Indicator*	*Colour change acid-alkali*
3.0–4.4	Methyl orange	Red- orange
4.4–6.3	Methyl red	Red-yellow
6.0–7.6	Bromothymol blue	Yellow-blue
6.0–8.0	Litmus	Red-blue
8.2–10.0	Phenolphthalein	Colourless-red

It is quite obvious from Figure 77 that any of these indicators will give a sharp end-point if a strong alkali is being titrated with a strong acid, because the addition of only one or two drops of acid causes the pH value to move over the whole of the vertical portion of the curve.

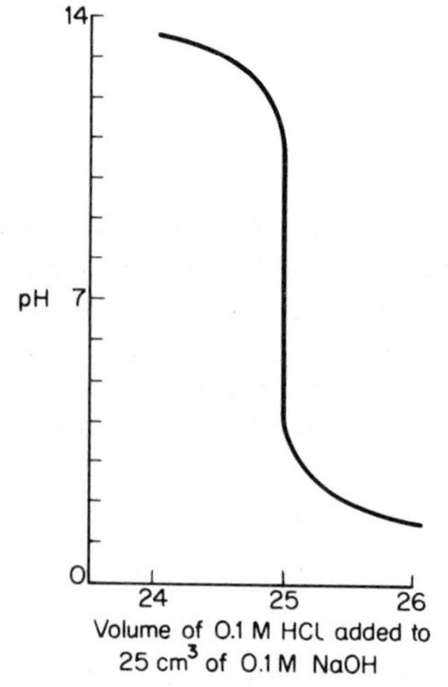

FIG. 77

ESSENTIAL CHARACTERISTICS OF A GOOD INDICATOR

1. The colour change of the indicator must be clear and sharp, i.e. it must be sensitive. Thus it would be useless if 2 or 3 cm^3 of the reagent were necessary to bring about the colour change.

2. The pH range over which the colour change takes place must be such as to indicate when the reaction (as shown by the equation) is complete.

CHOICE OF INDICATOR

(a) FOR TITRATION OF WEAK ACIDS WITH STRONG ALKALIS

We have already seen, p. 288, that a weak acid is only slightly dissociated, and hence its hydrogen ion concentration is low. Now a strong acid in 0.1 M solution will have a pH value of about 1, and we will suppose that weak acid has in 0.1 M solution a pH value of 4. It would still be definitely on the acid side of the neutral point. Now if this weak acid is titrated with 0.1 M alkali (pH value 13) it is clear that, no matter how much acid is added, a solution of pH value 3 cannot be obtained.

In other words the range of pH value of the solutions as the two are mixed cannot be outside the limits 4 to 13. For this reaction it would be useless to choose an indicator which changed at pH = 3. It would show no change. Furthermore it is not advisable to choose an indicator which changed at pH value 5 because the pure acid

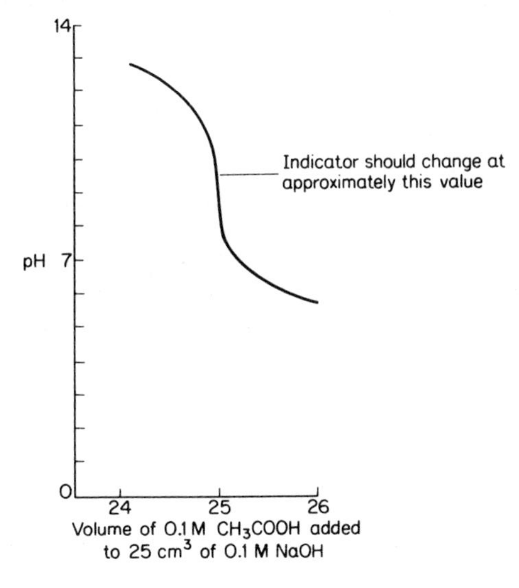

FIG. 78

itself would have to be present in a concentration somewhere near 0.01 M to produce this effect. The best indicator to use would be one which changed at about pH value 9. The graph, Figure 78, shows the pH change as a weak acid (acetic) is added to a strong alkali (sodium hydroxide). For this titration the best indicator is phenolphthalein.

Hence when titrating solutions of

oxalic (ethanedioic) acetic (ethanoic) succinic (butanedioic) formic (methanoic) boric (boric(III)) carbonic (carbonic(IV)) sulphurous (sulphuric(IV)) or any weak acid	against a strong alkali, use phenolphthalein as an indicator.

If the acid is being run into the alkaline solution, the disappearance of the pink coloration gives the correct end-point. Carbon dioxide of the air may dissolve sufficiently to cause the pink colour of the phenolphthalein to disappear if the flask is shaken too vigorously.

(b) FOR TITRATION OF WEAK ALKALIS WITH STRONG ACIDS

The case is reversed when considering the indicator to use if a strong acid and a weak alkali are to be titrated. The pH value of the weak alkali in the concentration used may not exceed 10 (i.e. the solution could not have a high concentration of hydroxide ions). The pH value cannot vary outside the limits 1 to 10 (for, say, 0.1 M solutions) and comparatively large concentrations of the alkali might be necessary to obtain a pH value anywhere approaching 10. Figure 79 shows the pH change during a titration of a strong acid (hydrochloric) against a weak alkali (ammonia). For this type of reaction an indicator changing at a pH value of about 4 is essential. That used in practice is methyl orange or methyl red.

Hence when titrating solutions of

ammonia sodium carbonate sodium borate or any weak alkali	against a strong acid, use methyl orange or methyl red.

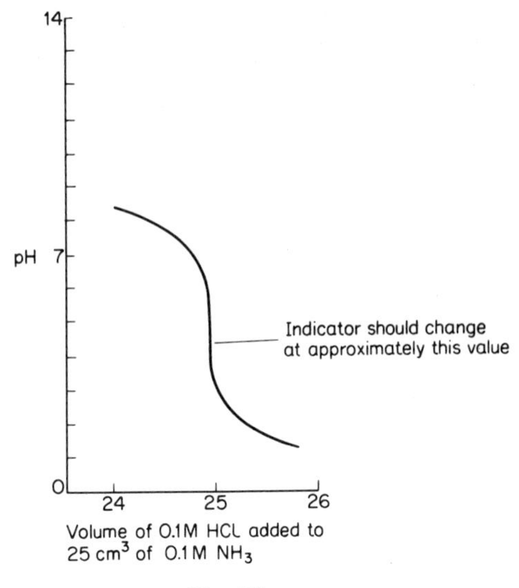

FIG. 79

It is impossible to follow the neutralization of a weak alkali by a weak acid as there is no point in the titration at which a rapid change of pH value takes place thus showing a sharp colour change of the indicator.

NOTES ON SOME OF THE INDICATORS IN GENERAL USE

LITMUS

This contains several dyes and the colour change is therefore not easy to follow if very accurate work is being done. Furthermore, the solution does not keep well out of contact with air. Litmus contains azolitmin, and the latter can be substituted for litmus.

Litmus solution. [Colour change red-blue pH 6.0–8.0.] Digest 10 g of commercial litmus with about 500 g cm^3 of warm water and allow to stand for some time. Filter off and add to the filtrate a few drops of dilute nitric acid until the purple colour is obtained. Keep in a bottle with a loose cork. Use two or three drops for 25 cm^3 of solution to be titrated.

Azolitmin. [Colour change red-blue pH 5.0–8.0.] Dissolve about 5 g of azolitmin in 500 g cm^3 of water to which a little sodium hydroxide solution has been added. Add dilute nitric acid until the purple colour is obtained. Use two or three drops for 25 cm^3 of solution to be titrated.

METHYL ORANGE

[Colour change red-orange pH 3.0 to 4.4.] This is an excellent indicator for use as described above if the concentration of the solutions is greater than 0.2 M. In 0.1 M solutions the end-point is not too easy to determine.

Methyl orange. Dissolve 1 g of methyl orange in 500 cm^3 water. Use two or three drops for 25 cm^3 of solution to be titrated.

METHYL ORANGE MODIFIED WITH XYLENE CYANOL F.F.

[Colour change magenta—blue-grey—green.] Hickman and Linstead advocate mixing methyl orange with xylene cyanol F.F., which has the effect of producing a blue-grey tint at the pH value 3.8, corresponding with the end-point given by methyl orange.

Screened methyl orange. Mix 1 g of methyl orange with 1.4 g xylene cyanol, F.F. and 500 cm^3 of 50% ethanol and water. Use two or three drops for 25 cm^3 of solution to be titrated.

METHYL RED

[Colour change red-yellow pH 4.4 to 6.3.] This is a sensitive indicator and very useful for titrating weak alkalis (ammonia and organic bases in solution) with strong acids.

Methyl red. Dissolve 1 g of the dye in 500 cm^3 of 60% ethanol. Use one or two drops only for 25 cm^3 of liquid to be titrated.

PHENOLPHTHALEIN

[Colour change colourless-red pH 8.2 to 10.0.] This is a useful indicator for titrating strong alkalis with weak acids.

Phenolphthalein. Dissolve 1 g in 500 cm^3 of 50% ethanol. Use one or two drops for 25 cm^3 of solution to be titrated.

33

Acids and Alkalis

Before estimations involving acids or alkalis can be carried out on given substances or mixtures, it is necessary to obtain accurately standard acidic and alkaline solutions. The common acids and alkalis cannot be employed directly for making standard solutions because they are variable in composition for the reasons given.

Hydrochloric acid. Hydrogen chloride is very volatile in high concentrations.

Sulphuric acid. This is hygroscopic.

Nitric acid. Hydrogen nitrate is volatile and subject to decomposition.

Furthermore, the so-called 'pure' mineral acids of the laboratory are not pure substances: they contain varying quantities of water.

Sodium and potassium hydroxides. These are deliquescent and react with carbon dioxide from the air, e.g.

$$2NaOH + CO_2 \rightarrow Na_2CO_3 + H_2O$$

Calcium hydroxide. This is insufficiently soluble and also reacts with carbon dioxide from the air.

$$Ca(OH)_2 + CO_2 \rightarrow CaCO_3 + H_2O$$

Ammonia solution. Ammonia is a volatile solute, and gives a solution of variable concentration.

Characteristics of a good standardizing reagent:

(*a*) It should be obtainable in a high degree of purity and it should be easy to test for impurities.

(*b*) It should be stable and unaffected by the atmosphere. It should not be deliquescent or efflorescent, so that it may easily be weighed out accurately. It should be stable to heat so that it may be dried.

(*c*) It should be fairly cheap.

(*d*) Its relative molecular mass should be such that a solution of concentration about 0.01 to 0.5 M can be prepared easily.

For standardization of acids the materials commonly used are:

1. Pure sodium carbonate prepared by heating sodium hydrogencarbonate (or analytical quality anhydrous sodium carbonate).
2. Pure borax $Na_2B_4O_7 \cdot 10H_2O$, disodium tetraborate.
3. Iceland spar, pure calcium carbonate.

Alkaline solutions may be standardized by using solid crystalline organic acids such as oxalic (ethanedioic) acid or succinic (butanedioic) acid, which can be obtained in a high state of purity.

Experiment 273 Standardization of hydrochloric acid by sodium carbonate

Material: concentrated hydrochloric acid, sodium hydrogencarbonate.

Concentrated hydrochloric acid is roughly 11 M. Pour out into a measuring cylinder about 2 cm^3 of concentrated hydrochloric acid. Transfer it to a 250 cm^3 flask and make up to the mark with water. Shake well. Put some pure sodium hydrogencarbonate or anhydrous sodium carbonate into an evaporating dish and heat gently over a low flame for about fifteen minutes, stirring continuously. Take care not to heat the mass too strongly or fusion may take place which will seriously retard solution and also cause slight decomposition. It will be obvious when carbon dioxide is being evolved as the mass appears unusually light as it is being stirred. Allow the dish to cool in a desiccator because anhydrous sodium carbonate absorbs moisture to form the monohydrate. Weigh, then heat again for five minutes, cool and reweigh. Repeat this process until the mass is constant.

$$Na_2CO_3 + 2HCl \rightarrow 2NaCl + H_2O + CO_2$$

Relative molecular mass of anhydrous sodium carbonate is 106. Mass of sodium carbonate for 250 cm^3 0.05 M solution is

$$0.25 \times 0.05 \times 106 = 1.325 \text{ g}$$

Weigh a clean, dry weighing bottle and weigh out exactly 1.325 g of the pure sodium carbonate into it. Transfer the carbonate to a beaker containing a little hot water (shaking gently as the carbonate comes in contact with the water) and wash the weighing bottle carefully by means of the wash-bottle, allowing the washings to drop into the beaker. Stir gently to ensure dissolution. Then cool to room temperature before proceeding. Smear an almost imperceptible amount of vaseline on the lip of the beaker and pour the solution down a glass rod into a funnel resting in the neck of a clean (but not necessarily dry) 250 cm^3 flask. Wash the beaker out with further small quantities of water, pouring all washings down the rod and funnel into the 250 cm^3 flask to ensure that no solution is left on the walls of the beaker. It is as well to remember that once you have weighed out the sodium carbonate into the weighing bottle, every particle of it, whether as solid or as a solution, must be transferred into the 250 cm^3 flask. Make up to the mark with water, adding the last few drops from a pipette, and shake well or pour into a large dry beaker in order that the solution may become homogeneous.

By pipette take 25 cm^3 of 0.05 M sodium carbonate solution, run it into a conical flask, add two or three drops of screened methyl orange solution, make a note of the burette reading and run in the acid from a burette. The first 15 to 20 cm^3 of the acid may be run in 5 cm^3 at a time without fear of overshooting the end-point. Then run in 1 cm^3 at a time, shaking after each addition until the colour changes from green to blue-grey. This is a trial titration and its accuracy may not warrant its use in the calculation. The titration should now be repeated with further portions of 25 cm^3 of the carbonate solution until two readings are obtained which agree to 0.1 cm^3 or three readings which do not show any trend in value, up or down.

Readings of burette in cm^3

	Trial	1	2	3
Final	23.4	45.4	25.5	47.5
Initial	1.2	23.4	3.4	25.5
Difference	22.2	22.0	22.1	22.0

Average accurate titre = 22.0 cm^3.
i.e. 22.0 cm^3 HCl of unknown concentration reacted with 25.0 cm^3 of 0.05 M Na_2CO_3. Hence if m mol dm^{-3} is the concentration of the acid, using the equation for the reaction

$$\frac{22 \times m}{25 \times 0.05} = \frac{2}{1}$$

$$m = 0.1136$$

The diluted hydrochloric acid is 0.114 M. The relative molecular mass is 36.5, so the concentration of hydrogen chloride is $0.1136 \times 36.5 = 41.5$ g dm^{-3}.

If the dilution ratio is known accurately then the concentration of the original acid can be calculated.

If any other mass of sodium carbonate is taken then the concentration of the solution can be calculated from

$$\frac{4 \times \text{mass of solute in 250 cm}^3 \text{ solution}}{106 \text{ g}}$$

Experiment 274 Standardization of hydrochloric acid by disodium tetraborate

Material: approximately 0.1 M hydrochloric acid; disodium tetraborate-10-water (borax).

It is best not to rely on one standardization to determine the concentration of a solution.

Borax ($Na_2B_4O_7 \cdot 10H_2O$) is not very soluble in cold water: it is used as a 0.05 M solution.

$$Na_2B_4O_7 \cdot 10H_2O + 2HCl \rightarrow 2NaCl + \underset{\text{boric acid}}{4H_3BO_3} + 5H_2O$$

Relative molecular mass of borax = 381. Hence for 250 cm^3 0.05 M solution, $381 \times 0.25 \times 0.05$ g of borax is required, i.e. 4.775 g.

Weigh out about 4.8 g of pure borax accurately, and make up to 250 cm^3. Take 25 cm^3 of this solution, add a few drops of screened methyl orange and titrate against the 0.1 M acid. The boric acid liberated does not affect the end-point if methyl orange is used as an indicator. This method can be used for nitric and hydrochloric acids, but with sulphuric acid the end-point is not satisfactorily defined.

First calculate the concentration of the borax solution that has been made up, then, using the titration results, calculate the concentration of the acid.

Experiment 275 Standardization of sodium hydroxide solution by oxalic acid

Material: stick sodium hydroxide; hydrated oxalic (ethanedioic) acid.

$$2NaOH + H_2C_2O_4 \cdot 2H_2O^{1} \rightarrow Na_2C_2O_4 + 4H_2O$$

2NaOH	$H_2C_2O_4 \cdot 2H_2O$
Relative molecular mass 40. For 250 cm^3 0.1 M solution use 1 g.	For 250 cm^3 0.05 M solution use 1.575 g.

Quickly weigh out on to a watch-glass about 1.3 g of stick sodium hydroxide (the purest form available for purchase). Put it into a beaker, add a little of water, swill round for a few moments and discard the solution. (This will have removed most of the surface layer of carbonate.) Dissolve the remainder in water and make up to about 250 cm^3 in a flask. Weigh out about 1.6 g of pure oxalic acid

[1] This water of crystallization takes no part in the reaction. It is included here because the crystals contain it.

accurately in a weighing bottle. (Analytical quality should be used—oxalic acid frequently contains both calcium and potassium oxalates as impurity.) Dissolve the acid in water in a beaker and transfer the solution completely to a 250 cm^3 flask and make up to the mark, shake well or pour into a large flask and swirl. Fill a burette with the acid and titrate with 25 cm^3 of the alkali using phenolphthalein as the indicator (one or two drops should be sufficient to colour the solution pink). Run the acid in until the solution becomes colourless. Perform two or three accurate titrations.

First calculate the concentration of the oxalic acid solution that has been made up, then, using the titration results, calculate the concentration of the alkali.

Experiment 276 Standardization of sodium hydroxide solution by succinic acid

Material: sodium hydroxide; succinic (butanedioic) acid.

Prepare a 0.1 M solution of sodium hydroxide as indicated in Experiment 275.

These hydrogen atoms
do not ionize.
↓

$$(CH_2COOH)_2 + 2NaOH \rightarrow (CH_2COONa)_2 + 2H_2O$$

Relative molecular
mass 118. For
250 cm^3 0.05 M
solution use 1.475 g.

Weigh out accurately in a weighing bottle about 1.5 g of succinic acid and transfer it into a beaker. Warm the beaker in order to dissolve the acid. Cool the solution to room temperature and pour it, together with the rinsings, into a 250 cm^3 measuring flask. Make it up to the mark. To 25 cm^3 of the acid solution add one or two drops of phenolphthalein and run in the alkali from the burette, proceeding more cautiously as you are approaching the end-point. Shake well after each addition and note the reading at the first permanent pink tinge. Perform several accurate titrations.

First calculate the concentration of the succinic acid solution that has been made up, then, using the titration results, calculate the concentration of the alkali.

N.B. Use a rubber stopper for the bottle in which you are keeping the sodium hydroxide solution and wash the burette out with water and then with very dilute acid and again with water, to prevent the tap of the burette from sticking in its socket. This procedure should be adopted whenever an alkaline solution is placed in the burette.

Experiment 277 Standardization of nitric acid by the Iceland Spar method

Material: concentrated nitric acid; Iceland Spar—a crystalline form of calcium carbonate.

This method depends on determining the mass of pure calcium carbonate dissolved by a known volume of the acid. It is suitable as a check, particularly when the acid solution is fairly concentrated. Note that it cannot be used to standardize sulphuric acid because calcium sulphate is only sparingly soluble.

$$CaCO_3 + 2HNO_3 \rightarrow Ca(NO_3)_2 + H_2O + CO_2$$

$$100 \text{ g of calcium carbonate} \equiv 2 \text{ dm}^3 \text{ 1 M } HNO_3$$

$$\therefore 1.25 \text{ g of calcium carbonate} \equiv 25 \text{ cm}^3 \text{ 1 M } HNO_3$$

Dilute some concentrated nitric acid by putting 25 cm^3 in about 125 cm^3 water in a measuring flask and then making the total volume up to 250 cm^3. Shake well. Weigh out accurately, on a watch-glass, a crystal of Iceland spar (2–3 g), and introduce it carefully into a conical flask and add about 50 cm^3 of water. Run into this water exactly 25 cm^3 of the nitric acid solution, place a funnel in the neck and leave the flask in a safe place until the next day (or until effervescence ceases).

Pour away the solution surrounding the crystal, and wash the crystal several times with distilled water. Transfer the crystal to a watch-glass, dry it thoroughly in a steam-oven and weigh it.

Suppose loss in mass of crytal is 1.340 g. From the equation

$$25 \text{ cm}^3 \text{ 1 M } HNO_3 \equiv 1.25 \text{ g of } CaCO_3$$

$\therefore$ the concentration of the acid is

$$\frac{1.340}{1.25} \times 1 \text{ M} = 1.072 \text{ M}$$

Hence the concentration of the original acid is 10.7 M.

Experiment 278 Determination of the relative molecular mass of calcium carbonate by back titration

Material: calcium carbonate (precipitated); 1 M hydrochloric acid; 0.1 M alkali.

Calcium carbonate is insoluble in water and difficulties would arise in titrating a standard acid against it. These difficulties can be avoided by dissolving a known quantity of the carbonate in excess of acid and determining the amount of excess by titration with alkali.

The concentration of the latter should be less than that of the acid used, as the amount of acid left will be considerably less than the amount taken up.

Weigh accurately a weighing bottle containing not more than 1.5 g of pure dry precipitated calcium carbonate. Empty this carefully into a funnel resting in a 250 cm^3 flask containing 50 cm^3 of 1 M hydrochloric acid and reweigh the weighing bottle accurately. Wash the carbonate through into the flask, and, when effervescence has ceased, make up to the mark with water and shake well. Withdraw 25 cm^3 and titrate against alkali. Use methyl orange or methyl red as an indicator.

$$CaCO_3 + 2HCl \rightarrow CaCl_2 + H_2O + CO_2$$

$$1 \text{ mole } CaCO_3 \equiv 2000 \text{ cm}^3 \ 1 \text{ M HCl}$$

Suppose the mass of calcium carbonate taken = 1.47 g; and that 20.6 cm^3 0.1 M alkali neutralized 25 cm^3 of residual solution.

$\therefore$ 20.6 cm^3 of 1 M alkali would neutralize 250 cm^3.

Volume of 1 M acid originally present

$$= 50 \text{ cm}^3.$$

$\therefore$ $(50-20.6)$ cm^3, i.e. 29.4 cm^3 is the volume of 1 M acid used up.

Hence 29.4 cm^3 of 1 M acid $\equiv$ 1.47 g calcium carbonate.

$$\therefore 2000 \text{ cm}^3 \text{ of 1 M acid} \equiv \frac{1.47}{29.4} \times 2000 \text{ g of calcium carbonate}$$

$$= 100 \text{ g}$$

$\therefore$ The relative molecular mass is 100.

In the determination of the relative molecular mass of an unknown compound on these lines, it is necessary, before weighings are made, to determine the acidity of the base and hence the form of the equation. Then, after making up the solution to 250 cm^3, it is necessary to verify that the solution is still acid, and if not, a further 25 cm^3 of acid should be added. If, on back-titrating, an 0.1 M solution of alkali were found to be unsuitable, a more concentrated solution could be used.

Experiment 279 Determination of the number of molecules of water of crystallization in hydrated sodium carbonate

Material: 0.2 M hydrochloric acid; hydrated sodium carbonate (washing soda).

$$Na_2CO_3 + 2HCl \rightarrow 2NaCl + H_2O + CO_2$$

Relative molecular mass 106 (of the anhydrous substance).
$\therefore$ 106 g of sodium carbonate $\equiv$ 2 dm^3 1 M acid.

Weigh out accurately about 4 g of soda crystals (this is assuming that about a third of the mass taken is water of crystallization) being careful to pick translucent crystals. Transfer these to a 250 cm^3 flask, add water, shaking after each addition, and make up to the mark. Shake well. Take 25 cm^3 of this solution in a conical flask, add a few drops of methyl orange and run in the hydrochloric acid from the burette until the first permanent orange colour is observed. Repeat to obtain two or three accurate results.

Suppose the mass of soda crystals taken = 5.35 g.
Suppose 25 cm^3 of solution required 18.7 cm^3 0.2 M acid.

$$\therefore \text{Concentration of solution} = \frac{18.7}{25} \times \frac{0.2}{2} \text{ M}$$

$$\therefore \text{Concentration in terms of anhydrous sodium carbonate} = \frac{18.7}{25} \times 0.1 \times 106 \text{ g dm}^{-3}$$

$$\therefore \text{Mass of anhydrous sodium carbonate in 250 cm}^3 \text{ solution} = \frac{18.7 \times 0.1 \times 106}{25 \times 4} \text{ g}$$

$$= 1.982 \text{ g}$$

Let the formula be $Na_2CO_3 \cdot xH_2O$. Then

$$\frac{Na_2CO_3}{Na_2CO_3 \cdot xH_2O} = \frac{1.982}{5.35} = \frac{106}{106 + 18x}$$

Hence $x = 9.9$

$\therefore$ The number of molecules of water of crystallization = 10 (because x is likely to be a whole number and the result is subject to the usual experimental error).

N.B. It is advisable to work with sodium carbonate solution in concentrations not less than 0.1 M. In 0.05 M solution the end-point using methyl orange is scarcely sharp enough.

Experiment 280 Determination of the proportions of sodium carbonate and sodium hydroxide in a mixture (double indicator method)

Material: mixture of about 5 g each of sodium hydroxide and sodium carbonate in 1000 cm^3 solution; 0.2 M hydrochloric acid.

		Indicator to exhibit completion of reaction
I.	$NaOH + HCl \rightarrow NaCl + H_2O$	Any indicator.
IIa.	$Na_2CO_3 + HCl \rightarrow NaHCO_3 + HCl$	Phenolphthalein.
IIb.	$NaHCO_3 + HCl \rightarrow NaCl + H_2O + CO_2$	Methyl orange.

As we have already seen, the neutralization of a strong alkali by means of a strong acid can be followed by the use of any indicator. Sodium hydrogencarbonate solution is slightly acid to phenolphthalein (you should verify this experimentally) whereas sodium carbonate solution is definitely alkaline to it. If acid is added to a mixture of sodium hydroxide and sodium carbonate in solution, using phenolphthalein as indicator, then the pink colour of the indicator is discharged when reactions I and IIa are complete. If methyl orange is now added and a further quantity of acid is also added, the amount required will be that necessary to complete reaction IIb. But one mole of sodium hydrogencarbonate formed from one mole of sodium carbonate and hence the amounts of acid required for reactions IIa and IIb will be the same.

Suppose the volume of acid needed to reach the end-point as indicated by the phenolphthalein is a cm^3 and the *additional* volume of acid to reach the end-point as indicated by the methyl orange is b cm^3. Then volume of acid reacting with sodium carbonate is $2b$ cm^3 and volume of acid reacting with sodium hydroxide is

$$(a+b)-2b \text{ cm}^3 = (a-b) \text{ cm}^3$$

Take 25 cm^3 of the solution containing the two alkalis and add one or two drops of phenolphthalein solution, and run in 0.2 M acid until the pink colour is just discharged. Note the burette reading and add a few drops of methyl orange and a further quantity of acid until the yellow colour of the methyl orange changes to orange.

Suppose 25 cm^3 of mixture required 21.3 cm^3 of 0.2 M HCl (phenolphthalein) and that the same 25 cm^3 of mixture required a further 7.9 cm^3 of 0.2 M HCl (methyl orange).

Total volume of 0.2 M HCl used = 29.2 cm^3.

The complete neutralization of the sodium carbonate thus requires $2 \times 7.9 = 15.8$ cm^3 of 0.2 M HCl.

∴ The sodium hydroxide required $(29.2-15.8)$ cm^3
or 13.4 cm^3 of 0.2 M acid.

∴ Concentration of the sodium carbonate solution is

$$\frac{15.8 \times 0.2}{25} \times \frac{1}{2} \text{ M} = \frac{15.8}{25} \times 0.1 \times 106 \text{ g dm}^{-3}$$

$$= 6.7 \text{ g dm}^{-3}$$

Concentration of the sodium hydroxide solution is

$$\frac{13.4}{25} \times 0.2 = \frac{13.4}{25} \times 0.2 \times 40$$
$$= 4.29 \text{ g dm}^{-3}$$

Hence the percentage of each alkali in the mixture can be calculated.

Note: The above exercise is a useful one on mixed indicators and one requiring thought. Unless the conditions are carefully controlled, however, and end-points correctly determined, its accuracy leaves much to be desired. As an aid to the memory the student is reminded that:

(*a*) the reaction between sodium hydroxide and hydrochloric acid cannot take place in two stages as can the reaction of sodium carbonate with hydrochloric acid;

(*b*) the phenolphthalein must obviously be the indicator first added since it is colourless in acid solution and will not mask the colour of the methyl orange added later.

Experiment 281 Determination of the proportions of sodium carbonate and sodium hydroxide in a mixture (Winkler)

Material: 0.1 M hydrochloric acid; mixture of 2–3 g each, sodium hydroxide and sodium carbonate in 1000 cm^3 solution; barium chloride solution.

Where accuracy is of prime importance the method due to Winkler is more satisfactory. The solutions used may be much more dilute without loss in accuracy.

The total alkali present (carbonate and hydroxide) is determined by titration with 0.1 M acid using methyl orange as an indicator. To a second portion excess barium chloride solution is added,

$$Na_2CO_3 + BaCl_2 \rightarrow 2NaCl + BaCO_3\downarrow$$

i.e.

$$Ba^{2+} + CO_3^{2-} \rightarrow BaCO_3\downarrow$$

Thus the carbonate is removed as insoluble barium carbonate. The hydroxide ions from the alkali remain in solution. The concentration of these is determined in the usual way, using phenolphthalein as an indicator.

Make up a solution of the mixture. To 25 cm^3 of this solution add methyl orange and titrate with 0.1 M hydrochloric acid until the yellow colour changes to orange. To a further 25 cm^3 of the original solution add about an equal volume of 0.1 M barium chloride solution and add one or two drops of phenolphthalein and titrate with 0.1 M hydrochloric acid, noting the burette reading when the solution is decolorized. Take care to run in the acid slowly otherwise some of the barium carbonate may be acted upon before the end-point is reached.

Suppose 25 cm^3 of mixed sodium hydroxide and carbonate solution required a cm^3 0.1 M HCl (methyl orange) and a further 25 cm^3 after treatment with $BaCl_2$ required b cm^3 0.1 M HCl (phenolphthalein). Hence

Volume of acid required to react with NaOH $= b$ cm^3
Volume of acid required to react with Na_2CO_3 $= (a-b)$ cm^3

$$\therefore \text{Concentration of NaOH} = \frac{b}{25} \times 0.1 \text{ M}$$

$$= \frac{b \times 0.1 \times 40}{25} \text{ g dm}^{-3}$$

$$\text{Concentration of Na}_2\text{CO}_3 = \frac{(a-b)}{25} \times 0.1 \times \frac{1}{2} \text{ M}$$

$$= \frac{(a-b) \times 0.1 \times 106}{25 \times 2} \text{ g dm}^{-3}$$

Hence the percentage of each alkali in the mixture can be calculated.

Experiment 282 Estimation of ammonia in ammonium sulphate by the indirect method

Material: 1 M sodium hydroxide; 0.1 M acid; ammonium sulphate.

The ammonium salt (as solid or solution) is boiled with excess sodium hydroxide and the excess determined by back-titration with acid.

$$\underset{132 \text{ g}}{(NH_4)_2SO_4} + \underset{\substack{2 \text{ dm}^3 \text{ 1 M} \\ \text{solution}}}{2NaOH} \rightarrow Na_2SO_4 + \underset{34 \text{ g}}{2NH_3} + 2H_2O$$

$$\therefore 1.6 \text{ g pure ammonium sulphate} \equiv \frac{2000 \times 1.6}{132} \text{ cm}^3 \text{ 1 M NaOH}$$

$$\equiv 24.2$$

Weigh accurately a weighing bottle containing about 1.6 g of pure ammonium sulphate and transfer this to a conical flask. Reweigh the bottle. Run exactly 50 cm^3 of the alkali into the flask and insert in the flask a rubber stopper into which a reflux condenser tube is fitted—even a glass funnel will do, with care—this prevents loss of solution. Boil the contents of the flask for about fifteen minutes or until no smell of ammonia is perceptible. Pour a little water down the condenser tube and so wash any alkali solution back into the flask. Pour the whole solution when cold into a 250 cm^3 flask, adding several washings from the conical flask, and make up to the mark.

Shake well. Titrate portions of 25 cm^3 of this solution against the acid using methyl orange as an indicator.

Suppose *a* g of ammonium sulphate were taken and that 25 cm^3 of solution (made up to 250 cm^3) required *b* cm^3 of 0.1 M acid.

$\therefore$ 250 cm^3 would require *b* cm^3 acid.
$\therefore$ $(50-b)$ cm^3 of 1 M NaOH were used up.
But 2000 cm^3 1 M NaOH $\equiv$ 34 g NH_3.

$$\therefore (50-b)\ \text{cm}^3\ 1\ \text{M NaOH} \equiv \frac{34}{2000} \times (50-b)\ \text{g NH}_3$$

$$\therefore \%\ \text{NH}_3 \text{ in sample} = \frac{34}{2000}\,\frac{(50-b)}{a} \times 100$$

Experiment 283 Estimation of ammonia in an ammonium salt by the direct method

Material: 0.5 M sulphuric acid; 0.1 M alkali; ammonium sulphate.

Ammonium sulphate can be obtained in a very pure state, and this method is used in industry to standardize acid solutions.

A weighed amount of the ammonium sulphate is boiled with excess sodium hydroxide solution. The ammonia which is driven off is absorbed in excess of standard acid, and the volume of the latter used up is found by back-titration.

Weigh accurately a weighing bottle containing about 1.6 g of ammonium sulphate (see p. 309) and empty this carefully into a funnel in a round-bottomed flask which should contain a few pieces of porous pot. Wash the sulphate through with distilled water. Re-weigh the weighing bottle. Run in exactly 50 cm^3 of 0.5 M sulphuric acid (or 1 M hydrochloric acid) into a conical flask standing in cold water. Fit up the apparatus shown in Figure 80. The trap T is to prevent droplets of alkali from being mechanically carried over into the standard acid.

Pour about 25 cm^3 (a large excess) of concentrated sodium hydroxide solution into the flask containing the ammonium sulphate. Warm the contents of the flask gently and finally boil for about ten minutes. At the end of that time open the tap and smell the vapour. If the ammonia can be detected, the boiling must be continued for a few more minutes and the test repeated until no ammonia can be smelt. Remove the burner from underneath the solution, raise the pipette out of the acid solution, and wash back into the conical flask any acid solution clinging to it inside or outside. Make up the residual acid to 250 cm^3, taking care to include all washings, and titrate 25 cm^3 of this against 0.1 M sodium hydroxide using methyl orange as an indicator. Calculate the percentage of ammonia from the fact that 1 dm^3 of 1 M H_2SO_4 $\equiv$ 34 g NH_3.

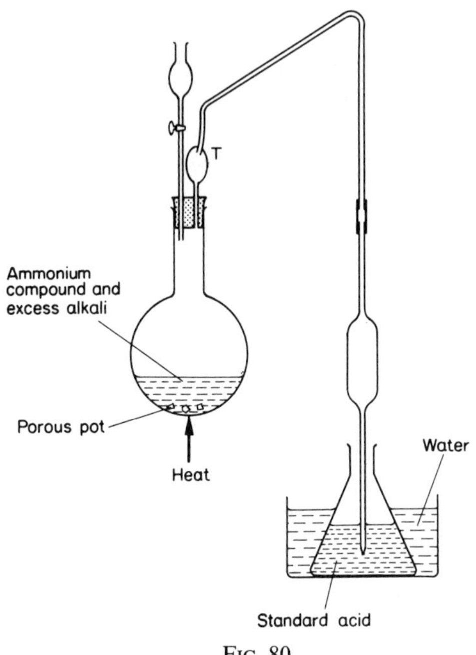

FIG. 80

Experiment 284 Determination of the degree of temporary hardness in water

Material: 0.1 M hydrochloric acid; temporarily hard water.

Prepare a temporarily hard water (if one is not available) by bubbling a rapid stream of carbon dioxide through calcium hydroxide solution (there is no need to wait until the solution is clear—any precipitate can be filtered off). Take 100 cm^3 of the solution, add a few drops of methyl orange and run in the standard hydrochloric acid until the yellow colour changes to orange. (If the water is known to be very hard because it comes from a limestone or chalk region, the volume taken for titration need only be 25 or 50 cm^3.)

$$Ca(HCO_3)_2 + 2HCl \rightarrow CaCl_2 + 2H_2O + 2CO_2$$

162 g   2 dm^3 1 M solution

Specimen results. Suppose a cm^3 of 0.104 M acid are required for 100 cm^3 of the temporarily hard water, then the mass of $Ca(HCO_3)_2$ in 100 cm^3 solution is

$$\frac{162 \times a \times 0.104}{2000 \times 1} \text{ g}$$

But 162 g of calcium hydrogencarbonate would form 100 g calcium carbonate if heated, so the mass of calcium carbonate from 100 cm^3 solution is

$$\frac{100 \times a \times 0.104}{2000 \times 1} \text{ g}$$

Hence the hardness can be calculated in parts per million.

Experiment 285 Determination of the relative molecular mass of amidosulphuric acid

Material: amidosulphuric (sulphamic) acid; 0.1 M sodium hydroxide.

Weigh accurately about 2.5 g of amidosulphuric acid. Dissolve it in water in a beaker, transfer the solution to a 250 cm^3 measuring flask and make up to the mark. Mix well to ensure uniform concentration.

Take 25 cm^3 of the solution by pipette, transfer to a conical flask, add two or three drops of methyl red or methyl orange and titrate against the sodium hydroxide. Repeat to obtain consistent readings.

$$NaOH + NH_2SO_2OH \rightarrow NH_2SO_2ONa + H_2O$$

Let the mass of amidosulphuric acid be 2.47 g. This mass in 250 cm^3 is equivalent to

$$4 \times 2.47 = 9.88 \text{ g dm}^{-3}$$

Let 25 cm^3 of solution require 25.9 cm^3 of 0.1 M alkali. Then the concentration of the amidosulphuric acid solution is

$$\frac{25.9 \times 0.1}{25} \text{ M}$$

$$\text{Molar mass} = \frac{\text{concentration in g dm}^{-3}}{\text{concentration in mol dm}^{-3}}$$

$$= \frac{9.88}{0.1036} = 95.4 \text{ g}$$

Theoretical value of the relative molecular mass of amidosulphuric acid = 97.1 See also Experiment 161, p. 145.

Experiment 286 Determination of the proportion of sodium carbonate and sodium hydrogencarbonate in a mixture

Material: mixture of A.R. quality sodium carbonate and sodium hydrogencarbonate; 0.1 M hydrochloric acid.

A solution of sodium carbonate has a pH of about 10 and turns phenolphthalein red. When an acid is run into the solution, the pH moves downwards and when it reaches pH 8, half the volume of acid required by the equation

$$Na_2CO_3 + 2HCl \rightarrow 2NaCl + CO_2 + H_2O$$

has been run in. At pH 8, the sodium carbonate has become sodium hydrogencarbonate and the phenolphthalein has become faintly pink. If methyl orange is then added and the titration continued, the end point is reached when a volume equal to that used previously has been added. The total titration for sodium carbonate can, therefore, be arrived at in two ways,

(i) Use phenolphthalein as indicator, note the volume of acid used and take double that figure, or

(ii) Use methyl orange as the sole indicator. See also Experiment 280, p. 306.

When acid is run into a solution containing sodium carbonate and sodium hydrogencarbonate, using phenolphthalein as indicator, only the carbonate is affected. Double the titration figure obtained is a measure of the concentration of sodium carbonate.

When methyl orange is then added and the titration completed to the end-point of the indicator, the acid used in this stage is a total of two volumes, that is the second half titration of the carbonate (which is equal to the phenolphthalein titration) and the acid used by the original hydrogencarbonate.

Let the volume of acid used with phenolphthalein as indicator be a cm^3 and the volume used after adding methyl orange as indicator be b cm^3. Then

$$\text{acid used by sodium carbonate} = 2 \times a \text{ cm}^3$$
$$\text{acid used by sodium hydrogencarbonate} = (b - a) \text{ cm}^3$$

Weigh accurately about 2 g of the mixture. Dissolve in distilled water and make up to 250 cm^3. If w g are taken, this is a concentration of $4w$ g dm^{-3}.

Take 25 cm^3 of the solution, add a drop of phenolphthalein, and titrate until the pink colour is just discharged. Then add two drops of methyl orange and continue the titration until the colour of the indicator is orange.

$$NaHCO_3 + HCl \rightarrow NaCl + H_2O + CO_2$$

If, using phenolphthalein, 25 cm^3 of solution needed 10 cm^3 of 0.1 M acid, the volume of acid needed for full neutralization of the carbonate = 20 cm^3.

$$\text{Concentration of the sodium carbonate} = \frac{20 \times 0.1}{25 \times 2} = 0.04 \text{ M}$$

If, after adding methyl orange, a further 15 cm^3 of acid were needed, the acid used by the original hydrogencarbonate = 15 − 10 cm^3 = 5 cm^3.

$$\text{Concentration of sodium hydrogencarbonate} = \frac{5 \times 0.1}{20} = 0.025 \text{ M}$$

$$\therefore \text{ sodium carbonate in the mixture} = \frac{0.04 \times 106}{4w} \times 100\%$$

$$\therefore \text{ sodium hydrogencarbonate in the mixture} = \frac{0.025 \times 84}{4w} \times 100\%$$

Experiment 287 Estimation of ammonia in an ammonium salt (formaldehyde method)

Material: 4% formaldehyde (methanal, made by diluting formalin ten times); 0.1 M sodium hydroxide; ammonium salt.

Weigh accurately about 1.5 g of the ammonium salt (say ammonium chloride), dissolve in water and make up to 250 cm^3. Neutralize about 100 cm^3 of the formaldehyde solution by adding a drop of phenolphthalein and then sodium hydroxide dropwise until the colour is faintly pink. This is called *neutral formaldehyde.*

Take 25 cm^3 of the ammonium salt solution and add 10 cm^3 of the neutral formaldehyde (by measuring cylinder). Allow the mixture to stand for two minutes. Add a drop of phenolphthalein. Titrate the liberated acid against the sodium hydroxide. Repeat to obtain consistent readings.

$$6HCHO + 4NH_4Cl \rightarrow (CH_2)_6N_4 + 6H_2O + 4HCl$$

If a g of ammonium chloride is dissolved in 250 cm^3 of solution, the concentration of ammonium chloride is $4a$ g dm^{-3}.

If 25 cm^3 of this solution give acid sufficient to neutralize b cm^3 of 0.1 M alkali, the solution is

$$\frac{b \times 0.1}{25} \text{ M}$$

Therefore, the solution contains

$$\frac{b \times 0.1 \times 18}{25} \text{ g dm}^{-3} \text{ of ammonium ion} = \frac{18b}{250} \text{ g dm}^{-3}$$

Also, percentage of ammonium ion in ammonium chloride

$$\frac{b \times 0.1 \times 18 \times 100}{25 \times 4a}$$

Expressed as ammonia, because $NH_3 \equiv NH_4^+$ and relative molecular mass of ammonia $= 17$,

$$\text{percentage of ammonia in ammonium chloride} = \frac{b \times 0.1 \times 17 \times 100}{25 \times 4a}$$

Experiment 288 Analysis of an indigestion tablet

Material: sodamint tablets; 0.1 M hydrochloric acid.

Sodamint tablets are mostly sodium hydrogencarbonate. Weigh out a tablet and crush it in 25 cm^3 of distilled water, add a few drops of methyl orange and titrate the solution with 0.1 M hydrochloric acid. Calculate the percentage of sodium hydrogencarbonate in the tablet.

$$NaHCO_3 + HCl \rightarrow NaCl + H_2O + CO_2$$

Experiment 289 Analysis of lemon squash

Material: lemon squash; 0.1 M sodium hydroxide.

When citric acid (2-hydroxypropane-1,2,3-tricarboxylic acid-1-water) is titrated with sodium hydroxide, the acid is two-thirds neutralized when phenolphthalein changes colour. Titrate a 25 cm^3 portion of the squash with 0.1 M alkali and calculate the concentration and hence the percentage m/V of citric acid in the squash.

$$H_3(C_6H_5O_7) \cdot H_2O + 2NaOH \rightarrow Na_2H(C_6H_5O_7) + 3H_2O$$

Experiment 290 Analysis of a toilet cleaner

Material: toilet cleaning crystals; 0.1 M sodium hydroxide.

The major constituent of many lavatory cleaners, e.g. Harpic, is the acid salt, sodium hydrogensulphate; this titration ascribes all the acidity as being due to this salt (a suitable concentration of solution to study is 12 g dm^{-3}). Titrate 25 cm^3 portions of 0.1 M sodium hydroxide with the 'Harpic' solution using phenolphthalein as the indicator. Calculate the concentration and hence the percentage of the salt in the cleaner.

$$NaHSO_4 + NaOH \rightarrow Na_2SO_4 + H_2O$$

Experiment 291 Analysis of a mixture of group IIA carbonates

Material: a suitable mixture to study can be made by dissolving

2 and 4 g respectively of calcium and barium carbonates in 1 dm^3 of 0.2 M hydrochloric acid (solution A); 0.1 M hydrochloric acid and 0.075 M sodium carbonate are also required.

Titrate 25 cm^3 of solution A with the standard sodium carbonate solution using methyl orange as the indicator and calculate the concentration of the excess acid in A.

Take 25 cm^3 of solution A, add 50 cm^3 of the standard sodium carbonate solution, heat the mixture to the boiling point and then allow it to cool slightly before filtering. Wash the precipitate thoroughly and add the washings to the filtrate; titrate the excess carbonate in the filtrate with the standard hydrochloric acid, using methyl orange as the indicator.

Calculate the concentration of the original acid in solution A and hence the masses of the two carbonates dissolved in 1 dm^3 of the original acid.

EXERCISES

1. *Material:* formaldehyde (methanal) solution, 4% solution; 1 M solutions of hydrochloric acid, ammonium chloride and sodium hydroxide.

   Investigate the reaction between ammonia and formaldehyde in aqueous solution. The following experiments are suggested.

   (*a*) Mix excess (75 cm^3) of the formaldehyde solution with 25 cm^3 of the ammonium chloride; add 50 cm^3 of the sodium hydroxide, and leave the liquid fifteen minutes before titrating with hydrochloric acid using phenolphthalein.

   (*b*) Mix excess (75 cm^3) of the ammonium chloride with 25 cm^3 of the formaldehyde solution, add 50 cm^3 of the sodium hydroxide, and leave for fifteen minutes before titrating with 1 M hydrochloric acid. Litmus may be used, but bromothymol blue will be found more satisfactory. Repeat the experiment but omit the formaldehyde.

   State briefly the conclusions at which you arrive. (Cambridge Schol.)

2. *Material:* 1 M ammonia; 0.1 M copper(II) sulphate; 0.025 M sulphuric acid; chloroform (trichloromethane).

   Determine the composition of the complex amminecopper(II) ion formed when excess ammonia is added to a copper sulphate solution. Mix 25 cm^3 ammonia with 25 cm^3 0.1 M copper sulphate. Shake the resulting liquid with 75 cm^3 of chloroform for ten minutes. Separate off 50 cm^3 of the chloroform and determine the amount of ammonia in it by titration with the standard acid provided. Assume that free ammonia distributes itself between the aqueous solution and chloroform in the ratio

$$\frac{\text{concentration in aqueous solution}}{\text{concentration in chloroform}} = 26$$

   Note: Use a measuring cylinder or a pipette with a suction device for the chloroform. (Cambridge Schol.)

3. *Material:* impure calcium carbonate; 1 M hydrochloric acid; 0.1 M sodium hydroxide.

The substance A is a mixture of calcium carbonate and silicon dioxide (silicon(IV) oxide). Find its percentage composition by mass.

Weigh out accurately about 1 g of the mixture and transfer it to a graduated 250 cm^3 flask. Add 50 cm^3 of hydrochloric acid. When the reaction is complete make the liquid up to the mark with distilled water, shake well to ensure thorough mixing and titrate 25 cm^3 portions with sodium hydroxide solution using methyl orange as indicator. (J.)

4. *Material:* sodium hydrogensulphate; sodium hydrogencarbonate.

Determine the purity of the laboratory specimen of sodium hydrogen-sulphate. You are supplied with pure sodium hydrogencarbonate.

5. *Material:* sodium hydroxide and ammonium chloride solution; 0.1 M hydrochloric acid.

You are provided with a solution containing ammonium chloride and sodium hydroxide in such proportions that there is less than one mole of ammonium chloride for every one of sodium hydroxide. Find the concentration of each (in g dm^{-3}) using the standard acid.

6. *Material:* 0.05 M sodium carbonate.

Estimate the concentration of the barium chloride solution in the reagent bottles. You are supplied with standard sodium carbonate solution. Phenolphthalein will show the end-point of the reaction:

$$BaCl_2 + Na_2CO_3 \rightarrow BaCO_3\downarrow + 2NaCl$$

7. *Determination of distribution ratio of succinic acid between water and diethyl ether.*

Material: succinic (butanedioic) acid, diethyl ether (ethoxyethane), 0.5 M sodium hydroxide; 0.1 M sodium hydroxide.

Grind a little succinic acid in a mortar and weigh out quantities of 1.5 g, 1.0 g and 0.5 g (these weighings need only be roughly done). Pour into a separating funnel about 40 cm^3 of water and add the 1 g of succinic acid. Add about 40 cm^3 of ether, shake until the acid has dissolved and allow to stand for a few minutes. Run off nearly all the water layer and titrate 25 cm^3 of this against 0.5 M sodium hydroxide using phenolphthalein as an indicator (about 30 cm^3 will be required). Discard the boundary portion. Take 25 cm^3 of the ether layer by means of a pipette filler (fairly accurate results can be obtained by using a measuring cylinder) and titrate this against the 0.1 M sodium hydroxide. Shake well after each addition of alkali. (The volume of the 0.1 M solution required will be approximately the same as the amount of 0.5 M solution required for the water layer.)

Obtain the distribution ratio:

$$K = \frac{5 \times [\text{vol. of 0.5 M NaOH for 25 cm}^3 \text{ aqueous layer}]}{\text{vol. of 0.1 M NaOH for 25 cm}^3 \text{ ether layer}}$$

Repeat the process with the other quantities of the succinic acid.

8. *Quantitative verification of solubility product* (*F. G. Mee*).

Material: 0.1 M sodium hydroxide; 0.1 M hydrochloric acid.

Calcium hydroxide should have a smaller solubility in a solution of sodium hydroxide than in water, and it is the object of this experiment to verify that the decrease of solubility that occurs is that which would be expected from the equilibrium constant for the electrolytic dissociation.

$$Ca(OH)_2 \rightleftharpoons Ca^{2+} + 2OH^-$$

Extra OH^- ions are provided from some other source—sodium hydroxide in this case—the concentration of Ca^{2+} ions decreases in such a way that the product of it and the square of the concentration of OH^- ions remains a constant, if the solution is saturated.

To verify this solutions of calcium hydroxide are made up in (*a*) water, (*b*) 0.025 M NaOH, (*c*) 0.05 M NaOH, (*d*) 0.1 M NaOH. The alkali should be as free as possible from carbonate and the solutions can be made by adding a quantity of calcium hydroxide to each of the sodium hydroxide solutions and allowing to stand, with shaking at intervals. Each of these solutions is titrated with 0.1 M HCl, which will, in each case, give a measure of the total OH^- ion concentration present. The OH^- ion concentration due to the sodium hydroxide is known, assuming this to be fully ionized also. Subtraction will yield the concentration of OH^- ions due to dissolved calcium hydroxide, and the concentration Ca^{2+} ions is exactly half of this. Titration thus gives all required to calculate the solubility product.

A typical set of results is the following:—

25 cm³ of each of the four solutions were titrated with 0.1 M HCl, using phenolphthalein indicator, with results shown in the second column of the table shown below.

The constancy of the product is striking. The effect of decreasing the solubility due to the sodium hydroxide appears during the experiment, before the detailed calculation is made. Calcium hydroxide in water gave a titre 11.9 cm³; 0.05 M NaOH alone would give 12.5 cm³. Yet calcium hydroxide in 0.05 M NaOH gives but 17.7 cm³, showing that the solubility has been reduced to about one-half. The result with the 0.1 M

Solution in	*Titration* (cm³)	*Ionic concentration* (mol dm^{-3})				*Solubility product*
		Total $[OH^-]$ present	$[OH^-]$ due to NaOH	$[OH^-]$ due to $Ca(OH)_2$	$[Ca^{2+}]$	$[Ca^{2+}] \times [OH^-]^2$ $(\times 10^5)$ mol³ dm^{-9}
Water	11.9	0.048	—	0.048	0.0237	5.4
0.025 M NaOH	14.5	0.058	0.025	0.033	0.0165	5.6
0.05 M NaOH	17.7	0.071	0.050	0.021	0.0105	5.3
0.1 M NaOH	27.3	0.109	0.100	0.009	0.0045	5.4

solution cannot be expected to be very accurate, since at this concentration the calcium hydroxide contributes so very little to the total OH^- ion concentration. If more than four results are desired, they should therefore be taken with sodium hydroxide solutions between the above values, and not more concentrated than 0.1 M.

(From the Science Master's Book published by Messrs. John Murray.)

34
Potassium Permanganate

Potassium permanganate (manganate(VII)), $KMnO_4$, is a powerful oxidizing agent and is used for the estimation of many reducing agents, especially compounds of iron, and oxalic (ethanedioic) acid and its salts.

CONDITION OF USE OF POTASSIUM PERMANGANATE

In acid solution two moles of potassium permanganate react with, for example, ten moles of iron(II) sulphate in the presence of excess dilute sulphuric acid.

$$2KMnO_4 + 8H_2SO_4 + 10FeSO_4 \rightarrow K_2SO_4 + 2MnSO_4 + 5Fe_2(SO_4)_3 + 8H_2O$$

or, in ionic terms,

$$2MnO_4^- + 16H^+ + 10Fe^{2+} \rightarrow 2Mn^{2+} + 10Fe^{3+} + 8H_2O$$

These equations can be considered from the point of view of the oxidizing agent, the potassium permanganate, which gains electrons to be

$$MnO_4^- + 8H^+ + 5e^- \rightarrow Mn^{2+} + 4H_2O$$

and of the reducing agent, the iron(II) salt, which supplies the electrons

$$Fe^{2+} \rightarrow Fe^{3+} + e^-$$

In alkaline solution, potassium permanganate by a different reaction yields manganese(IV) oxide as a brown precipitate.

$$MnO_4^- + 2H_2O + 3e^- \rightarrow MnO_2\downarrow + 4OH^-$$

Consideration of these facts makes it clear at once that for quantitative work, potassium permanganate must be used in conditions which exclude entirely one of these reactions. In practice, potassium permanganate is almost always used to titrate solutions which are sufficiently acidic to exclude altogether the formation of manganese(IV) oxide.

Of the three mineral acids available, only sulphuric acid is suitable for use with potassium permanganate because the latter reacts with hydrochloric acid,

$$2KMnO_4 + 16HCl \rightarrow 2KCl + 2MnCl_2 + 8H_2O + 5Cl_2$$

while nitric acid is itself a vigorous oxidizing agent and might interfere with the oxidizing action of the permanganate. The solution which is in process of titration with potassium permanganate must be sufficiently acidic to prevent the formation of any precipitate of manganese(IV) oxide. As dilute sulphuric acid is usually 1 M and potassium permanganate solution about 0.02 M, a bulk of acid equal to one half that of the solution to be titrated will usually provide a sufficient excess of acid.

INDICATOR AND END-POINT

As the titration proceeds, manganese(II) ions accumulate, but at the dilution used give a colourless solution. As soon as potassium permanganate is in excess, the solution becomes pink and therefore potassium permanganate acts as its own indicator, the end-point being the first permanent pink coloration.

STANDARD SOLUTION OF POTASSIUM PERMANGANATE

The equation derived in the preceding paragraph is

$$MnO_4^- + 8H^+ + 5e^- \rightarrow Mn^{2+} + 4H_2O$$

Therefore solutions of potassium permanganate are usually made having a concentration of one-fifth of 0.1 M, i.e. 0.02 M; 1 dm^3 of 0.02 M potassium permanganate solution contains 3.16 g of the salt.

An accurately 0.02 M solution of potassium permanganate cannot be made up directly from the solid because this may be reduced by organic matter from the atmosphere and so rendered impure; further, organic matter present in the water in which the salt is dissolved may reduce it. It is therefore desirable to make up a solution slightly more concentrated (say about 3.25 g dm^{-3}) and allow it to stand several days. It may then be standardized by methods described below. A potassium permanganate solution slowly decomposes and should be protected from light and standardized again at intervals. It may be standardized by a pure iron(II) salt or a pure oxalate (ethanedioate).

Experiment 292 Standardization of potassium permanganate solution by an iron(II) salt

Material: approximately 0.02 M $KMnO_4$; ammonium iron(II) sulphate.

The oxidation of an iron(II) salt by potassium permanganate may be expressed ionically thus:

$$Fe^{2+} \rightarrow Fe^{3+} + e^-$$

or using iron(II) sulphate, the oxidation may be written in molecular terms:

$$10FeSO_4 + 2KMnO_4 + 8H_2SO_4 \rightarrow 5Fe_2(SO_4)_3 + K_2SO_4 + 2MnSO_4 + 8H_2O$$

Iron(II) sulphate crystals, $FeSO_4 \cdot 7H_2O$, cannot be used for standardization because they are rendered impure by efflorescence and by atmospheric oxidation to form a brown basic sulphate as a result of a reaction of the type

$$12FeSO_4 + 3O_2 + 6H_2O \rightarrow 4[Fe(OH)_3 \cdot Fe_2(SO_4)_3].$$

The salt, ammonium iron(II) sulphate, $Fe(NH_4)_2(SO_4)_2 \cdot 6H_2O$, is free from these disadvantages and can be obtained in a high state of purity. (It is prepared by dissolving iron(II) and ammonium sulphates in the calculated quantities in hot water containing sulphuric acid and allowing the solution to crystallize.) In solution, it reacts freely as iron(II) ions, sulphate ions and ammonium ions. Only the first named react with the permanganate.

One mole of ammonium iron(II) sulphate contains one mole of iron(II) ions and the partial equation given above shows that this loses one mole of electrons, thus the mass of the salt to be taken for 250 cm^3 0.1 M solution is $0.25 \times 0.1 \times 392$ g, i.e. 9.8 g.

This amount should be weighed out in a weighing bottle. To prevent oxidation, the salt should now be dissolved in diluted sulphuric acid, which has been boiled to remove air and then cooled. Make up 250 cm^3 of solution in this way: it is then accurately 0.1 M.

Measure out 25 cm^3 of the solution into a conical flask. Add about 15 cm^3 of dilute sulphuric acid and titrate with the potassium permanganate solution from a burette until the first permanent pink coloration is observed. No brown precipitate should appear. Repeat the titration twice.

Suppose that 25 cm^3 of the iron(II) solution is oxidized by an average volume of 23.4 cm^3 of the permanganate solution. Then, because of the iron(II) salt solution is exactly 0.1 M the concentration of the permanganate solution must be $\frac{25 \times 0.1}{23.4 \times 5}$ M or 0.0214 M, and hence contains $0.0214 \times 158 = 3.38$ g dm^{-3}.

Experiment 293 Standardization of potassium permanganate solution by sodium oxalate

Material: 0.02 M potassium permanganate; sodium oxalate (ethanedioate).

Care! Oxalates are poisonous so use a pipette and filler for measuring out the solution.

An acidified solution of an oxalate is, for purposes of titration with potassium permanganate solution, equivalent to a solution of oxalic acid itself.

$$Na_2C_2O_4 \text{ is } 2Na^+ + C_2O_4^{2-}$$
$$C_2O_4^{2-} + \underset{\text{from acid}}{2H^+} \rightleftharpoons H_2C_2O_4$$

Sodium oxalate is used for standardization because it can be obtained in a pure state more easily than can oxalic acid.

The oxidation of an oxalate is represented essentially by the equation

$$C_2O_4^{2-} \rightarrow 2CO_2 + 2e^-$$

The molecular equation is

$$2KMnO_4 + 5H_2C_2O_4 + 3H_2SO_4 \rightarrow K_2SO_4 + 2MnSO_4 + 8H_2O + 10CO_2$$

250 cm^3 0.05 M sodium oxalate solution will contain $0.25 \times 0.05 \times 134 = 1.675$ g of the salt. Weigh out this amount and make up to 250 cm^3 of solution with distilled water.

Potassium permanganate does not oxidize oxalates in cold solution; a temperature of 70 °C is necessary to cause the reaction to occur rapidly.

To 25 cm^3 of the 0.05 M oxalate solution in a conical flask add about 15 cm^3 of dilute sulphuric acid and heat the mixture to 70 °C. This temperature can be estimated accurately enough by testing with the palm of the hand. When the bottom of the flask is just too hot to hold, the temperature of the liquid is approximately correct. Titrate with potassium permanganate, heating again as the liquid cools, till a permanent pink coloration is observed. Manganese(II) sulphate formed during the reaction has a catalytic effect, but side reactions are prevented if the temperature is maintained at about 60 °C or more. Repeat the titration twice with further portions of 25 cm^3 of the oxalate.

Calculate the concentration of the permanganate solution.

Experiment 294 Determination of the number of molecules of water of crystallization in a molecule of hydrated iron(II) sulphate

Material: 0.02 M potassium permanganate; iron(II) sulphate crystals.

$$10(FeSO_4 \cdot xH_2O) + 2KMnO_4 + 8H_2SO_4 \rightarrow 5Fe_2(SO_4)_3 + K_2SO_4 + 2MnSO_4 + (10x+8)H_2O$$

Relative molecular mass of hydrated iron(II) sulphate crystals is $152 + 18x$ where x is the number of molecules of water of crystallization for each iron(II) and sulphate ion.

For 250 cm^3 0.1 M solution use $(152 + 18x) \times 0.25 \times 0.1$ g of iron(II) sulphate crystals, i.e. $(3.8 + 0.45x)$ g.

x will probably be between 1 and 10, i.e. the mass of crystals should be between 4.3 g and 8.3 g. Weigh out about 7 g of crystals and make up to 250 cm^3 of acidified solution, as described for ammonium iron(II) sulphate (p. 321). Titrate against 0.02 M potassium permanganate solution.

Suppose a g of crystals were made up to 250 cm^3 of solution. Hence the concentration (m M) of the iron(II) solution and the mass of crystals in solution are related by the equation

$$a = (3.8+0.45x)\frac{m}{0.1}$$

Solve for x.

Experiment 295 Estimation of the percentage by mass of iron in iron wire

Material: 0.02 M potassium permanganate; iron wire; flask and Bunsen valve.

A suitable mass of iron is converted to iron(II) sulphate solution, which is then titrated with potassium permanganate.

To obtain 250 cm^3 0.1 M iron(II) sulphate solution $56 \times 0.25 \times 0.1$ g pure iron (i.e. 1.4 g) would be required. Weigh accurately 1.3 to 1.5 g of the wire.

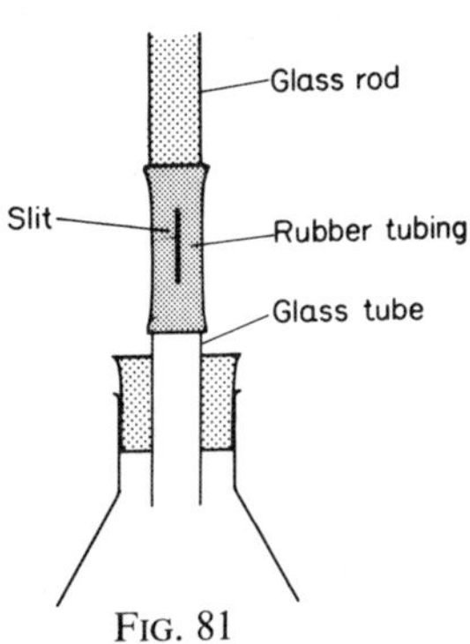

FIG. 81

The iron is treated with dilute sulphuric acid, heat being necessary to secure a sufficiently rapid reaction. Conditions must be such that iron(III) compounds are not formed. While the reaction is actually proceeding, the hydrogen generated acts as a reducing agent, but when solution is complete and the liquid is still hot, entry of air might bring about oxidation. To prevent this, the Bunsen valve is used. It consists of a piece of rubber tubing, carrying a longitudinal slit for part of its length, closed at the upper end by glass rod and connected by a stopper and glass tubing to the reaction flask (Figure 81). The slit opens outwards only; it allows the escape of hydrogen, but when the flask is cooling air cannot be drawn in because the valve closes. When the solution is cool, the risk of oxidation is negligible.

Fit a Bunsen valve to a small flask and in it warm the weighed iron wire with dilute sulphuric acid till all the iron is dissolved.

$$Fe + H_2SO_4 \rightarrow FeSO_4 + H_2$$

The solution will probably be cloudy. This effect is caused by precipitation of fine particles of carbon, an impurity in the iron. They will not interfere. When the reaction is complete and the flask has cooled, transfer the solution to a 250 cm^3 flask, wash out the reaction flask several times with air-free distilled water and add the washings to the 250 cm^3 flask. Make up to the mark with air-free distilled water and shake well. Titrate against the potassium permanganate.

Calculate the concentration (m M) of the iron(II) solution and hence the mass of iron. From the latter the percentage purity can be evaluated.

Experiment 296 Estimation of iron in ammonium iron(III) sulphate

Material: 0.02 M potassium permanganate; ammonium iron(III) sulphate (iron alum); zinc.

Estimation of iron in the iron(III) state is carried out by first reducing it quantitatively to the iron(II) state and then titrating the resulting iron(II) solution with potassium permanganate. The relative molecular mass of the alum is 482, corresponding to a formula $FeNH_4(SO_4)_2 \cdot 12H_2O$. Thus for 250 cm^3 of 0.1 M ammonium iron(III) sulphate solution $482 \times 0.25 \times 0.1$ g, i.e. 12.05 g, of alum are required. Weigh out about 12 g of the alum and make up this mass to 250 cm^3 of solution, acidifying with dilute sulphuric acid to prevent hydrolysis.

Measure out 25 cm^3 of the alum solution in a conical flask, add dilute sulphuric acid and several pieces of zinc. (The zinc must be free from iron.) Allow effervescence to proceed for forty minutes (set up two more similar reduction flasks in the meantime). The mixture may then be tested for completion of the reduction.

$$Zn + 2Fe^{3+} \rightarrow Zn^{2+} + 2Fe^{2+}$$

Place on a white tile a drop of solution of potassium thiocyanate, KSCN. Dip a glass rod into the reduction flask and allow a drop of liquid from it to mix with the drop of solution on the tile. If a blood-red tinge appears, an iron(III) salt still remains and reduction is not complete.

$$Fe^{3+} + SCN^- \rightarrow FeSCN^{2+}$$

Continue the reduction until the test is negative, i.e. the mixture remains colourless or is only very faintly coloured.

Cool the contents of the flask, filter through glass-wool (for speed), and wash the zinc and flask well with air-free distilled water. Add the washings to the main solution and titrate against potassium permanganate. (Acidify further if necessary.) Repeat with the other two flasks, but delay the test for completion until reduction has proceeded for at least as long as was necessary before. This prevents loss of solution in repeated testing.

Calculate the concentration and mass of iron in the alum solution. From the latter result the percentage of iron in the alum can be calculated.

Experiment 297 Other methods of reducing the iron(III) salt: use of zinc amalgam

The method of reduction given above is rather slow. A more rapid result is achieved by using zinc amalgam. To prepare it, weigh out 200 g of mercury and 5 g of zinc foil. Put the mercury into a dish and heat it on the steam bath. Add a few drops of dilute sulphuric acid and stir into the mercury a thin strip of the zinc until all the zinc has dissolved. If the amalgam solidifies when cold, warm it and add a little more mercury. Add a small quantity of dilute sulphuric acid to the amalgam in a stoppered conical flask. Run in 25 cm^3 of the iron(III) solution, insert the stopper and shake gently for a few minutes. No gas is evolved. The reduction should be complete in about five minutes. Decant off the reduced solution (test as described in Experiment 296) into a flask, wash the amalgam several times with small quantities of distilled water, adding the washings to the reduced solution. Titrate against potassium permanganate. The amalgam may now be used to reduce a further portion of 25 cm^3 of the iron(III) solution.

Experiment 298 Estimation of oxalic acid and one of its soluble salts in a mixture of the two

Material: 0.02 M potassium permanganate; 0.1 M sodium hydroxide; mixture of oxalic (ethanedioic) acid and sodium oxalate (ethanedioate).

Care! Oxalates are poisonous so use a pipette and filler for measuring out the solution.

The oxalic acid is determined separately by titration with 0.1 M sodium hydroxide; the total oxalate is determined by titration with 0.02 M potassium permanganate.

$$2MnO_4^- + 5C_2O_4^{2-} + 16H^+ \rightarrow 2Mn^{2+} + 8H_2O + 10CO_2$$

Relative molecular mass of oxalic acid is 90.

For 250 cm^3 0.05 M solution $90 \times 0.25 \times 0.05$ g of anhydrous oxalic acid are needed, i.e. 1.125 g. Allowing for the fact that the given mixture will probably contain hydrated crystals of the acid ($H_2C_2O_4 \cdot 2H_2O$) and for the higher relative molecular mass of sodium oxalate ($Na_2C_2O_4$ = 134) about 2 g of the mixture will probably provide 250 cm^3 of a suitable solution. Weigh out accurately about 2 g of the mixture and make up to 250 cm^3 of solution.

(*a*) Titrate 25 cm^3 of the solution with 0.1 M sodium hydroxide using phenolphthalein as indicator.

(*b*) Titrate 25 cm^3 of the solution, acidified with dilute sulphuric acid and heated to 70 °C with 0.02 M potassium permanganate.

The sodium hydroxide solution reacted with the free oxalic acid only; the potassium permanganate solution reacted with the oxalic acid and sodium oxalate.

Both oxalic acid and sodium oxalate contain one oxalate ion in a molecule so the volume of potassium permanganate can be separated into the portions reacting with oxalic acid and sodium oxalate.

Suppose a g of the mixture were made up to 250 cm^3 solution and 25 cm^3 of this needed b cm^3 of 0.1 M sodium hydroxide and d cm^3 of 0.02 M potassium permanganate for titration.

From the titre in part (*a*) calculate the concentration (in mol dm^{-3}) and hence the mass of anhydrous oxalic acid in 250 cm^3 solution.

From the titre in part (*b*) calculate the concentration (in mol dm^{-3}) of the solution with respect to oxalate ions and subtract the concentration of oxalic acid. Hence calculate the mass of sodium oxalate in 250 cm^3 solution and the percentages of each oxalate in the mixture.

Experiment 299 Estimation of hydrogen peroxide

Material: 10- or 20-volume hydrogen peroxide; 0.02 M potassium permanganate.

An acidified solution of hydrogen peroxide reacts with potassium permanganate,[1] liberating oxygen, and can be estimated by this reaction.

$$5H_2O_2 + 2KMnO_4 + 3H_2SO_4 \rightarrow K_2SO_4 + 2MnSO_4 + 8H_2O + 5O_2$$

$$H_2O_2 \rightarrow O_2 + 2H^+ + 2e^-$$

The 10-volume and 20-volume solutions of hydrogen peroxide are both more concentrated than 0.05 M, but, as hydrogen

[1] Mannitol and other carbohydrates are often present in hydrogen peroxide solution to retard its decomposition; they will react giving a high reading.

peroxide decomposes slowly at ordinary temperature, the appropriate ratio of dilution for a given sample must be found by trial.

By means of a pipette measure out 1 cm^3 of the given solution of hydrogen peroxide into a conical flask, acidify it with dilute sulphuric acid and run in potassium permanganate from a burette 0.5 cm^3 at a time till the mixture is pink. No attempt should be made to obtain an accurate end-point. If d cm^3 of potassium permanganate solution are needed, the hydrogen peroxide solution must be diluted with distilled water so that the original and diluted volumes are in the ratio of $1:d$ approximately. Hence dilute a cm^3 of the hydrogen peroxide to 250 cm^3 accurately.

Repeat the titration with 25 cm^3 of the diluted hydrogen peroxide solution, acidify with dilute acid and obtain an accurate end-point.

Suppose 25 cm^3 of the diluted solution were titrated by p cm^3 0.02 M potassium permanganate.

$$2KMnO_4 \equiv 5H_2O_2$$

2 dm^3 1 M or 100 dm^3 0.02 M solution ≡ 5×34 g

Mass of hydrogen peroxide in 25 cm^3 diluted solution

$$= 5 \times 34 \times \frac{p}{100000} \text{ g}$$

∴ Mass of hydrogen peroxide in 1 dm^3 diluted solution

$$= 5 \times 34 \times \frac{p}{100000} \times \frac{1000}{25} \text{ g}$$

and hence the concentration of the original solution

$$= \frac{5 \times 34 \times p}{1000 \times 25} \times \frac{250}{a} \text{ g dm}^{-3}$$

The volume strength is calculated using the equation for thermal or catalytic decomposition,

$$2H_2O_2 \rightarrow 2H_2O + O_2$$

If the 2×34 g of hydrogen peroxide are present in 1 dm^3 of solution then it is a 22.4-volume solution. Calculate the volume strength of the original hydrogen peroxide.

Experiment 300 Estimation of the purity of commercial sodium nitrite

Material: sodium nitrite (nitrate(III)); 0.02 M potassium permanganate.

An acidified solution of sodium nitrite is oxidized by potassium permanganate to nitrate (nitrate(V)).

$$5NaNO_2 + 2KMnO_4 + 3H_2SO_4 \rightarrow 5NaNO_3 + K_2SO_4 + 2MnSO_4 + 3H_2O$$

Relative molecular mass of sodium nitrite is 69.

$$NO_2^- + H_2O \rightarrow NO_3^- + 2H^+ + 2e^-$$

The nitrite solution cannot be titrated with permanganate solution from a burette in the usual way, because, as soon as it is acidified, the nitrous acid formed begins to decompose.

$$2NaNO_2 + H_2SO_4 \rightarrow Na_2SO_4 + 2HNO_2$$
$$3HNO_2 \rightarrow HNO_3 + H_2O + 2NO$$

The nitrite solution is placed in the burette and is added, *slowly* and with constant stirring, to an acidified solution of potassium permanganate.

For 250 cm^3 0.05 M sodium nitrite solution, $69 \times 0.25 \times 0.05$ or 0.86 g would be required. To allow for impurity make up a solution containing about 1 g of sodium nitrite in 250 cm^3.

Pipette 25 cm^3 0.02 M potassium permanganate solution into a beaker, acidify with 15 cm^3 of dilute sulphuric acid, and from a burette add the sodium nitrite solution slowly and with continual stirring till the permanganate colour is just discharged. Repeat the titration to obtain two concordant results.

Suppose a g of sodium nitrite were used and that the titre was b cm^3.

Calculate the concentration in mol dm^{-3} and then the mass of pure sodium nitrate used.

Hence the percentage purity of sodium nitrite

$$= 5 \times 69 \times \frac{25}{100\,000} \times \frac{250}{b} \times \frac{100}{a}$$

Experiment 301 Determination of the number of molecules of water of crystallization in ammonium iron(II) sulphate crystals

Material: ammonium iron(II) sulphate; 0.02 M potassium permanganate.

$$10[Fe(NH_4)_2(SO_4)_2 \cdot xH_2O] \equiv 2KMnO_4$$

Making a reasonable assumption for a value of x, calculate the approximate mass of ammonium iron(II) sulphate solution needed for 250 cm^3 of 0.1 M solution. Make up this solution in dilute sulphuric acid (p. 321) and titrate with potassium permanganate.

Experiment 302 Analysis of a mixture of potassium sulphate and potassium permanganate

Material: mixture of potassium sulphate and potassium permanganate; 0.1 M ammonium iron(II) sulphate.

Make up the standard solution of ammonium iron(II) sulphate (p. 321) in dilute sulphuric acid. Assuming that the proportion of potassium permanganate in the given mixture is about 50 %, make up a solution of accurately known concentration and about 0.02 M. Titrate the solution with it: the potassium sulphate is not affected.

$$10[Fe(NH_4)_2(SO_4)_2 \cdot 6H_2O] \equiv 2KMnO_4$$

Experiment 303 Determination of the solubility of ammonium oxalate in water at room temperature

Material: ammonium oxalate (ethanedioate); 0.02 M potassium permanganate.

Prepare about 100 cm^3 of a saturated solution of ammonium oxalate at room temperature by heating the salt with distilled water at about 50 °C, adding so much of the salt that a small sample cooled in a test tube yields crystals. Cool the whole solution under the tap, determine its temperature and filter through dry glass-wool into a dry flask. Titrate 1 cm^3 of this filtrate, using the method of Experiment 299 (p. 326), with the potassium permanganate. On this evidence, dilute a known volume of the saturated solution to 250 cm^3 in a measuring flask to give approximately a 0.05 M solution.

Titrate portions of 25 cm^3 with 0.02 M potassium permanganate solution:

$$5(NH_4)_2C_2O_4 \equiv 2KMnO_4 \text{ (see p. 322)}$$
$$5 \times 124 \text{ g} \equiv 100 \text{ dm}^3 \text{ 0.02 M}$$

Experiment 304 A study of potassium hydrogenoxalate

Material: potassium hydrogenoxalate (hydrogenethanedioate) solution 8 g dm^{-3}; 0.1 M sodium hydroxide; 0.02 M potassium permanganate.

Care! Oxalates are poisonous.

Titrate 25 cm^3 portions of the oxalate solution with the alkali using phenolphthalein as the indicator. Continue with this portion of the oxalate solution, add 100 cm^3 of dilute sulphuric acid and after warming the solution to 70 °C titrate rapidly with the permanganate.

If the formula of the oxalate is $K_aH_b(C_2O_4)_c \cdot dH_2O$ calculate a, b, c and d. See also Experiment 298 (p. 325).

Experiment 305 Determine the percentage of manganese(IV) oxide in a sample of pyrolusite

Material: pyrolusite; 0.25 M oxalic acid (*Care!*); 0.02 M potassium permanganate. (See Experiment 308 for alternative method.)

Assume that manganese(IV) oxide reacts with oxalic acid in sulphuric acid solution according to the equation

$$MnO_2 + H_2C_2O_4 + H_2SO_4 \rightarrow MnSO_4 + 2H_2O + 2CO_2$$

Make up 250 cm^3 of a solution of oxalic acid which is about 0.25 M, i.e. 3.15 g dm^{-3} of crystals.

Transfer 100 cm^3 of it (accurately measured by pipette, using a pipette filler) to a conical flask and add 25 cm^3 dilute sulphuric acid. Weigh a weighing-bottle containing about 1.1 g of the powdered pyrolusite, add the powder to the acidified oxalic acid solution and obtain the mass of the powder by difference. Boil the mixture gently (with a funnel in the neck of the flask) until the remaining solid particles (silica) are white. Transfer the liquid after cooling to a 250 cm^3 measuring flask, wash the conical flask and funnel and add the washings to the bulk of the solution. Make up to the mark. Titrate with the potassium permanganate.

Dilute 100 cm^3 of the original oxalic acid to 250 cm^3 and titrate 25 cm^3 portions with permanganate.

The smaller volume of permanganate solution needed in the first case is due to oxidation of some of the oxalic acid by the manganese(IV) oxide. Both solutions were diluted in the same ratio of 100:250 so that the difference between the *total* volumes of potassium permanganate needed for titration measures the oxidizing action of the manganese(IV) oxide.

$$5MnO_2 \equiv 2KMnO_4$$

$$5(55+32)\ g \equiv 100\ dm^3\ 0.02\ M$$

EXERCISES

1. *Material:* solution of formic (methanoic) acid 1–2 g dm^{-3}; 0.02 M potassium permanganate; 0.1 M sodium hydroxide.

(*a*) Titrate the solution of formic acid (HCOOH) with the standard sodium hydroxide solution, and thereby determine its concentration in mol dm^{-3}.

(*b*) Add to 25 cm^3 of the formic acid solution an excess of sodium carbonate solution, heat almost to boiling and titrate while hot with the standard potassium permanganate solution until the clear liquid above the precipitate is coloured pink.

(*c*) Titrate 25 cm^3 of the solution of formic acid, heated to boiling but without neutralization with the sodium carbonate, with the potassium permanganate solution.

(*d*) Interpret your results as far as you can. (Oxford Schol.)

2. *Material:* 0.02 M potassium permanganate; 0.1 M ammonium iron(II) sulphate solution; 0.05 M hydrogen peroxide.

Check the relationship between the permanganate and the iron(II) salt by titrating 25 cm^3 of the latter with the former. Titrate a mixture of 25 cm^3 of the iron(II) salt and 10 cm^3 of the peroxide with the permanganate. Repeat the previous step using next 20 cm^3, then 30 cm^3 and finally 40 cm^3 of peroxide. Dilute sulphuric acid should be added as usual for each titration.

Plot a graph of the titres against the volume of peroxide added to the permanganate (0 to 40 cm^3). Comment upon the shape of the graph and deduce what you can.

35
Potassium Dichromate

Potassium dichromate (dichromate(VI)) is used as an oxidizing agent in reactions very similar to those described in the last chapter in which potassium permanganate was employed, especially for titration of iron(II) salts.

It will be remembered that potassium permanganate is not used with chlorides because (especially in the presence of iron(III) salts) it may be used up in oxidizing hydrochloric acid. Potassium dichromate is free from this disadvantage and may be used in the presence of hydrochloric acid or its salts. The great disadvantage of potassium dichromate is that the colour change is difficult to judge quantitatively unless an indicator is added. The chromium(III) salt produced is deep green in colour and effectively masks the colour of any excess of potassium dichromate, which cannot act (like potassium permanganate) as its own indicator.

STANDARD POTASSIUM DICHROMATE SOLUTION

In acid solutions each mole of potassium dichromate will react with six moles of electrons

$$Cr_2O_7^{2-} + 14H^+ + 6e^- \rightarrow 2Cr^{3+} + 7H_2O$$

or, in the molecular form with, for example, iron(II) chloride,

$$K_2Cr_2O_7 + 14HCl + 6FeCl_2 \rightarrow 2KCl + 2CrCl_3 + 6FeCl_3 + 7H_2O$$

Relative molecular mass of potassium dichromate is 294.

Hence for 1 dm^3 0.0167 M solution the mass of potassium dichromate is 294×0.0167 g $= 4.90$ g.

The solution may be made up directly from the pure salt if it is first melted in a porcelain dish, so that no water remains in it, and then ground to powder after cooling.

INDICATOR

The indicator necessary is diphenylamine, and this compound, which is used in solution in concentrated sulphuric acid, has no influence on the colour of the mixture, but, when oxidized by a slight excess of potassium dichromate, it produces an intensely coloured blue compound. It is necessary, however, to prevent iron(III) ions, which are formed in the course of titration of an iron(II) salt, from oxidizing the diphenylamine prematurely, i.e. before potassium dichromate is in excess, and for this purpose phosphoric acid is used. The iron(III) ions are then taken up, as fast as they are produced by oxidation, into an iron(III) phosphate complex which is almost undissociated, and the iron(III) ions are prevented from oxidizing the diphenylamine.

The titration is carried out in the following way: make a solution of 1 g of diphenylamine in 50 cm^3 of concentrated sulphuric acid. Dilute some syrupy phosphoric acid with twice its bulk of water. To 25 cm^3 of the solution to be titrated add one drop of the diphenylamine solution and 5 cm^3 of the diluted phosphoric acid. Titrate with potassium dichromate solution till the mixture (which is deep green in colour) just turns blue.

A superior indicator is N-phenylanthranilic (N-phenyl-2-aminobenzoic) acid, which changes from a green coloration to a violet-red coloration on adding excess of potassium dichromate solution. The indicator is prepared by dissolving 0.5 g of N-phenylanthranilic acid in 10 cm^3 of 0.5 M sodium carbonate solution and then making up the volume to 500 cm^3 with distilled water. To 25 cm^3 of the solution to be titrated, 20 cm^3 of dilute sulphuric acid and 1 cm^3 of the above indicator solution are added. Potassium dichromate solution is run in from a burette in the usual way.

Potassium dichromate solution may be used for the estimation of iron and its salts by any of the methods previously described employing potassium permanganate. The following estimation is instructive. Potassium permanganate could not be used because of the presence of chlorides.

Experiment 306 Estimation of the purity of metallic tin

Material: tin; iron(III) chloride solution; 0.0167 M potassium dichromate.

The tin is converted to tin(II) chloride solution, which is then used to reduce a portion of an iron(III) solution to the iron(II) state. The iron(II) solution is then oxidized back to the iron(III) condition by the potassium dichromate solution.

The reducing action of tin(II) chloride solution on an iron(III) salt may be expressed ionically in the form

$$Sn^{2+} + 2Fe^{3+} \rightarrow Sn^{4+} + 2Fe^{2+}$$

or as a partial equation

$$Sn^{2+} \rightarrow Sn^{4+} + 2e^-$$

Thus for 250 cm^3 of 0.05 M tin(II) chloride solution $119 \times 0.25 \times 0.05$ g or 1.49 g of tin are required.

Weigh out accurately about 1.5 g of tin, place in a conical flask and add about 50 cm^3 of concentrated hydrochloric acid diluted with its own volume of water. Put a funnel into the neck of the flask to prevent loss by splashing and when the tin is completely dissolved transfer the tin(II) chloride solution to a 250 cm^3 flask. Wash the funnel and beaker well with acid of the concentration previously used—excess hydrochloric acid is necessary to prevent precipitation of basic tin salts—and add the washings to the main solution, finally making up to the mark with distilled water.

Make up a solution containing about 9 g of iron(III) chloride in 250 cm^3 of dilute hydrochloric acid. Put about 25 cm^3 of it into a conical flask and add, by pipette, 25.0 cm^3 of the tin(II) chloride solution. Since the iron(III) solution has the higher concentration all the tin(II) ions are converted into tin(IV) ions. Titrate the mixture with potassium dichromate solution, using indicator and conditions as described on p. 333. Repeat to obtain concordant results.

Suppose a g of tin were converted into 250 cm^3 of tin(II) chloride solution and 25 cm^3 of this reduced the iron(III) chloride solution so that b cm^3 of 0.0167 M potassium dichromate were required for its oxidation.

$$Sn + 2HCl \rightarrow SnCl_2 + H_2$$
$$SnCl_2 + 2FeCl_3 \rightarrow SnCl_4 + 2FeCl_2$$
$$6FeCl_2 + K_2Cr_2O_7 + 14HCl \rightarrow 6FeCl_3 + 2KCl + 2CrCl_3 + 7H_2O$$

From these equations calculate the concentration and hence the mass of pure tin taken, then the percentage purity of the tin is

$$\frac{3 \times 119 \times 10b \times 100}{60\,000 \times a}$$

Experiment 307 Determination of the purity of potassium chromate

Material: potassium chromate; 0.1 M ammonium iron(II) sulphate.

Potassium chromate is converted by acids into potassium dichromate.

$$2CrO_4^{2-} + 2H^+ \rightarrow Cr_2O_7^{2-} + H_2O$$

or

$$2K_2CrO_4 + H_2SO_4 \rightarrow K_2SO_4 + K_2Cr_2O_7 + H_2O$$

Calculate the mass of potassium chromate needed to produce 250 cm^3 0.0167 M potassium dichromate when in acidified solution (p. 332). Weigh out this amount of potassium chromate (by difference) and, including in the solution 50 cm^3 of dilute sulphuric acid, make up to 250 cm^3 with distilled water.

Titrate this, from a burette, against the ammonium iron(II) sulphate solution (to which 5 cm^3 of phosphoric acid have been added), using diphenylamine as indicator. Calculate the percentage purity from the equations given above.

Experiment 308 Determination of the purity of potassium chlorate

Material: potassium chlorate; 0.2 M ammonium iron(II) sulphate; 0.0167 M potassium dichromate.

In hot acid solution, potassium chlorate[1] oxidizes iron(II) salts.

$$6Fe^{2+} + 6H^+ + ClO_3^- \rightarrow 6Fe^{3+} + Cl^- + 3H_2O$$
$$1 \text{ mole } KClO_3 \equiv 6 \text{ dm}^3 \text{ } 0.0167 \text{ M } K_2Cr_2O_7$$

Calculate the mass of potassium chlorate which is required for 250 cm^3 0.1 M solution and weigh out this amount. Make it up to 250 cm^3 of solution with distilled water.

Prepare 250 cm^3 of 0.2 M ammonium iron(II) sulphate solution (p. 321). To 25 cm^3 of this solution, add 25 cm^3 of the potassium chlorate solution, about 15 cm^3 of dilute sulphuric acid, and boil the mixture, with a funnel in the mouth of the flask, for about twenty minutes. Repeat this twice to obtain three results.

Then wash the funnels, cool the mixtures and titrate with the dichromate solution. Calculate the percentage purity from the equations given above.

Experiment 309 Determination of the percentage of iron in an iron(III) salt, reducing the iron by tin(II) chloride

Material: ammonium iron(III) sulphate; 0.0167 M potassium dichromate; tin(II) chloride.

Prepare a solution of ammonium iron(III) sulphate of accurately known concentration and roughly 0.1 M.

Prepare a solution of tin(II) chloride by dissolving about 3 g of the salt ($SnCl_2 \cdot 2H_2O$) in 50 cm^3 of concentrated hydrochloric acid and diluting to 250 cm^3.

Add 5 cm^3 of concentrated hydrochloric acid to 25 cm^3 of the iron solution in a conical flask and boil the liquid. Run in the tin(II) chloride solution drop by drop from a burette until the yellow colour of the iron(III) ions is discharged. Add two or three more drops of the tin(II) chloride solution to make certain that the reduction is completed. (Care should be taken not to add too great an excess of

[1] The percentage purity of manganese(IV) oxide, trilead tetraoxide (dilead(II) lead(IV) oxide) and many other oxidizing agents can be obtained by using these substances in the place of potassium chlorate in Experiment 308.

tin(II) chloride.) Now add 1 or 2 cm^3 of a saturated solution of mercury(II) chloride. Some white precipitate should be observed showing that sufficient tin(II) chloride had been added. Titrate this solution with the dichromate using diphenylamine as indicator.

Experiment 310 Determination of the proportion of iron in spathic iron ore (iron(II) carbonate)

Material: iron ore; zinc amalgam; 0.0167 M potassium dichromate.

Most of the iron in this ore is in the iron(II) but some is in the iron(III) condition.

The relative molecular mass of iron(II) carbonate is 116 so to obtain 250 cm^3 of an approximately 0.1 M iron(II) solution, weigh out about $116 \times 0.25 \times 0.1$ g of the ore (i.e. about 3 g).

Weighing by difference, put the ore into a conical flask and add about 50 cm^3 of dilute hydrochloric acid. Cover the mouth of the flask with a small watch-glass and, warming the mixture and adding more acid if necessary, allow the action to continue till no brown particles remain. (Silica will remain undissolved.)

Wash the watch-glass and reduce all the iron to the iron(II) state by zinc amalgam (p. 325). Make up the reduced solution to 250 cm^3 in a measuring flask. Titrate with dichromate solution, using diphenylamine as indicator. Calculate the percentage of iron in the ore.

$$6Fe^{2+} + Cr_2O_7^{2-} + 14H^+ \rightarrow 6Fe^{3+} + 2Cr^{3+} + 7H_2O$$

EXERCISES

1. *Material:* 0.1 M potassium permanganate; 0.1 M potassium dichromate; ammonium iron(II) sulphate; diphenylamine.

   Compare the oxidizing powers in acid solution of the permanganate and the dichromate by oxidation of the solution of ammonium iron(II) sulphate (this need not be made up accurately). Explain your result by means of equations.
2. *Material:* mixture containing potassium dichromate and potassium sulphate; ammonium iron(II) sulphate.

   Make a solution of the mixture to be approximately 0.016 M with respect to the dichromate (assume the solid to contain 50% of the dichromate). Make up a standard solution of ammonium iron(II) sulphate and titrate with the unknown solution using diphenylamine as an indicator.
3. *Material:* 0.0167 M potassium dichromate; 0.05 M iron(III) sulphate; 0.05 M titanium(III) sulphate; solution of hydroxylamine in sulphuric acid containing about 1.5 g dm^{-3}.

Hydroxylamine (NH_2OH) can be reduced to ammonia or oxidized to definite stages such as N_2, N_2O, HNO_2 or HNO_3. You are given standard solutions of potassium dichromate, iron(III) sulphate and titanium(III) sulphate and a solution of hydroxylamine in sulphuric acid. Given that titanium(III) sulphate reduces hydroxylamine quantitatively to ammonia determine the nitrogen compound to which it is oxidized by (*a*) iron(III) sulphate, (*b*) potassium dichromate. (Oxford Schol.)

4. *Material:* 0.1 M ammonium iron(II) sulphate solution; 0.016 M potassium dichromate solution.

   Titrate 25 cm^3 of the iron(II) solution to which about 15 cm^3 of dilute sulphuric acid and about 5 cm^3 of 1 M phosphoric acid have been added with the dichromate solution using diphenylamine or N-phenylanthranilic acid as indicator. Calculate the concentration of the iron(II) solution.

5. *Material:* as for exercise 4, plus copper powder.

   Weigh out accurately about 0.1 g of copper powder and put it into 50 cm^3 potassium dichromate solution mixed with 25 cm^3 dilute sulphuric acid. Warm the mixture for twenty minutes, cool and titrate the solution, in the presence of phosphoric acid as in exercise 4, with the iron(II) solution. Calculate the percentage purity of the copper powder.

36
Cerium Salts

Cerium(IV) salts are powerful oxidizing agents; N-phenylanthranilic acid can be used as an indicator as for dichromate titrations.

$$Ce^{4+} + e^- \rightarrow Ce^{3+}$$

The relative molecular mass of cerium(IV) sulphate-4-water is 404.

Experiment 311 Determination of the purity of cerium(IV) sulphate

Material: cerium(IV) sulphate-4-water; 0.1 M ammonium iron(II) sulphate.

Weigh out about 10 g of cerium(IV) sulphate accurately, dissolve it in about 50 cm^3 concentrated sulphuric acid and then make up the solution to 250 cm^3 in a measuring flask.

Titrate 10 cm^3 portions of the standard iron(II) solution mixed with 25 cm^3 dilute sulphuric acid with the cerium(IV) salt, using a few drops of N-phenylanthranilic acid as indicator.

Experiment 312 Determination of the purity of copper(I) chloride

Material: copper(I) chloride (made as in Experiment 111); ammonium iron(III) sulphate solution (5 g in 50 cm^3 dilute sulphuric acid); 0.03 M cerium(IV) sulphate.

Weigh out about 0.1 g of the copper(I) salt accurately and add it to about 10 cm^3 of the iron(III) salt solution; swirl the contents of the flask for a few minutes. Add a few drops of indicator and titrate the mixture with cerium(IV) salt solution.

$$Cu^+ + Fe^{3+} \rightarrow Cu^{2+} + Fe^{2+}$$

37
Iodine and Sodium Thiosulphate

Sodium thiosulphate (thiosulphate(VI)) reacts with iodine, producing sodium iodide and sodium tetrathionate

$$I_2 + 2Na_2S_2O_3 \rightarrow Na_2S_4O_6 + 2NaI$$

i.e.

$$I_2 + 2S_2O_3^{2-} \rightarrow S_4O_6^{2-} + 2I^-$$

This reaction may be used for the estimation of iodine or, indirectly, for the estimation of a substance which participates in a reaction in which iodine is liberated.

STANDARD SODIUM THIOSULPHATE SOLUTION

The sodium thiosulphate purchased for laboratory purposes is the hydrated salt $Na_2S_2O_3 \cdot 5H_2O$ (relative molecular mass 248).

From the above equation: $2S_2O_3^{2-} \rightarrow S_4O_6^{2-} + 2e^-$. Thus 1 dm^3 of 0.1 M sodium thiosulphate solution contains 24.8 g of the pure hydrated salt. An accurately standard solution of the salt cannot be made up directly from the salt as usually purchased because it is not sufficiently pure. Make up a roughly 0.1 M solution by dissolving 25 g of sodium thiosulphate crystals in warm distilled water and diluting to 1 dm^3 in a measuring flask. This solution must now be standardized. This is usually carried out by the use of either standard potassium permanganate (manganate(VII)) or potassium iodate (iodate(V)) solution.

INDICATOR AND END-POINT

The indicator used in titrating iodine solutions with sodium thiosulphate is starch solution. With free iodine it produces a deep blue coloration, the blue colour disappearing as soon as sufficient sodium thiosulphate has been added to react with all the iodine.

The procedure is as follows: sodium thiosulphate solution is run from a burette into the iodine solution until its original brown colour is changed to pale yellow. Then a few drops of starch solution are added producing a blue coloration. Further drop-by-drop addition of thiosulphate solution is continued until the blue coloration disappears.

PREPARATION OF THE STARCH INDICATOR

Mix 1 g of starch to a thin paste with water in an evaporating dish, then pour the paste into about 250 cm^3 of boiling water. Boil the mixture for two to three minutes then cool it for use. This solution will not keep; moulds will grow on it. The addition of about 0.5 g of salicylic (2-hydroxybenzoic) acid to the 250 cm^3 of water before boiling will prevent such growth and the starch indicator will then keep for a long time.

Experiment 313 Standardization of sodium thiosulphate solution by potassium permanganate

Material: 0.02 M potassium permanganate; approximately 0.1 M sodium thiosulphate; potassium iodide.

The permanganate is added to acidified potassium iodide solution. Iodine is liberated and is titrated with the thiosulphate solution.

$$2KMnO_4 + 10KI + 16HCl \rightarrow 12KCl + 2MnCl_2 + 8H_2O + 5I_2$$

ionically,

$$2MnO_4^- + 10I^- + 16H^+ \rightarrow 2Mn^{2+} + 8H_2O + 5I_2$$

Dissolve about 1 g of potassium iodide[1] in about 20 cm^3 distilled water in a conical flask and acidify with an equal volume of dilute sulphuric or hydrochloric acid. Add 25 cm^3 0.02 M potassium permanganate. From a burette, add the sodium thiosulphate solution until the mixture is pale yellow; add the starch indicator and continue titration till the blue colour is discharged. Repeat the titration two or three times.

Suppose 25 cm^3 of 0.02 M potassium permanganate liberate iodine requiring 24.7 cm^3 of the sodium thiosulphate solution.

Then the sodium thiosulphate solution is

$$\frac{25}{24.7} \times 0.02 \times 5 \text{ M}$$

or 0.101 M. The concentration of hydrated salt is 248×0.1012 g dm^{-3}.

---

[1] Relative molecular mass of potassium iodide is 166. For 25 cm^3 of 0.1 M solution, $166 \times (25/10000)$ g of potassium iodide, i.e. 0.4 g are needed. 1 g potassium iodide provides an excess for solution of the liberated iodine—alternatively use about 10 cm^3 of a 10% *m*/*V* solution of potassium iodide in each titration.

0.0167 M potassium dichromate solution may be substituted for 0.02 M permanganate, the estimation being otherwise identical.

Experiment 314 Standardization of sodium thiosulphate solution by potassium iodate

Material: approximately 0.1 M sodium thiosulphate; potassium iodate; potassium iodide.

Standard potassium iodate solution reacts with excess potassium iodide in acidified solution and the liberated iodine is titrated with the sodium thiosulphate solution,

ionically $$IO_3^- + 5I^- + 6H^+ \rightarrow 3H_2O + 3I_2$$

or $$KIO_3 + 5KI + 3H_2SO_4 \rightarrow 3K_2SO_4 + 3H_2O + 3I_2$$

Relative molecular mass of potassium iodate is 214.

For 250 cm^3 of 0.0167 M KIO_3 solution, $214 \times 0.25 \times 0.0167$ g or 0.892 g are needed.

Make up a solution of potassium iodate to contain 0.892 g in 250 cm^3. Pipette 25 cm^3 of it into a conical flask and add about 1 g of potassium iodide. Acidify the solution with about 15 cm^3 of dilute sulphuric acid and titrate the liberated iodine with the sodium thiosulphate solutions as described before. Repeat the titration two or three times. The calculation is similar to that given in Experiment 313.

Experiment 315 Preparation of a standard iodine solution

Material: iodine; potassium iodide; 0.1 M sodium thiosulphate.

Iodine is only very slightly soluble in water. It is, however, readily soluble in potassium iodide solution forming a brown solution which contains the compound KI_3. This compound liberates iodine so readily that the solution behaves as if the dissolved iodine were all free iodine.

$$I_2 + I^- \rightleftharpoons I_3^-$$

Iodine can be purchased sufficiently pure to justify direct preparation of a standard solution, but it is volatile enough to make it almost impossible to avoid loss while the solution is being prepared. Hence, standardization is necessary.

1 dm^3 0.05 M iodine solution contains 12.7 g of iodine, so weigh 13 g in a weighing bottle. Transfer it to a 1 dm^3 flask. Dissolve about 25 g of potassium iodide in 100 cm^3 of water and add about 80 cm^3 of it to the iodine, using the remaining 20 cm^3 for washing out the weighing bottle. When the iodine has dissolved, dilute the solution to 1 dm^3 and shake well.

Titrate 25 cm^3 of this solution with standard 0.1 M sodium thiosulphate solution. The calculation is similar to that given on p. 339.

An alternative way of obtaining a standard solution of iodine is to add a known quantity of an iodate to excess of an iodide—see Experiment 314.

Experiment 316 Estimation of available chlorine in bleaching powder

Material: bleaching powder; potassium iodide; acetic (ethanoic) acid; 0.1 M sodium thiosulphate.

Bleaching powder liberates chlorine when reacting with a dilute acid; this chlorine is available for bleaching and is known as 'available' chlorine.

Bleaching powder may deteriorate because:

(i) it is attacked by carbon dioxide of the air;

(ii) a hypochlorite (chlorate(I)) tends to decompose on standing.

Both these changes reduce the 'available' chlorine content and an old sample of bleaching powder may be almost worthless for bleaching. The following method estimates 'available' chlorine. Solutions of household bleaches can be estimated along similar lines, the dilution factor for titration purposes being about 10.

Bleaching powder is mixed with potassium iodide solution and the mixture is acidified. The liberated iodine is titrated by sodium thiosulphate solution. Acetic acid is used to liberate the chlorine, being quicker than atmospheric carbon dioxide.

A mineral acid, such as hydrochloric, would allow calcium chlorate in the bleaching powder to liberate iodine from the potassium iodide, giving a result which is too high.

$$2\text{–Cl} + 2KI \rightarrow 2KCl + I_2$$

or

$$2\text{–Cl} + 2I^- \rightarrow 2Cl^- + I_2$$

To allow for impurity in the bleaching powder and for deterioration in 'available' chlorine content, about 2.5 g should be used. Weigh accurately a weighing bottle containing about 2.5 g of bleaching powder which has been powdered as finely as possible. Prepare in a 250 cm^3 measuring flask a solution of 4 to 5 g of potassium iodide in 20 cm^3 of water. Transfer the bleaching powder to a clean mortar and weigh the bottle again. Rub the bleaching powder into a paste with a little water at a time, transferring the paste to the 250 cm^3 flask, until all the bleaching powder is in the flask. Acidify the mixture with acetic acid and make up the clear brown solution to 250 cm^3. Shake the flask well and titrate 25 cm^3 of the solution against the sodium thiosulphate solution as described previously, using starch indicator. Repeat the titration two or three times.

Suppose a g of bleaching powder were used and 25 cm^3 of the resulting iodine solution required b cm^3 0.1 M $Na_2S_2O_3$.

$$\underset{(2\times 35.5)\text{ g}}{Cl_2} \equiv I_2 \equiv \underset{\substack{2\text{ dm}^3\text{ 1 M}\\ \text{solution}}}{2Na_2S_2O_3}$$

$$\text{Mass of available chlorine} = 2\times 35.5\times\frac{b}{20000}\text{ g}$$

Mass of available chlorine per 250 cm^3 of solution

$$= 2\times 35.5\times\frac{10b}{20000}\text{ g}$$

Percentage of available chlorine in the bleaching powder

$$= 2\times 35.5\times\frac{10b}{20000}\times\frac{100}{a}$$

Experiment 317 Estimation of copper

Material: copper(II) sulphate crystals; potassium iodide; acetic (ethanoic) acid; 0.1 M sodium thiosulphate.

The copper salt solution must be free from anything but a trace of mineral acid (sulphuric acid present by hydrolysis). Otherwise the end-point is not accurate. This solution then liberates iodine from potassium iodide solution in accordance with the equation:

$$2Cu^{2+} + 4I^- \rightarrow \underset{\substack{\text{copper(I)}\\ \text{iodide}}}{2CuI\downarrow} + I_2$$

or $$Cu^{2+} + e^- \rightarrow Cu^+$$

The iodine is titrated by the sodium thiosulphate. From the equation

$$2Cu^{2+} \equiv I_2 \equiv 2Na_2S_2O_3$$

i.e. $$CuSO_4\cdot 5H_2O \equiv Na_2S_2O_3$$

Relative molecular mass of copper sulphate-5-water is 249.5.

To produce 250 cm^3 of 0.1 M solution, $249.5\times 0.25\times 0.1$ g of copper sulphate crystals are needed. Weigh accurately a weighing bottle containing about 6 g of copper sulphate crystals. Transfer the crystals to a 250 cm^3 flask and weigh the weighing bottle again. Add sodium carbonate solution till a *slight* permanent bluish precipitate of copper carbonate has formed. Acidify the mixture with a little acetic acid when a clear blue solution is obtained. Make the solution up to 250 cm^3 with distilled water and shake well. Pipette 25 cm^3 of the solution into a conical flask containing about 1.5 g potassium iodide dissolved in a little water. Titrate the liberated iodine against the sodium thiosulphate solution as previously described. Calculate the percentage purity of the copper sulphate-5-water crystals.

Experiment 318 Estimation of a sulphite

Material: 0.05 M iodine; 0.1 M sodium thiosulphate; potassium iodide; sodium sulphite (sulphate(IV)) crystals.

The sodium sulphite solution is added to an excess of standard iodine solution and the excess of iodine is estimated by sodium thiosulphate.

In the presence of sodium hydrogencarbonate, sodium sulphite is oxidized quantitatively by iodine to sodium sulphate.

$$Na_2SO_3 + I_2 + H_2O \rightarrow Na_2SO_4 + 2HI$$
$$2HI + 2NaHCO_3 \rightarrow 2NaI + 2H_2O + 2CO_2$$

From these,

$$SO_3^{2-} + H_2O \rightarrow SO_4^{2-} + 2H^+ + 2e^-$$

Relative molecular mass of sodium sulphite-7-water is 252, i.e. for 250 cm^3 0.05 M solution use $252 \times 0.25 \times 0.05$ g or 3.15 g crystals. Make up a solution containing about 3.4 g of the impure crystals.

Pipette 50 cm^3 of 0.05 M iodine solution into a conical flask and from a pipette run into it, slowly and with constant shaking, 25 cm^3 of the sodium sulphite solution. Add about 2 g of sodium hydrogen-carbonate, shake the flask and titrate the excess iodine with the sodium thiosulphate solution, using starch indicator as previously described. Repeat the estimation two or three times.

Suppose a g of sodium sulphite crystals were made up to 250 cm^3 of solution and, after 25 cm^3 of this solution had been added to 50 cm^3 0.05 M iodine solution, b cm^3 0.1 M sodium thiosulphate solution were needed to titrate excess iodine, i.e. b cm^3 0.05 M iodine solution were left in excess.

$$\underset{252\text{ g}}{Na_2SO_3 \cdot 7H_2O} \equiv \underset{\text{20 dm}^3\text{ 0.05 M solution}}{I_2}$$

25 cm^3 of the sodium sulphite solution reacted with $(50-b)$ cm^3 of 0.05 M iodine solution. To oxidize the whole of the sodium sulphite, $10(50-b)$ cm^3 iodine solution would be needed.

$\therefore$ Mass of sodium sulphite-7-water present

$$= \frac{252 \times 10(50-b)}{20\,000} \text{ g}$$

∴ Percentage of sodium sulphite-7-water in the crystals

$$= \frac{252 \times 10(50-b)}{20000} \times \frac{100}{a}$$

EXERCISES

1. The substance A is an impure specimen of copper carbonate. Estimate by the following method the percentage of copper which the substance contains.

   Material: A is malachite; 0.1 M sodium thiosulphate; potassium iodide.

   Weigh out accurately about 3 g of the specimen A, dissolve it with caution in the minimum amount of dilute hydrochloric acid, add sodium carbonate solution drop by drop until a faint precipitate appears and then add dilute acetic acid until this precipitate just dissolves. Make up the solution to 250 cm^3 with distilled water. To 25 cm^3 portions of this solution add a few crystals (not less than 1 g) of potassium iodide, and when these have dissolved titrate the liberated iodine with the sodium thiosulphate solution provided.

   $$2CuCl_2 + 4KI \rightarrow 2CuI\downarrow + 4KCl + I_2$$ (J.)

2. Estimate the concentration of potassium sulphate (K_2SO_4) of the given solution A.

   Material: unknown potassium sulphate solution; barium chromate (about 5 g dm^{-3}) in dilute hydrochloric acid; potassium iodide; 0.1 M sodium thiosulphate.

   You are provided with a solution of barium chromate in dilute hydrochloric acid. To 100 cm^3 of this solution at its boiling point add 25 cm^3 of the sulphate solution A. Subsequently, cool and then add ammonia solution cautiously to the liquid until the colour changes from orange to yellow. Filter, and treat the filtrate with hydrochloric acid and potassium iodide. Titrate the iodine produced with the standard solution of sodium thiosulphate provided. (Cambridge Schol.)

3. *Material:* unknown mixture of copper(II) sulphate and sodium chloride, about 35 g dm^{-3}; 0.05 M iodine; sodium thiosulphate approximately 0.1 M; potassium iodide 10% m/V solution.

   Using the given solution of iodine determine the concentration of the solution of sodium thiosulphate and use this solution to determine the proportion of copper(II) sulphate ($CuSO_4$) in the mixture which contains copper sulphate and sodium chloride. (L.)

4. *Material:* a solution of hydrogen peroxide (about 0.05 M); potassium iodide; ammonium molybdate; 0.1 M sodium thiosulphate.

   To 25 cm^3 of the peroxide solution add 50 cm^3 of dilute sulphuric acid, 15 cm^3 of potassium iodide solution and, as a catalyst to promote the release of iodine, three drops of ammonium molybdate solution. After waiting five minutes for the reaction to be completed, titrate the solution with 0.1 M sodium thiosulphate, using starch as the indicator. Calculate the concentration and the volume strength of the hydrogen peroxide solution.

38
Silver Nitrate

Silver nitrate is a non-hygroscopic solid capable of being prepared in a high degree of purity, and hence very satisfactory for use in volumetric analysis. Its use depends upon the insolubility of its halogen salts (except fluoride) and it is commonly used to estimate chlorides.

$$AgNO_3 + NaCl \rightarrow NaNO_3 + AgCl$$

or ionically,

$$Ag^+ + Cl^- \rightarrow AgCl$$

It can be used as its own indicator, for example, by running a solution of silver nitrate into a solution of sodium chloride until on allowing the precipitate of silver chloride to settle, a further drop produces no precipitate. This is the end-point. It is clear that this method would be tedious and difficult for a beginner.

If silver nitrate solution is added to a sodium chloride solution containing a few drops of potassium chromate solution then silver chloride is selectively precipitated before silver chromate.

The conditions for precipitation of silver chloride and silver chromate depend on their relative solubility products. For the chloride, $[Ag^+][Cl^-] = 1 \times 10^{-10}$ mol^2 dm^{-6}; thus in a saturated solution of silver chloride, $[Ag^+] = 10^{-5}$ mol dm^{-3}. For the chromate, $[Ag^+]^2[CrO_4^{2-}] = 1 \times 10^{-12}$ mol^3 dm^{-9}; in a saturated solution of silver chromate the concentration of silver ion is twice that of chromate ion, i.e. $[Ag^+] = 2[CrO_4^{2-}]$, and on substituting for this in the previous expression it becomes $[2CrO_4^{2-}]^2[CrO_4^{2-}] = 1 \times 10^{-12}$ or $4[CrO_4^{2-}]^3 = 1 \times 10^{-12}$. From this $[CrO_4^{2-}] = 0.63 \times 10^{-4}$ mol dm^{-3} and consequently, $[Ag^+] = 1.26 \times 10^{-4}$ mol dm^{-3}.

A lower concentration of silver ions is needed to give a precipitate of the chloride than to give a precipitate of the chromate: silver chloride is precipitated before sufficient silver nitrate has been added to give a precipitate of silver chromate. Silver nitrate solution in excess will produce sufficient silver ions to exceed the solubility product of the silver chromate, and hence that is precipitated as a brick-red substance, and the end-point is the first appearance in the mixture of a reddish tinge.

$$2Ag^+ + CrO_4^{2-} \rightleftharpoons Ag_2CrO_4\downarrow$$

LIMITATIONS OF USE OF POTASSIUM CHROMATE

Potassium chromate can only be used in neutral solutions because
(*a*) silver chromate is soluble in acids;
(*b*) an alkaline solution would react with the silver nitrate to form silver oxide.

$$2Ag^+ + 2OH^- \rightarrow Ag_2O\downarrow + H_2O$$

If an acidic solution of a chloride is to be estimated there are available the following alternatives:
(*a*) Neutralize the acid with excess of calcium carbonate (free from chloride) and use potassium chromate as an indicator. (See p. 348.)
(*b*) Add excess silver nitrate and estimate the excess by potassium thiocyanate. (See p. 358.)
(*c*) Use an adsorption indicator (see p. 360).

These titrations are very accurate and therefore silver nitrate is usually used in 0.1 M solution (or more dilute).

N.B. Solutions of silver nitrate must be made up with distilled water because tap water usually contains chloride ions.

Experiment 319 Standardization of silver nitrate solution

Material: silver nitrate; pure sodium chloride.

A standard solution of silver nitrate can be made up by weighing the silver nitrate directly. It may be necessary, however, to standardize a solution of silver nitrate of unknown concentration.

$$AgNO_3 + HCl \rightarrow AgCl + HNO_3$$

Relative molecular mass of silver nitrate is 170, hence 250 cm^3 of 0.1 M solution contains

$$170 \times 0.25 \times 0.1 = 4.25 \text{ g crystals}$$

Weigh accurately a weighing bottle containing about 4.25 g of silver nitrate. Transfer these to a 250 cm^3 measuring flask and weigh the bottle accurately. Add distilled water to the flask and shake after each addition. When all the silver nitrate has dissolved, add distilled water up to the mark and shake well.

$$NaCl + AgNO_3 \rightarrow AgCl + NaNO_3$$

Relative molecular mass of sodium chloride is 58.5, hence 250 cm^3 0.1 M solution should contain 1.46 g.

Proceed in exactly the same way as the above to make an approximately 0.1 M solution of sodium chloride, weighing out accurately about 1.5 g of pure dry sodium chloride.

Take 25 cm^3 of the sodium chloride solution, add 1 cm^3 of a 5% *m/V* solution of potassium chromate and run in the silver nitrate from the burette. Proceed slowly towards the end-point and take the reading when the first permanent reddish tinge is apparent. Repeat the experiment two or three times until concordant results are obtained. (An allowance may be made for the silver nitrate used to affect the indicator by performing a 'blank' experiment, using 1 cm^3 potassium chromate solution in 50 cm^3 of water.)

Suppose mass of sodium chloride taken is 1.52 g in 250 cm^3 and that 25 cm^3 of its solution required 26.2 cm^3 of the silver nitrate solution.

$$\text{Concentration of NaCl} = 1.52 \times 4 \text{ g dm}^{-3}$$
$$= 6.08 \text{ g dm}^{-3}$$

$$\therefore \text{Concentration of NaCl solution} = \frac{6.08}{58.5} \text{ M}$$

$$\therefore \text{Concentration of AgNO}_3 \text{ solution} = \frac{25}{26.2} \times \frac{6.08}{58.5} \times 0.1$$
$$= 0.0992 \text{ M}$$

$$\text{Concentration of AgNO}_3 = \frac{25}{26.2} \times \frac{6.08}{58.5} \times 170 \text{ g dm}^{-3}$$
$$= 16.9 \text{ g dm}^{-3}$$

Experiment 320 Standardization of hydrochloric acid by means of silver nitrate

Material: approximately 0.1 M hydrochloric acid; 0.1 M silver nitrate; calcium carbonate (precipitated).

The calcium carbonate will react with the acid (any unused carbonate remaining undissolved) leaving the chloride ions unaffected in a neutral solution.

$$CaCO_3 + 2HCl \rightarrow CaCl_2 + H_2O + CO_2$$

or

$$CO_3^{2-} + 2H^+ \rightarrow H_2O + CO_2\uparrow$$

Thus 100 g of $CaCO_3$ would neutralize 2 dm^3 1 M HCl, i.e. 25 cm^3 0.1 M HCl require 0.125 g of $CaCO_3$.

This need not be weighed out, use a liberal excess, say about 1 g.

Run 25 cm^3 of the hydrochloric acid from a pipette into a conical flask and add about 1 g of calcium carbonate. Add 1 cm^3 of 5%

potassium chromate solution and run in the silver nitrate from the burette until the first permanent reddish tinge is observed. Repeat twice or until two concordant results are obtained.

$$CaCl_2 + 2AgNO_3 \rightarrow 2AgCl + Ca(NO_3)_2$$

$$\therefore HCl \equiv AgNO_3$$

Suppose 25 cm^3 of the acid required 24.3 cm^3 of 0.0993 M $AgNO_3$.

$$\begin{aligned} \therefore \text{Concentration of HCl solution} &= \frac{24.3}{25} \times 0.0993 \text{ M} \\ &= 0.0965 \text{ M} \\ &= 0.0965 \times 36.5 \text{ g dm}^{-3} \\ &= 3.52 \text{ g dm}^{-3} \end{aligned}$$

Experiment 321 Determination of the proportions of sodium and potassium chloride in a mixture

Material: mixture[1] of potassium and sodium chlorides (about 3 g of each to make 1 dm^3 of solution); 0.1 M silver nitrate.

$$KCl + AgNO_3 \rightarrow AgCl + KNO_3$$

Relative molecular mass of potassium chloride is 74.5.

$$NaCl + AgNO_3 \rightarrow AgCl + NaNO_3$$

Relative molecular mass of sodium chloride is 58.5.

In both cases,

$$Ag^+ + Cl^- \rightarrow AgCl\downarrow$$

Assuming the mixture to consist of approximately equal proportions of sodium and potassium chloride, 1 dm^3 1 M solution would contain about 65 g. Hence the amount to be dissolved in 250 cm^3 to make a 0.1 M solution is $65 \times 0.25 \times 0.1$ g = 1.6 g (approx.).

Weigh accurately a weighing bottle containing about 1.6 g of the mixture. Transfer the solid to a 250 cm^3 flask and again weigh the bottle. Make up the solution to 250 cm^3 in distilled water and shake well. Titrate 25 cm^3 of the solution with 0.1 M $AgNO_3$ using 1 cm^3 of 5% potassium chromate solution as an indicator.

Take the burette reading at the first permanent reddish tinge and repeat the experiment two or three times.

Suppose mass of mixed chlorides = 1.550 g and that 25 cm^3 of

[1] After making up the mixture, it is advisable to heat it gently for some time and allow it to cool in a desiccator.

solution required 23.1 cm^3 0.1 M $AgNO_3$. Let the 1.55 g of mixture contain x g of sodium chloride. From the equation above,

58.5 g of NaCl would require 10000 cm^3 0.1 M $AgNO_3$.

$\therefore$ x g of NaCl would require $\frac{x}{58.5} \times 10000$ cm^3 $AgNO_3$.

Similarly $(1.55 - x)$ g of KCl would require

$$\frac{(1.55-x)}{74.5} \times 10000 \text{ cm}^3 \text{ 0.1 M AgNO}_3$$

But 25 cm^3 of the solution required 23.1 cm^3 $AgNO_3$.
$\therefore$ 250 cm^3 of the solution would require 231 cm^3 $AgNO_3$.

$$\text{Hence } \frac{x}{58.5} \times 10000 + \frac{(1.55-x)}{74.5} \times 10000 = 231.$$

Then $x = 0.623$ g so mass of NaCl is 0.623 g and mass of KCl is

$(1.55 - x) = 0.927$ g
$\therefore$ % of NaCl by mass = 40. % of KCl by mass = 60.

N.B. The results from this experiment cannot be expected to show the same accuracy as an ordinary estimation, for example, of the amount of sodium chloride in a solution. An error of 0.1 cm^3 in a titration would produce an error of about 5% in the above estimation of the amount of sodium chloride in the mixture, but less than 0.5% in the estimation of the amount of sodium chloride on its own in a solution.

Experiment 322 Determination of the number of molecules of water of crystallization in hydrated barium chloride

Material: 0.1 M silver nitrate; hydrated barium chloride; sodium sulphate.

If this experiment is to be performed using potassium chromate solution as indicator then the barium ions must first be removed from the solution, because the reaction

$$Ba^{2+} + CrO_4^{2-} \rightarrow BaCrO_4\downarrow$$

would remove the indicator from the solution. The barium ions can be removed by adding excess sodium sulphate, which removes the barium as insoluble barium sulphate.

$$Ba^{2+} + SO_4^{2-} \rightarrow BaSO_4\downarrow$$

The result is, in effect, that an equivalent solution of sodium chloride is titrated instead of the barium chloride solution.

$$BaCl_2 + Na_2SO_4 \rightarrow BaSO_4 + 2NaCl$$

Relative molecular masses 208 142

$$NaCl + AgNO_3 \rightarrow AgCl + NaNO_3$$

From these equations it is evident that

208 g $BaCl_2$ require
- 142 g Na_2SO_4 to precipitate the barium ions.
- 2 dm^3 1 M $AgNO_3$ to precipitate the chloride ions.

Hence mass of *anhydrous* barium chloride in 250 cm^3 of 0.05 M solution is

$$208 \times 0.25 \times 0.05 \text{ g} = 2.60 \text{ g}$$

and 25 cm^3 of this solution would require

$$142 \times \frac{25}{20\,000} \text{ g} = 0.18 \text{ g of sodium sulphate}$$

Hence about 1 g of the anhydrous solid (or 2 g of the crystals $Na_2SO_4 \cdot 10H_2O$) would provide ample excess.

Allowing for the presence of water of crystallization a suitable mass of barium chloride to use would be about 3 g.

Weigh out accurately (by means of a weighing bottle) about 3 g of barium chloride crystals and make up to 250 cm^3 with distilled water. Shake well. Take 25 cm^3 of this solution by means of a pipette, transfer to a conical flask, add about 1 g of anhydrous sodium sulphate and shake. Add about 1 cm^3 of 5% potassium chromate solution and run in silver nitrate from a burette until a permanent reddish tinge is observed. Repeat the experiment two or three times.

Suppose the mass of barium chloride crystals made up to 250 cm^3 solution was a g and that 25 cm^3 of this solution required b cm^3 of silver nitrate solution.

Calculate the concentration of anhydrous barium chloride and hence the mass of it in 250 cm^3 solution. Hence if x is the number of moles of water for each mole of barium chloride

$$\frac{BaCl_2}{BaCl_2 \cdot xH_2O} = \frac{b \times 0.1 \times 208}{25 \times 2 \times 4a}$$

$$\frac{208}{208 + 18x} = \frac{b \times 0.1 \times 208}{25 \times 2 \times 4a}$$

Experiment 323 Estimation of chloride and alkali in a solution containing both

Material: 0.1 M hydrochloric acid; 0.1 M silver nitrate. An alkaline sodium chloride solution: a solution of suitable concentration can be quickly made by putting a known mass (about 3 g) of sodium chloride in water in a 1000 cm^3 flask and adding 500 cm^3 of 0.1 M sodium hydroxide solution before making up to the mark.

Silver nitrate cannot be used in alkaline solution with potassium chromate as an indicator. Hence the alkali is neutralized with standard hydrochloric acid and the total chloride present is estimated with standard silver nitrate solution.

$$NaOH + HCl \rightarrow NaCl + H_2O$$

$$NaCl + AgNO_3 \rightarrow NaNO_3 + AgCl$$

Take 25 cm^3 of the alkaline sodium chloride solution and add a few drops of phenolphthalein solution. It will turn red. Run hydrochloric acid into this solution until the pink colour is just discharged. Note the reading. Add 1 cm^3 of 5% solution of potassium chromate to this neutral solution and run in silver nitrate solution until the first reddish tinge is observed. Note the reading and repeat the process twice.

Suppose 25 cm^3 of alkaline sodium chloride solution required a cm^3 0.1 M HCl. Suppose the same neutral solution required b cm^3 0.1 M $AgNO_3$. Then the concentration of sodium hydroxide is

$$\frac{a}{25} \times 0.1 \text{ M} = \frac{a}{25} \times \frac{40}{10} \text{ g dm}^{-3}$$

The concentration of sodium chloride is

$$\frac{b-a}{25} \times 0.1 \text{ M} = \frac{b-a}{25} \times \frac{53.5}{10} \text{ g dm}^{-3}$$

Experiment 324 Alternative method for Experiment 323

Material: alkaline sodium chloride solution; 0.1 M hydrochloric acid; 0.1 M silver nitrate; calcium carbonate (precipitated chalk).

To determine the concentration of the solution with respect to the alkali titrate with 0.1 M acid as indicated above.

To determine the chloride content take a separate 25 cm^3 of the solution, add 3 cm^3 of nitric acid (1 M excess) and then calcium carbonate until some is in excess. If no effervescence is observed on adding the chalk, then more nitric acid is added. The solution is now neutral and can be titrated with silver nitrate, using potassium chromate as an indicator. In this case the silver nitrate reading gives the chloride content directly because no hydrochloric acid has been added.

Experiment 325 Estimation of chloride and acid in a solution containing both

Material: acidified chloride solution: this solution can be conveniently prepared by putting about 3 g of pure sodium chloride in a 1000 cm^3 flask, adding 500 cm^3 0.1 M hydrochloric acid and making up to the mark; 0.1 M sodium hydroxide; 0.1 M silver nitrate; calcium carbonate (precipitated).

Silver nitrate cannot be used to titrate an acid solution of a chloride using potassium chromate as an indicator. The concentration of the acid is first determined by means of standard alkali. A further quantity of solution is made neutral with calcium carbonate and the chloride content determined from this. Titrate 25 cm^3 of the original solution against sodium hydroxide using phenolphthalein as an indicator. To a further 25 cm^3 add excess calcium carbonate (see p. 348) and 1 cm^3 of 5% potassium chromate solution as an indicator, and titrate with silver nitrate. The concentration of the acid and chloride are calculated from the volumes of alkali and silver nitrate respectively.

Experiment 326 Estimation of potassium chlorate in a mixture of potassium chlorate and potassium sulphate

Material: mixture of pure potassium chlorate and potassium sulphate; 0.1 M silver nitrate.

Potassium chlorate will react with concentrated hydrochloric acid, liberating chlorine, and is reduced to the chloride which is estimated in the usual way with silver nitrate.

$$KClO_3 + 6HCl \rightarrow KCl + 3H_2O + 3Cl_2$$

Relative molecular mass of potassium chlorate is 122.5.

Weigh out accurately about 3 g of the mixture and place the solid in a large evaporating basin. Add about 20 to 30 cm^3 of concentrated hydrochloric acid, cover the dish with a watch-glass and heat gently in a fume chamber for twenty to thirty minutes. There should now be no smell of chlorine. Wash any splashed liquid from the watch-glass into the dish, place on a water-bath and evaporate carefully down to dryness. Make up the solid left to 250 cm^3 of solution and titrate with the silver nitrate.

Experiment 327 Estimation of ammonium chloride and ammonium sulphate in a mixture

Material: 4% formaldehyde (methanal); 0.1 M sodium hydroxide; 0.1 M silver nitrate; mixture of ammonium chloride and ammonium sulphate.

The total ammonium ion concentration is found by the method described in Experiment 283, p. 310. The chloride ion concentration is found by titration with standard silver nitrate.

Weigh accurately about 1.5 g of the mixture, dissolve in water and make up to 250 cm^3. Prepare about 100 cm^3 of neutral formaldehyde (see Experiment 287, p. 314).

(*a*) Take 25 cm^3 of the ammonium salt solution and add 10 cm^3 of the neutral formaldehyde (by measuring cylinder). Allow to stand for two minutes. Add a drop of phenolphthalein and titrate the liberated acids against the alkali. Repeat to obtain consistent results.

(*b*) Take 25 cm^3 of the ammonium salt solution, add 1 cm^3 of potassium chromate solution and titrate against the silver nitrate. Repeat to obtain consistent results.

Suppose that the mass of mixture taken is 1.525 g in 250 cm^3. Then the concentration of the mixture in solution is 6.100 g dm^{-3}.

(i) If 25 cm^3 of the ammonium salt solution gives acid sufficient to neutralize 26.9 cm^3 of 0.1 M alkali, then the solution is

$$\frac{26.9}{25} \times 0.1 = 0.1076 \text{ M}$$

This is the total concentration of the ammonium ion.

(ii) If 25 cm^3 of the ammonium salt solution requires 20.0 cm^3 of 0.1 M silver nitrate, then the chloride concentration is

$$\frac{20.0}{25} \times 0.1 = 0.0800 \text{ M}$$

From (i) and (ii), the concentration of the sulphate is

$$0.5(0.1076 - 0.0800) = 0.0138 \text{ M}$$

The concentration of ammonium chloride in the solution is 0.0800 M. The relative molecular mass of ammonium chloride, NH_4Cl, is 53.5.

$$\text{Concentration of ammonium chloride} = 0.0800 \times 53.5 \text{ g dm}^{-3}$$
$$= 4.28 \text{ g dm}^{-3}$$

The percentage of ammonium chloride in the mixture is

$$\frac{4.28}{6.10} \times 100 = 70.2$$

The concentration of ammonium sulphate in the solution is 0.0138 M. The relative molecular mass of ammonium sulphate, $(NH_4)_2SO_4$, is 132. Therefore, concentration of ammonium sulphate is $132 \times 0.0138 = 1.82\ g\ dm^{-3}$.

The percentage of ammonium sulphate in the mixture is

$$\frac{1.82}{6.10} \times 100 = 29.9$$

VOLHARD'S METHOD

By the use of a solution of either potassium or ammonium thiocyanate the concentration of a silver nitrate solution can be found. This can be done in acid solution which is sometimes very advantageous (see p. 347). If the solution is neutral it is necessary to add nitric acid to prevent hydrolysis of the iron(III) salt used as indicator. The reaction depends upon the insolubility of silver thiocyanate.

$$Ag^+ + SCN^- \rightarrow AgSCN\downarrow$$

A solution of an iron(III) salt (usually the alum) is the indicator. If a thiocyanate solution is added to an acidic silver nitrate solution containing iron(III) sulphate the silver thiocyanate is selectively precipitated before the production of any iron(III) thiocyanate because of the very small solubility product of the silver thiocyanate.

As soon as the thiocyanate ions are in excess they react with the iron(III) ions present to produce the blood red coloration.

$$Fe^{3+} + SCN^- \rightarrow FeSCN^{2+}$$

Volhard's method can be applied to chlorides in acid solution by precipitating all the chloride by the addition of excess silver nitrate solution and determining the excess silver nitrate by titration with potassium thiocyanate.

Note that in this method it is not necessary for the silver nitrate to be in the burette. The thiocyanate solution is run into the silver nitrate solution. The reason for this is that the precipitated silver thiocyanate can adsorb thiocyanate ions, and this would take place to some extent if the potassium or ammonium thiocyanate were in the conical flask and silver nitrate solution were added from the burette.

Experiment 328 Standardization of potassium thiocyanate solution

Material: potassium thiocyanate; 0.1 M silver nitrate; ammonium iron(III) sulphate (10% *m/V* solution).

Potassium thiocyanate (also ammonium thiocyanate) is a very deliquescent solid and it is impossible to weigh out exactly the amount required for any particular solution.

$$KSCN + AgNO_3 \rightarrow AgSCN + KNO_3$$

Relative molecular mass of potassium thiocyanate is 97 and hence 1 dm^3 of 0.1 M solution will contain 9.7 g of anhydrous potassium thiocyanate.

Weigh out about 12 g of the crystals on a watch-glass and transfer the crystals to a measuring flask, add distilled water to make up about 1 dm^3 and shake well. Measure out 25 cm^3 of silver nitrate solution (its concentration must be accurately known) into a conical flask, add 2 or 3 cm^3 of the ammonium iron(III) sulphate and a little dilute nitric acid. The acid prevents the hydrolysis of the iron(III) sulphate, which would otherwise cause the solution to be brownish in colour and so prevent a clear indication of the end-point. Put the thiocyanate solution into a burette and run in the thiocyanate until the first permanent reddish tinge is observed. Repeat the process two or three times.

Suppose 25 cm^3 of 0.104 M $AgNO_3$ required 23.7 cm^3 of KSCN solution.

$$\begin{aligned}\text{Concentration of KSCN solution} &= \frac{25}{23.7} \times 0.104 \text{ M} \\ &= 0.1097 \text{ M} \\ &= 0.1097 \times 97 \text{ g dm}^{-3} \\ &= 10.64 \text{ g dm}^{-3}\end{aligned}$$

Experiment 329 Estimation of the purity of sodium chloride (Volhard)

Material: sodium chloride; 0.1 M silver nitrate; 0.05 M potassium thiocyanate; ammonium iron(III) sulphate.

The sodium chloride is dissolved in water, excess silver nitrate added; the precipitated silver chloride is removed and the excess of silver nitrate is determined by back titration.

Weigh accurately a weighing bottle containing not more than 0.4 g of sodium chloride. Transfer this to a 250 cm^3 flask, adding a little distilled water to dissolve the sodium chloride. Reweigh the weighing bottle. Run in 150 cm^3 silver nitrate solution, add 2 to 3 cm^3 of concentrated nitric acid, replace the stopper and shake until the precipitated silver chloride settles, leaving a clear supernatant liquid. Add distilled water up to the mark and shake thoroughly.

Next the solution must be filtered because silver chloride is more soluble than the silver thiocyanate and tends to react with the iron(III) thiocyanate; filter through a dry filter paper. Measure out 50 cm^3 of the filtrate by means of a pipette, add 2 or 3 cm^3 of the iron(III) solution and run in the potassium thiocyanate solution until the first permanent reddish tinge is observed. Repeat the titration two or three times.

Suppose the mass of sodium chloride is 0.376 g. To this 150 cm^3 0.1 M $AgNO_3$ were added and the solution was made up to 250 cm^3. Suppose 50 cm^3 of this solution required 34.8 cm^3 0.05 M KSCN

$$NaCl + \underset{\substack{10\ dm^3\ 0.1\ M \\ solution}}{AgNO_3} \rightarrow AgCl + NaNO_3$$

$$\therefore\ 50\ cm^3 \text{ of the solution} \equiv 17.4\ cm^3\ 0.1\ M\ KSCN$$

$$\therefore\ 250\ cm^3 \text{ of the solution} \equiv 87\ cm^3\ 0.1\ M\ KSCN$$

$(150-87)$ cm^3 of 0.1 M $AgNO_3$ were used up by the chloride, i.e. 63.0 cm^3 0.1 M $AgNO_3 \equiv$ 0.376 g of the impure sodium chloride.

But 10 dm^3 0.1 M $AgNO_3 \equiv$ 58.5 g NaCl

$$\therefore\ 63.0\ cm^3\ 0.1\ M\ AgNO_3 \equiv \frac{58.5}{10\,000} \times 63\ g\ NaCl$$

$$= 0.3686\ g\ NaCl$$

$$\text{Percentage purity NaCl} = \frac{0.3686}{0.376} \times 100$$

$$= 98$$

Experiment 330 Determination of the percentage of silver in an alloy

Material: silver alloy; 0.1 M potassium thiocyanate; ammonium iron(III) sulphate.

For this purpose the alloy should not contain too high a proportion of copper. If too much copper is present the colour of the copper ions interferes. The pre-1920 silver coinage is excellent for this purpose as it contains 92.5% of silver. The 1920–47 coin is not very suitable as it contains 50% Ag; 40% Cu; 5% Ni; 5% Zn. The current silver coinage contains no silver whatsoever.

Weigh accurately a piece of silver of mass about 2 g and introduce the metal carefully into a 250 cm^3 beaker. Add about 5 cm^3 of distilled water followed by 5 cm^3 of concentrated nitric acid. Put the beaker in a fume chamber and allow the metal to dissolve. When the metal has dissolved boil the solution to decompose nitrous acid and then add distilled water, washing the sides of the beaker. Transfer the solution to a measuring flask and add the rinsings from the beaker so that no silver nitrate solution is lost. Make up to the mark and shake well. Measure out 25 cm^3 of this solution with a pipette, add 2 or 3 cm^3 of the iron(III) solution and titrate with the thiocyanate solution. Note the burette reading at the first permanent reddish tinge, and repeat the titration two or three times.

$$Ag + 2HNO_3 \rightarrow AgNO_3 + H_2O + NO_2$$
$$AgNO_3 + KSCN \rightarrow AgSCN + KNO_3$$

Hence $\quad Ag \equiv KSCN$

i.e. $\quad 108\ g\ Ag \equiv 10\ dm^3\ 0.1\ M\ KSCN$ solution

Experiment 331 Standardization of hydrochloric acid

Material: approximately 0.1 M hydrochloric acid; 0.1 M silver nitrate; 0.1 M potassium thiocyanate; nitrobenzene; ammonium iron(III) sulphate.

The acid should be approximately 0.1 M. Put 25 cm^3 of 0.1 M silver nitrate in a conical flask and add 10 cm^3 of the acid, then about 4 cm^3 of dilute nitric acid and finally about 2 cm^3 of nitrobenzene which coats the precipitate and stops it redissolving; 0.5 cm^3 of an iron(III) solution acts as indicator. Titrate the mixture with a standard thiocyanate solution until a red coloration is obtained; thorough shaking is essential.

$$\underset{\text{(in excess)}}{AgNO_3} + HCl \rightarrow AgCl + HNO_3$$
$$AgNO_3 + KSCN \rightarrow AgSCN + KNO_3$$

This is another example of back titration (see Experiments 278, p. 304, and 318, p. 344). If the concentrated hydrochloric acid supplied to the laboratory is diluted accurately one hundred times this method may be used for its standardization.

ADSORPTION INDICATORS

Adsorption indicators play an important part in volumetric analysis: certain dyes can act as indicators for many types of reaction. These adsorption indicators are very sensitive, enabling dilute solutions to be employed. The following example gives an idea of the mechanism of an adsorption indicator.

CHARGE ON COLLOIDAL PARTICLE ALTERS AS THE REACTION IS COMPLETED

Many precipitates are colloidal in nature and the colloidal particles can adsorb ions from solution. Ions are not adsorbed by the colloidal particles indiscriminately, for the latter show preference for certain ions over others. The colloidal particles of a silver bromide precipitate will adsorb either bromide or silver ions from solutions containing either of these. Thus silver bromide in potassium bromide solution will adsorb bromide ions but not potassium ions, and silver bromide in silver nitrate solution will adsorb silver ions but not nitrate ions.

Consider the action which takes place when silver nitrate is run into a solution of potassium bromide. So long as the latter is in excess (i.e. before the reaction is completed) there are present potassium ions, bromide ions, nitrate ions and silver bromide particles, i.e. assuming excess potassium bromide.

$$2K^+ + 2Br^- + Ag^+ + NO_3^- \rightarrow AgBr + 2K^+ + NO_3^- + Br^-$$

The colloidal silver bromide adsorbs the bromide ions, forming a *negatively* charged complex which we may regard as $[AgBr]Br^-$, i.e. a negatively charged particle. As soon as the reaction is completed there will be no bromide ions in excess and a further drop of silver nitrate will produce a solution containing potassium ions, nitrate ions, *silver ions* and silver bromide particles, i.e. assuming excess silver nitrate,

$$K^+ + Br^- + 2Ag^+ + 2NO_3^- \rightarrow K^+ + 2NO_3^- + Ag^+ + AgBr$$

The silver bromide will now adsorb silver ions to form a complex such as $[AgBr]Ag^+$ which will be *positively* charged. The important point to note is that when the reaction is completed and a further quantity of the reagent (in this case silver nitrate) is added, the colloidal particle changes in sign from negative to positive.

MECHANISM OF THE INDICATOR

We will consider eosin to have the formula NaEo to indicate that the solution in water contains sodium cations and eosinate anions. These latter impart a pink colour to the solution. (Red ink is a solution of sodium eosinate in water.)

$$\underset{\text{colourless}}{NaEo} \rightleftharpoons Na^+ + \underset{\text{pink}}{Eo^-}$$

Whilst the bromide adsorbed complex $[AgBr]Br^-$ is present there is no tendency for the colloidal particle to attract to itself the coloured eosinate ions, which therefore *colour the solution but not the precipitate.* When the silver adsorbed complex $[AgBr]Ag^+$ is formed some of the eosinate ions are at once attracted to it by reason of their opposite charge and the precipitate becomes coloured by the eosinate ions. Furthermore, the colour of the eosinate ions when adsorbed by the colloidal particle is not exactly the same as the colour observed when the eosinate ion is in solution (the eosinate ion may combine with the silver ion to form silver eosinate). Hence quite a definite change in colour is observed, and it is the precipitate which becomes coloured. This is an important point in the use of adsorption indicators. It is the precipitate which must be observed and not the solution.

It is also often noticed that the precipitate coagulates as the end-point is reached because this is also the isoelectric point where there is no charge on the colloid particles.

If difficulty is experienced in noticing the end-point, the following method will help. To two conical flasks add some of the reagent, e.g. hydrochloric acid and sufficient calcium carbonate to render the solution neutral, and a few drops of fluorescein. Into flask A run an amount of silver nitrate obviously insufficient to cause any alteration of colour and into flask B sufficient to give a definite change, i.e. an amount greater than that necessary to attain the end-point.

Now place the solution to be titrated by the side of these and run the silver nitrate into it. The first permanent change will be then quite obvious.

The following examples are simple and the indicators used are either eosin (red ink will do) or fluorescein.

Experiment 332 Standardization of hydrochloric acid using fluorescein as indicator

Material: 0.1 M hydrochloric acid; 0.1 M silver nitrate; calcium carbonate.

Although many adsorption indicators can be used in acid solution fluorescein can only be used in neutral solution.

Measure out from a pipette 25 cm^3 of the hydrochloric acid solution to be standardized and add about 1 g of precipitated calcium carbonate (see Experiment 320, p. 348, for theoretical considerations): a little calcium carbonate should remain undissolved.

Add two or three drops of fluorescein (which will colour the solution green) and run in the silver nitrate solution from the burette until the precipitate (which is yellowish at first) turns pink. Repeat the experiment two or three times. Calculate the concentration of the acid from the equation

$$\underset{\substack{36.5\text{ g or}\\10\text{ dm}^3\text{ 0.1 M}\\\text{solution}}}{HCl} + \underset{\substack{10\text{ dm}^3\text{ 0.1 M}\\\text{solution}}}{AgNO_3} \rightarrow AgCl + HNO_3$$

N.B. This equation does not represent the actual reaction which took place. The silver nitrate is titrated against an equivalent solution containing chloride ions.

$$2HCl + CaCO_3 \rightarrow CaCl_2 + H_2O + CO_2$$
$$2H^+ + CO_3^{2-} \rightarrow H_2O + CO_2$$

Experiment 333 Determination of the relative molecular mass of potassium bromide

Material: 0.01 M silver nitrate; potassium bromide; eosin.

Weigh accurately a weighing bottle containing about 0.3 g of pure potassium bromide and transfer this to a 250 cm^3 flask, then reweigh the bottle. The weighing must be accurate, for a small actual error in weighing this small amount would create a large percentage error. Dissolve the bromide in distilled water, make the solution up to the mark, and shake well. Measure out 25 cm^3 of this solution from a pipette, add two or three drops of eosin and titrate with silver nitrate from a burette. Note the reading when the precipitate becomes salmon-pink in colour. Repeat two or three times.

Suppose a g of KBr were made up to 250 cm^3 of solution and 25 cm^3 of this solution required b cm^3 of 0.01 M $AgNO_3$.

$$b\ \mathrm{cm^3\ of\ 0.01\ M\ AgNO_3} \equiv 25\ \mathrm{cm^3\ of\ KBr\ solution}$$

$$\therefore\ 10b\ \mathrm{cm^3\ 0.01\ M\ AgNO_3} \equiv 250\ \mathrm{cm^3\ of\ KBr\ solution}$$

$$\equiv a\ \mathrm{g\ of\ KBr}$$

$$1\ \mathrm{cm^3\ 0.01\ M\ AgNO_3} \equiv \frac{a}{10b}\ \mathrm{g\ KBr}$$

100 000 cm^3 0.01 M $AgNO_3$,

i.e.

$$1\ \mathrm{dm^3\ 1\ M\ AgNO_3} \equiv \frac{a}{10b} \times 100\,000\ \mathrm{g\ KBr}$$

$$= \frac{10\,000 \times a}{b}\ \mathrm{g\ KBr}$$

which is the molar mass of potassium bromide because the equation for the reaction is

$$KBr + AgNO_3 \rightarrow AgBr + KNO_3$$

Experiment 334 Estimation of the purity of lead nitrate crystals by titration with sodium hydroxide

Material: 0.1 M sodium hydroxide; lead(II) nitrate; fluorescein.

$$\underset{331\ \mathrm{g}}{Pb(NO_3)_2} + \underset{2\ \mathrm{dm^3\ 0.1\ M\ solution}}{2NaOH} \rightarrow Pb(OH)_2 + 2NaNO_3$$

Weigh accurately a weighing bottle containing about 4 g of pure lead nitrate crystals and transfer it to a 250 cm^3 flask. Reweigh the weighing bottle. Add distilled water to the crystals, shake, and when dissolved make up to the mark, shake and allow to stand. Fill the burette with the lead nitrate solution. Measure out 25 cm^3 of the sodium hydroxide into a conical flask and add two to three drops of a solution of fluorescein. This will colour the solution green. As the lead nitrate solution is run in, the precipitate becomes yellowish in colour. Run in the lead nitrate solution until there is the first permanent pink coloration on the precipitate. Repeat the experiment two or three times.

Experiment 335 Determination of the concentration of a potassium thiocyanate solution

Material: 0.1 M silver nitrate; approximately 0.1 M potassium thiocyanate; fluorescein.

The solution of potassium thiocyanate should be approximately 0.1 M because in silver nitrate titrations with adsorption indicators it is necessary to proceed rapidly; otherwise the action of light upon the precipitate may obscure the change. Measure out 25 cm^3 of the potassium thiocyanate solution, add a few drops of fluorescein and rapidly run in silver nitrate solution until the precipitate coagulates and turns pink. Make two or three more accurate determinations, proceeding more slowly towards the end of the titration.

$$AgNO_3 + KSCN \rightarrow AgSCN + KNO_3$$

From the equation

$$1\ dm^3\ 0.1\ M\ AgNO_3 \equiv 9.7\ g\ KSCN$$

calculate the concentration of the potassium thiocyanate in g dm^{-3}.

EXERCISES

1. *Material:* solutions of sodium carbonate and sodium chloride; 0.1 M hydrochloric acid; 0.1 M silver nitrate.

   The solution contains only sodium carbonate and sodium chloride. Estimate the separate amounts of each salt in g dm^{-3}.
2. *Material:* solution containing sodium chloride, sodium hydroxide and sodium carbonate; 0.1 M hydrochloric acid; 0.1 M silver nitrate.

   The solution contains sodium chloride, sodium hydroxide and sodium carbonate. Estimate the mass of each compound in 1 dm^3 solution.

   (In the estimation of sodium hydroxide the sodium carbonate should first be precipitated by an excess of barium chloride and advantage taken of the fact that the concentration of hydrogen ion required to decompose barium carbonate is greater than that to decolorize phenolphthalein.)

   (Cambridge Schol.)

3. *Material:* solutions A and B; 0.02 M silver nitrate; 0.02 M potassium permanganate (manganate(VII)).

Determine the percentages of potassium chlorate and potassium chloride in a mixture X of these two compounds.

Solution A contains in each 1000 cm^3 the residue left on ignition (to constant mass) of 13.57 g of X.

Solution B contains in each 1000 cm^3 the product of boiling 13.57 g of X dissolved in sulphuric acid with 60 g iron(II) sulphate, the iron(II) being in excess of that required for the reaction:

$$ClO_3^- + 6Fe^{2+} + 6H^+ \rightarrow Cl^- + 6Fe^{3+} + 3H_2O$$ (Oxford Schol.)

4. To determine the solubility of barium chloride at the temperature of the laboratory.

Material: barium chloride-2-water; 0.1 M silver nitrate.

Crush some crystals of barium chloride and add to water in a boiling tube. Shake vigorously, add more crystals if necessary, and allow to stand with intermittent shakings for about twenty minutes. Calculate the mass of solution that on dilution to 250 cm^3 will give one of roughly 0.05 M: the solubility is of the order of 400 g dm^{-3} water at the temperature of the laboratory. Weigh in a weighing bottle empty, filter off about the calculated mass into the weighing bottle and weigh again. (If you use a pipette (*with a filler*) it must be washed out with the saturated solution.) Transfer completely the contents of the weighing bottle to a 250 cm^3 flask, empty all washings from the bottle into the flask and make up to 250 cm^3. Proceed as in Experiment 322 and calculate the result in g dm^{-3} solvent.

5. To determine the percentage purity of a sample of potassium bromide.

Material: potassium bromide; 0.1 M silver nitrate.

$$AgNO_3 + KBr = AgBr + KNO_3$$

Relative molecular mass of potassium bromide 119.

Hence 0.1 M potassium bromide will contain 11.9 g dm^{-3}. Weigh out accurately about 3 g of potassium bromide. Make up to 250 cm^3 with distilled water. Titrate against silver nitrate using potassium chromate solution as an indicator. Calculate the percentage purity of the bromide.

6. *Material:* solution A contains 5 g of a mixture of potassium nitrate and silver nitrate made up to 250 cm^3 of solution; 0.1 M potassium thiocyanate. Determine the concentrations of the two salts in g dm^{-3}.

7. Estimation of percentage purity of potassium bromide.

Material: potassium bromide; 0.1 M silver nitrate; 0.05 M potassium thiocyanate.

Repeat Experiment 329, using potassium bromide instead of sodium chloride. Weigh out about 0.7 g of the bromide. As silver bromide is less soluble than silver thiocyanate there is no need to filter off the precipitated silver bromide.

8. Estimation of the concentration of mercury(II) nitrate solution.

Material: mercury(II) nitrate; 0.1 M potassium thiocyanate; sodium nitroprusside (pentacyanonitrosylferrate(II)).

Sodium nitroprusside gives a permanent precipitate with mercury(II) nitrate but not with potassium thiocyanate. Make up a solution of mercury nitrate to be about 0.1 M. Run this from a burette into 25 cm^3

of potassium thiocyanate solution to which a few drops of a 10% solution of sodium nitroprusside have been added. Take as the end-point the first permanent turbidity.

$$Hg(NO_3)_2 + 2KSCN \rightarrow Hg(SCN)_2\downarrow + 2KNO_3$$

9. *Material:* sodium chloride solution containing 5.6 g dm^{-3}; unknown silver nitrate; unknown ammonium thiocyanate.

   By means of the standard sodium chloride solution determine the concentration of the solution of silver nitrate, and use this solution to determine the purity of ammonium thiocyanate (NH_4SCN). (L.)

10. *Material:* lead(II) acetate (ethanoate); sodium hydroxide; dichlorofluorescein.

    Investigate the possibility of using dichlorofluorescein to indicate the end-point of the reaction

$$Pb(CH_3COO)_2 + 2NaOH \rightarrow Pb(OH)_2 + 2CH_3COONa$$

    Use solutions which are approximately 0.1 M and allow the acetate to run into the sodium hydroxide solution.

11. Estimation of the concentration of sodium oxalate solution.

    Material: unknown sodium oxalate.

    The lead acetate solution prepared above may be used to estimate the concentration of a solution of sodium oxalate (or any normal oxalate) using fluorescein as adsorption indicator. The precipitate becomes permanently pink in colour at the end-point.

12. Estimation of bromide and iodide in a mixture.

    Material: mixture of potassium bromide and iodide; diiodofluorescein; eosin; 0.02 M silver nitrate.

    Weigh out accurately about 1 g of the mixture and dissolve in distilled water, and make up to 250 cm^3. Titrate 25 cm^3 of the solution,

    (*a*) with silver nitrate using diiodofluorescein (yellow to pink) which indicates the end-point when the iodide only has been precipitated.

    (*b*) with silver nitrate using eosin which indicates the end-point when the whole of the halide has been precipitated.

    Estimate the proportions of the iodide and bromide in the mixture.

39
Complexometric Titrations

Edta or ethylenediaminetetraacetic acid is usually employed as its disodium salt; it is also called diaminoethanetetraacetic acid, ethane-diaminetetraacetic acid and bis[di(carboxymethyl)amino] ethane. The formula of the acid is

$$\begin{array}{l} CH_2N(CH_2COOH)_2 \\ | \\ CH_2N(CH_2COOH)_2 \end{array}$$

and its anhydrous sodium salt, of relative molecular mass 336, is sometimes designated as Na_2H_2Y. It forms very stable complexes with metal ions in solution usually in a 1:1 molar ratio. The indicator for its titrations is Eriochrome Black T, a complex organic substance that is itself a weak complexing agent. The principles of the titration are as for acid-alkali titrations and if a graph is plotted of the negative logarithm (numerical value of the metal ion concentration in mol dm^{-3}) against the volume of edta, a graph is obtained of the same shape as in Figure 77.

Experiment 336 Determination of the total hardness of water

Material: 0.01 M edta; a buffer solution made from 57 cm^3 of concentrated ammonia solution and 7 g of ammonium chloride made up with distilled water to 100 cm^3.

$$H_2Y^{2-} + M^{2+} \rightarrow YM^{2-} + 2H^+$$

Titrate 50 cm^3 of tap water to which 1 cm^3 of buffer solution and 0.5 cm^3 of indicator have been added with 0.01 M edta solution: the colour change is from wine red to pure blue at the end-point. Calculate the hardness of water in parts per million (mg kg^{-1}) of calcium (as carbonate).

Experiment 337 Determination of the number of molecules of water of crystallization in hydrated aluminium sulphate

Material: a solution which is 0.01 M with respect to aluminium ions—either hydrated aluminium sulphate or hydrated aluminium potassium sulphate; 0.01 M zinc sulphate; 0.01 M edta.

Aluminium sulphate crystals are highly hydrated and have a relative molecular mass of about 650. Make up 100 cm^3 of a solution which is about 0.01 M with respect to aluminium ions. The solution of zinc sulphate is required for back titration of the excess edta.

$$Al^{3+} + H_2Y^{2-} \rightarrow AlY^{-} + 2H^{+}$$

To 10 cm^3 of the aluminium solution add 50 cm^3 of edta. Adjust the pH to between 7 and 8 by adding some dilute ammonia solution and then boil the solution for ten minutes to ensure completion of complex ion formation by the aluminium. Cool the solution to room temperature, add the indicator and titrate rapidly with the standard zinc solution until the colour goes from blue to wine red.

PART V

Qualitative Analysis

40
Introduction

The purpose of an analysis is to determine the composition of a compound or of a mixture of compounds. *Qualitative* analysis is the process of determining what elements are present without regard to metrical quantities. Qualitative analysis must precede *quantitative* analysis of an unknown material.

As a result of a qualitative analysis it can be stated which metallic and acidic radicals are in the sample, but, where the sample is a mixture of compounds, identification of the compounds which were used to make the mixture is not to be expected, and indeed, could only be arrived at by inference. Thus a mixture of copper(II) sulphate and sodium nitrate would give the same qualitative analytical result as a mixture of copper(II) nitrate and sodium sulphate. It is usual, therefore, to present the result of an analysis as a list of cations and anions.

Apart from information afforded by visual observation of a sample, an analysis usually includes attempted decomposition of the sample by heat, reaction with acids and alkalis, a flame test, and reactions of a *solution* of the sample with reagents to give either precipitates or colorations by which elements or groups of elements may readily be recognized. For reliable results the tests must be performed in the order suggested.

The scheme for the identification of cations is an application of solubility products. A solution of the sample, which may contain one or more compounds, is treated with specified reagents in a definite order, the purpose being to precipitate the cations present as insoluble, or sparingly soluble, compounds in groups, each group of precipitates being removed completely before the next precipitating reagent is added to the remaining solution. This 'group separation' depends on the different solubilities of the chlorides, sulphides, hydroxides and carbonates of the metals present in the original solution as ions. Of all the metal ions likely to be present only silver, lead and mercury(I) ions form precipitates of their chlorides on the addition of hydrochloric acid. After the removal of this group of precipitates, the remaining solution on treatment with hydrogen sulphide forms precipitates of sulphides with the ions of mercury(II), bismuth, copper(II), cadmium, lead(II), antimony(III), tin(II) and tin(IV). This rather large group is later divided into sub-groups depending on the comparative solubilities of the sulphides in ammonium sulphide. An outline of the whole scheme of separation into groups is as follows:

Analytical group	*Ions of solution precipitated*	*Precipitating reagent*	*Nature of precipitate*
I	Ag^+, Pb^{2+}, Hg_2^{2+}	Hydrochloric acid.	Chlorides.
II	(*a*) Hg^{2+}, Cu^{2+}, Cd^{2+}, Bi^{3+}, (Pb^{2+})	Hydrogen sulphide, in presence of hydrogen ions left in solution from the acid added in Group I.	Sulphides, insoluble in ammonium sulphide.
	(*b*) Sb^{3+}, Sn^{2+}, Sn^{4+}		Sulphides, soluble in ammonium sulphide.
III	Fe^{3+}, Al^{3+}, Cr^{3+}	Ammonia solution, in presence of ammonium chloride.	Hydroxides.
IV	Zn^{2+}, Mn^{2+}, Ni^{2+}, Co^{2+}	Hydrogen sulphide, in presence of hydroxide ions left in the buffered solution from Group III.	Sulphides.
V	Ba^{2+}, Ca^{2+}	Ammonium carbonate, in presence of buffered solution from Groups III and IV.	Carbonates.
VI	Na^+, K^+, Mg^{2+}	These cations are not precipitated collectively; they are identified individually by appropriate reagents.	——

The group precipitate is then analysed to find the particular cation (or cations) present. The ammonium ion (NH_4^+) is not included in the above list; its presence in the sample is indicated during preliminary tests.

The full-scale method makes use of test-tubes of approximately 100 × 16 mm, the small-scale method needs tubes of the size 75 × 10 mm; where in the former method reagent solutions are added in cubic centimetre quantities, in the latter method they are added in drops. The use of such small total volumes of liquids makes it impracticable to separate solid from liquid by filtration because the liquid would be almost completely lost by absorption in the filter paper. The essential technique is to centrifuge the mixture of suspension and liquid in order to gather the solid into a compact mass at the bottom of the tube, and then to draw off the liquid with a teat pipette. Another consequence of small volumes in small tubes is that the usual method of direct heating in a Bunsen flame is more difficult. For gentle warming, direct heating is possible if care is exercised, but in general, indirect heating on a metal block is preferable.

APPARATUS

For each student:

6 test-tubes (75 × 10 or 100 × 16 mm)
2 ignition tubes (50 × 10 mm)
1 boiling tube (75 × 12 or 150 × 25 mm)
A test-tube rack, consisting of a wooden block, about 150 mm long, 75 mm wide and 25 mm deep, bored with 10 mm holes to take the tubes and the boiling tube (or larger for full-scale apparatus), larger holes to take the beaker and crucible, and provided with a groove to accommodate the teat-pipette, spatula and forceps
A heating block, obtainable from suppliers, and consisting of an aluminium cylinder, mounted on legs and bored to take a tube or boiling tube and a crucible or beaker
A beaker of 5–10 cm^3 capacity
A crucible of 5–10 cm^3 capacity, lid and tongs
A standard Bunsen and a micro-burner
A small-scale spatula of flattened nickel, approximately 120 mm long and 3 mm broad, with one end angled to give a surface of 4 × 3 mm;
A watch-glass of approximately 25 mm diameter
A teat pipette, made from 5 mm glass tubing, tapered to about a quarter of its diameter to form a nozzle and fitted with a teat of stout rubber at the other end, giving an overall length of about 125 mm
A pair of forceps or a clothes peg or a piece of doweling fitted with a 'Terry' clip
A beaker, tall type, about 250 cm^3 capacity
A wash bottle (polythene 'squeeze' type) containing distilled water
A glass rod, about 4 mm diameter and 125 mm long, having one end fitted with a short length of rubber tubing (this end serving as a test tube 'brush')
A nichrome wire; cobalt blue glass.

For each pair of students:

A set of the more frequently used reagents (as Set A on p. 373)

For each group of four students:

A centrifuge, preferably placed on a side bench and conveniently near to the working bench

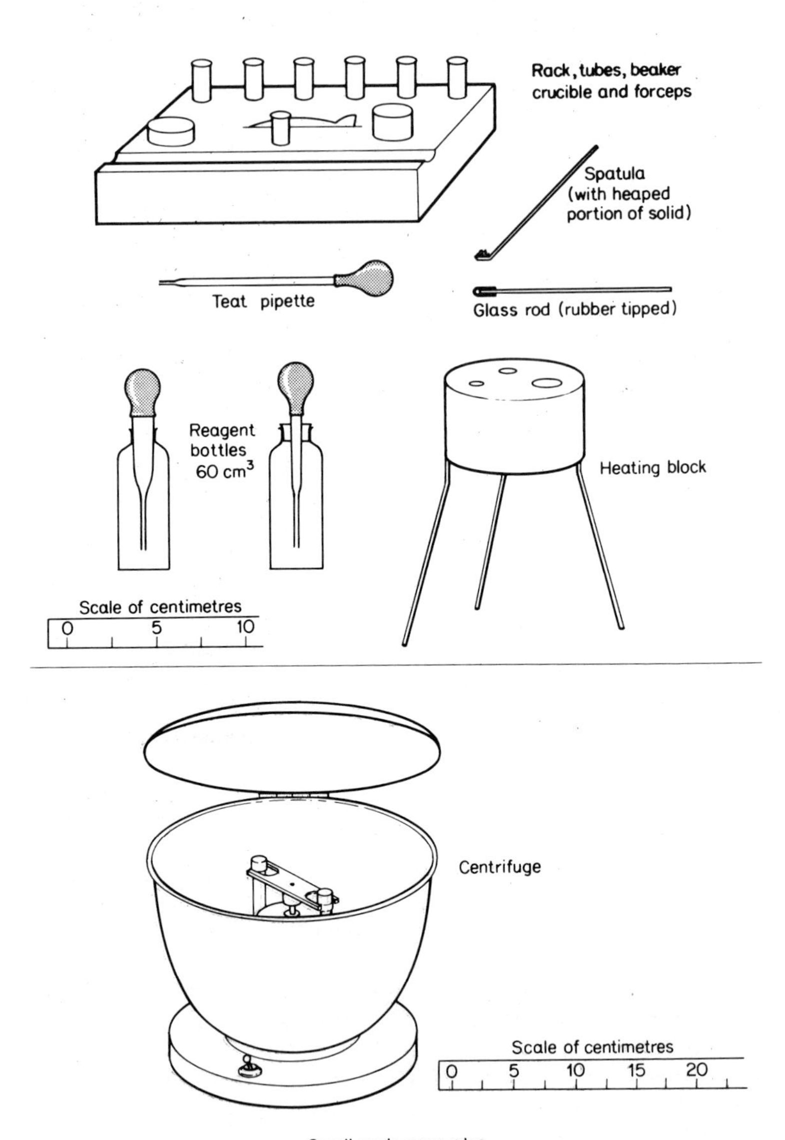

Small scale apparatus

FIG. 82

A set of less frequently used reagents placed centrally to give ready access, and given in Set B below

For common use:

A set of reagents, and other material, for use on a more limited scale, on a readily accessible shelf or side bench (Set C. p. 374)

REAGENTS

Reagents may be conveniently arranged in three sets: a set of the more frequently used reagents for the use of each individual student (Set A), a set of less frequently used reagents which may be placed in the centre of the bench for the use of a group of four students (Set B) and a third set of reagents and material, mainly for confirmatory tests and placed on a side bench or shelf (Set C).

SET A

Dilute hydrochloric acid, ammonia solution, ammonium carbonate, 1 or 2 M

Saturated solution of hydrogen sulphide in moist acetone (0.5 cm^3 water to 100 cm^3 of acetone (propanone.))

Ammonium sulphide, yellow ammonium sulphide diluted with an equal volume of water

Concentrated nitric acid

Solid ammonium chloride

SET B

Dilute nitric acid, acetic (ethanoic) acid, sodium hydroxide, sulphuric acid, 1 M

Hydrogen peroxide, 10 volume strength (1 M)

Calcium hydroxide, ammonium oxalate (ethanedioate), saturated solutions 1 M

Silver nitrate, 0.1 M

Barium chloride, 0.1 M

Potassium chromate (chromate(VI)), 0.1 M

Potassium permanganate (manganate(VII), 0.02 M

Concentrated hydrochloric acid), 0.02 M

Lead acetate (ethanoate) 1 M

Litmus solution (20% approx. and filtered)

Concentrated ammonia solution

Potassium hexacyanoferrate(II), 0.1 M

Disodium hydrogenphosphate, 1 M

Concentrated sulphuric acid

Solid anhydrous sodium carbonate (high quality)

Solid potassium hexacyanoferrate(III) (for preparing approximately 5% solution as required)

Test papers . . . red and blue litmus, lead acetate (ethanoate), in stoppered bottles

SET C

Thiourea, 1 M

Titan yellow, 0.1% solution, or Magneson, 0.001% solution in 1 M sodium hydroxide

Dimethylglyoxime (butanedione dioxime), 1% solution in ethanol

Mercury(II) chloride, 0.1 M

Cobalt(II) nitrate, 0.1 M

Tin(II) chloride, 5 g tin(II) chloride crystals boiled in 10 cm^3 concentrated hydrochloric acid to dissolve and diluted to 100 cm^3

Ammonium molybdate, 15 cm^3 concentrated ammonia diluted with an equal volume of water, 10 g ammonium molybdate dissolved in the solution. The solution added gradually with stirring, to 100 cm^3 concentrated nitric acid, kept cold. When settled, decant and dilute the solution with an equal volume of water

Ethanol

Diethyl ether (ethoxyethane)

Solid sodium hexanitrocobaltate(III), iron(II) sulphate, lead(IV) oxide, ammonium acetate (ethanoate), potassium chlorate (chlorate(V)), sodium bismuthate(V), disodium tetraborate

Zinc foil.

PROCEDURE

(a) CLEANLINESS OF APPARATUS

It is particularly important in small-scale analysis to exclude traces of impurities. The teat pipette is a frequently used piece of apparatus and must be kept clean between operations. For this purpose a 250 cm^3 beaker containing water is provided and the pipette is thoroughly washed by drawing up water into it, expelling the washings into the sink and repeating this once or twice; it is then left standing in the water until required. Before re-use the pipette should, of course, be reasonably dry to avoid dilution of reagents.

Test-tubes should be cleaned immediately their contents are no longer required; a small-scale 'test-tube brush' consisting of a piece of glass rod tipped with rubber tubing is a useful alternative and may be preferred.

(b) USING REAGENTS

Drops of reagents are added, as required, by means of the dropping tube with which the reagent bottle is provided. The tip of the dropping tube must be kept clear of the wall of the test-tube to avoid contamination of the reagent in the bottle.

When a reagent gives a precipitate, it is essential to ensure that sufficient has been added to complete the reaction. In the scheme for separation of metallic radicals it should become a routine procedure to add a further drop of the reagent to the liquid left after centrifuging before proceeding to the next group, carefully examining for any further precipitation.

At all times it is important to ensure that a solution is in the condition intended, e.g. if ammonia solution is to be added to make the solution alkaline it is insufficient to rely on a sense of smell or to test the upper layer of the solution with litmus paper; the correct procedure is to mix thoroughly either by using a glass rod or by decanting into another tube, and *then* to test a drop of the solution with litmus paper. Much trouble can be avoided, particularly from Group III onwards, by a proper regard to this operation.

(c) TESTING FOR GASES

The detection of gases which are evolved during the **ignition** of a sample follows the usual procedure. An ignition tube is filled about one-third full of the sample and heated, gently at first and later to completion, and appropriate reagents and test papers are applied at the mouth of the tube.

The procedure for the testing of gases in small-scale analysis evolved **by the action of acid or alkali** also follows, in general, that of the full-scale method, but practical difficulties arise from the reduced scale of quantities of the sample used and from the size of the tubes. These difficulties are overcome by using apparatus other than analysis tubes and by variation of technique. The boiling-tube, although of about the same capacity as an analysis tube, has a wider cross section and may be used for the generation of, and the testing for, gases without difficulty provided that suitable quantities of the sample and reagents are used. An alternative method, which is particularly suitable for the detection of carbon dioxide, is to put a quantity of the sample into the crucible, add acid, cover with the watch-glass from which on the under-surface hangs a drop of calcium hydroxide solution, and observe the action of the gas on the solution. Whilst this method may be adapted for detection of other gases it is generally found that the simpler boiling tube method is adequate.

(d) PREPARATION OF A SOLUTION

The sample must be in finely powdered form and, if a mixture, it must appear to be homogeneous.

The concentration of the solution should be as low as efficient working permits; it should be within the limits of 1–5 %, preferably near to the lower limit.

The 75 × 10 mm test-tube holds approximately 3.5 cm^3 when completely full (the 100 × 16 mm tube about 12 cm^3). The small, angled end of the spatula is a useful measure for the solid sample: trial tests show that, for a variety of mixtures, a moderately heaped portion on the spatula end weighs between 10 and 30 mg. Since 20 mg of solid dissolved in 2 cm^3 gives a 1 % solution, a moderately heaped portion in half a tubeful of the solvent gives a solution of about the required concentration.

Before attempting to prepare the solution for analysis, it is first requisite to find a suitable solvent. In many cases distilled water will dissolve the sample, but in others acid may be necessary. In these preliminary trials it is advisable to use small test portions, about a quarter of a spatula measure in a tube half full of the solvent. The solvents should be tried in the order

(i) distilled water,

(ii) dilute hydrochloric acid,

(iii) concentrated hydrochloric acid.

When testing with water as the solvent, the mixture should be shaken vigorously. Heat should be used only when there is some doubt about solubility in the cold, and the solution should then be cooled to room temperature and poured into another tube. The appearance of white solid may be due to lead(II) chloride which is soluble in hot water and sparingly soluble in cold water. Should this occur, the solution prepared for analysis must be centrifuged and the solid examined for lead in Group I.

If a solution is made with dilute hydrochloric acid, the sample is obviously clear of lead(II), silver or mercury(I) ions and is ready for treatment with hydrogen sulphide in Group II. If concentrated hydrochloric acid is the solvent, the solution must be diluted to five times its volume before use in analysis; if a milkiness occurs during dilution, hydrolysis of antimony(III) chloride has probably occurred, and the mixture should be boiled, treated with hydrogen sulphide, and the procedure of Group II followed. It is unlikely that samples requiring drastic methods to take them into solution will be encountered at this level, and it may be expected that in the majority of cases either water or dilute hydrochloric acid will prove suitable.

Having ascertained the suitable solvent, half a tubeful of solution containing a moderately heaped spatula portion of sample is made and divided into two tubes; one part is used for the group analysis, whilst the other part is available for testing for some anions.

(e) USE OF THE CENTRIFUGE

As damage will result if the head of the centrifuge is not balanced, each of the two buckets must always contain a test-tube and be of approximately the same mass before the head is allowed to rotate. A tube is filled to the same depth with water as the tube being used for the analysis is filled with solution so that the head can move smoothly. When in the course of an analysis a precipitate is formed, the tube is placed in the bucket, the level of water in the counter-balancing tube checked, the lid closed and the switch pressed for 10–15 seconds, by which time the solid should have settled. Difficulty in settling sometimes occurs, particularly with sulphide precipitates, and may be overcome by warming the mixture and recentrifuging; the addition of one or two drops of ethanol helps to prevent 'creeping' of the suspension on the side of the tube. Instructions, given in the scheme, to 'centrifuge' are intended to include all the procedure given in this paragraph. Centrifuging must be done to separate the precipitate and solution in small-scale analysis; it is optional in full-scale analysis.

(f) ISOLATING AND WASHING A PRECIPITATE

After centrifuging, the supernatant liquid is removed from the tube by means of a teat pipette. The test-tube is held at an angle of about 45°, the pipette is dipped into the liquid while the teat is kept pressed and with its tip just clear of the solid, and the pressure is slowly released to draw the liquid into the stem; the liquid is then transferred to another tube. This procedure is implied in the instruction to 'draw off the liquid'.

To wash the precipitate, add to the solid sufficient water from the wash bottle to half-fill the tube. Break up the solid with a glass rod, then centrifuge. Draw off the liquid and discard it. It is not necessary to attempt to remove all the liquid. Repeat the procedure to leave the solid with just a covering of water. In the instructions this is referred to as 'washing the precipitate'.

(g) ADDITION OF HYDROGEN SULPHIDE

A solution containing a high concentration of hydrogen sulphide is readily prepared by saturating acetone (propanone) with the gas.

About 0.5 cm^3 of water is added to 100 cm^3 of acetone and then saturated with hydrogen sulphide. This solution retains its efficiency for a reasonably long time, but it is advisable to empty the dropping bottles at three-weekly intervals and refill with a freshly prepared solution. Because of the high concentration of gas in the solution, three or four drops are usually enough to complete precipitation and this method of addition of hydrogen sulphide has much to recommend it. The reagent is readily eliminated, when it is necessary to do so, by boiling the solution on the heating block; the appearance of a blue flame on the block is due to acetone vapour and should be ignored. An alternative method is to use a hydrogen sulphide generator of a type designed for small-scale analysis. When using a generator care should be taken to see that the tip of the delivery pipette is just below the surface of the liquid while the tap is turned, then, when the desired flow of gas has been adjusted, the test-tube is raised enough to bring the tip of the pipette to the lower portion of the liquid. Where this method is used, the instruction in the scheme to 'add drops of hydrogen sulphide solution' should be considered as equivalent to 'saturate with hydrogen sulphide'.

COURSE OF AN ANALYSIS

The stages of an analysis are considered under the following headings:

	Average time (minutes)
1. Preliminary tests	10
2. Tests for acidic radicals	10
3. Metallic radicals: separation into groups	15
4. Metallic radicals: examination of group precipitates	10

Observations made in the course of the analysis should be written down immediately and presented in a clear manner as a final report. A suggested form of report is given opposite. An alternative is to rule three columns, and head them 'Test', 'Observation' and 'Inference' respectively.

The examination of single salts is more often set in examinations than that of four radical mixtures: if the latter is set, only one metal is present in any one group. The following is a suggested form of report:

REPORT ON A SAMPLE

Operations	*Inferences*
Preliminary tests	
Appearance. Faintly green powder.	Cr^{3+}, Cu^{2+}, Fe^{2+}, Ni^{2+}?
Ignition. Water.	Water of cyst.?
Calcium hydroxide solution milky: carbon dioxide.	Carbonate or hydrogen carbonate
Potassium permanganate solution decolorized: sulphur dioxide.	Sulphate?
Black residue: oxide of copper, cobalt, nickel or iron?	Copper etc. compound
Flame test: indefinite.	
Ammonium radical: absent.	Warm with sodium hydroxide solution: no ammonia/absent
Tests for acidic radicals	
Gases evolved by treatment with hydrochloric acid: carbon dioxide.	Carbonate or hydrogen carbonate
Using prepared solution: barium chloride gave precipitate.	Sulphate
[Using prepared solution: potassium shown to be absent.]	
Group separation	
Sample soluble in dilute hydrochloric acid.	
Group I absent.	
Group II: brown precipitate.	Group II present
Group III: absent.	
Group IV: black precipitate.	Group IV present
Group V: absent.	
Examination of group precipitates	
Group II. Insoluble in ammonium sulphide . . . IIa. Soluble in hydrochloric acid. Milkiness on dilution. Thiourea test positive.	Bismuth
Group IV. Soluble in concentrated hydrochloric acid. Yellow solid on evaporation. Red coloured precipitate with dimethylglyoxime (butanedione dioxime).	Nickel
Group VI solution: magnesium absent, trace only of sodium.	

Result

Bi^{3+} CO_3^{2-}

Ni^{2+} SO_4^{2-}

and water (probably water of crystallization)

A FEW POINTS OF GENERAL APPLICATION FOR THE PERFORMING AND WRITING UP OF AN ANALYSIS

Learn your tables thoroughly, letting them rest upon first principles as much as possible. Make yourself familiar with any new properties of compounds of which you were previously unaware. This applies even if you are allowed to take books into the practical examination.

Be methodical. Do not empty the solution from Group IV down the sink in mistake for an unwanted solution. You are not likely to have time to go through the main group separation more than once. Label any tube or precipitate which you must leave temporarily. Keep your bench clean and tidy.

Record your observations honestly. Do not be biased by a preliminary test, for interference is often possible. Put down exactly what you see. It may seem wrong at that moment, but it may be most illuminating in the light of further knowledge.

Make yourself familiar with precipitates which are not absolutely convincing, e.g.

Sulphate does not always come down as a copious white precipitate.

Chloride can usually be shown to be present in tap water, due to impurity.

Iron often comes from a frothing over of the Kipp's apparatus, if a trap is not used.

Aluminium hydroxide, zinc sulphide and magnesium ammonium phosphate are often missed because of lack of care in observation.

Also sodium is often present as an impurity.

Remember an examiner has only what you have written by which to judge the success of your experiments. Do not leave the writing up until the very last minute. Do it as you go along.

Interpret your result in the light of reason. Do not return barium and a sulphate if the original powder was soluble in water. Do not return copper or chromium if the original substance dissolved in water to form a colourless solution.

41

Preliminary Tests

These tests should be limited both in scope and in the time allowed for them. Test papers, reagents and apparatus should be conveniently placed to allow smooth and efficient transfer of attention from one test to another. Careful observation can afford useful information about the nature of the sample.

(a) GENERAL CHARACTERISTICS

Observe the appearance and colour of the sample and test it for acidity or alkalinity with damp litmus paper.

If a substance is white (even when moistened) it cannot contain such coloured materials as:

hydrated copper(II) salts	blue or green
hydrated nickel(II) salts	bright green
hydrated iron(II) salts	pale green
hydrated chromium(III) salts	dark green
hydrated iron(III) salts	yellow
hydrated cobalt(II) salts	crimson
manganese(IV) oxide copper(II) oxide etc.	black

After testing the substance with litmus, find a suitable solvent for performing the separation of the cations in Groups I to VI: water and hydrochloric acid (dilute or concentrated) should suffice at this stage (see p. 376).

(b) ACTION OF HEAT

Fill an ignition tube about one-third full of the sample, support in the forceps, and heat directly in the Bunsen flame, gently at first then gradually raising the temperature to the maximum. Note changes as they occur, test for gases evolved and observe the colour and nature of the residue. When testing for gases evolved, test first for carbon dioxide by resting the lip of the ignition tube over the lip of an analysis tube, the tube being about one-quarter full of calcium hydroxide solution. After a reasonable time shake the analysis tube. If this test is prolonged, it is advisable to start testing for the remaining gases with a fresh portion of the sample.

Observation	*Inference*
Moisture	Water of crystallization, basic salts, acid salts, hydroxides, hydrogencarbonates.
Sublimate: white	Ammonium salts (test with damp litmus paper), antimony oxide, mercury(II) chloride.
yellow	Sulphur, mercury(II) iodide.
grey	Mercury.
violet vapour and black deposit	Iodine.
Gases evolved: brown (coloured)	Nitrogen dioxide (from a heavy metal nitrate), bromine (from bromide by oxidation).
yellow-green	Chlorine (from chloride by oxidation). (Test with damp litmus paper).
Gases evolved: hydrogen sulphide (smell)	Sulphide (test with lead acetate paper).
sulphur dioxide	Sulphate, sulphite (test with potassium permanganate on filter paper).
ammonia	Ammonium salt.
Gases evolved: carbon dioxide (colourless, odourless)	Carbonate, hydrogencarbonate.
oxygen	Oxy-salt, easily decomposed oxide (an oxide of mercury probably).
steam	Hygroscopic substance or one with water of crystallization, some basic hydroxides and carbonates.
Residues: yellow hot, white cold	Zinc oxide, tin(IV) oxide.
yellow, melts	Lead(II) oxide, bismuth(III) oxide.
black	Oxides of copper, cobalt, nickel, iron.
red-brown	Iron(III) oxide.

Melting and decrepitation (noisy fusion) are too common to give a useful indication.

(c) FLAME TESTS

Flame tests are mainly of value as confirmatory tests after examination of group precipitates, but they also have an indication value at this earlier stage.

Take three or four drops of concentrated hydrochloric acid on the watch-glass. Make the nichrome wire red hot and dip it into the acid; again heat the wire and repeat this treatment until there is no noticeable colour when the wire is in the flame. Clean the watch-glass. Take a similar small quantity of the acid on the glass and place a small portion of the sample near to the acid. Moisten the wire with acid, touch it against the sample, then heat, first in the edge of the flame and finally in the hottest part. Observe the flame colour, viewing it directly and then through blue glass.

Viewed directly	*Through blue glass*	*Inference*
Yellow (intense)	——	Sodium
Lilac	Crimson	Potassium
Brick-red	Light green	Calcium
Light green	——	Barium
Blue-green	——	Copper
Blue	——	Lead, antimony

(d) TEST FOR AMMONIUM RADICAL

Take enough of the sample to give a depth of about 5 mm in the boiling-tube and add sodium hydroxide solution to cover the solid. Warm gently. Test by smell and by damp red litmus paper. A positive result indicates that the *ammonium* ion is present in the sample.

42

Tests for Acidic Radicals

Some acidic radicals may have been detected in the preliminary tests. A sample of the solid material is treated with dilute hydrochloric acid and then, if no result has been obtained, a new sample is carefully treated with concentrated sulphuric acid.

(a) ACTION WITH DILUTE HYDROCHLORIC ACID (heat if necessary)

Gas evolved	*Indication*
Carbon dioxide	carbonate or hydrogencarbonate
Hydrogen	some free metals
Sulphur dioxide	sulphite (sulphate(IV)) or hydrogensulphite
Sulphur dioxide and sulphur	thiosulphate
Hydrogen sulphide	sulphide
Nitrogen dioxide	nitrite (nitrate(III))
Chlorine	hypochlorite (chlorate(I)) or oxidizing agent

Carbonate. Take enough of the sample to give a depth of about 5 mm in the crucible and add dilute hydrochloric or nitric acid until the vessel is about a quarter full. If effervescence occurs cover the crucible with the watch-glass, on the underside of which a drop of calcium hydroxide solution has been placed; milkiness in the hanging drop indicates carbon dioxide and hence a *carbonate.*

Sulphite, thiosulphate, sulphide, nitrite. Take enough of the sample in the boiling-tube to give a depth of about 5 mm; add dilute hydrochloric acid until the tube is about a quarter full. Warm gently. The immediate appearance of light-brown fumes indicates a *nitrite*; a slightly blue colour may also be seen in the solution. Test for sulphur dioxide and hydrogen sulphide respectively by smell and by appropriate test papers; a spot of potassium dichromate solution on filter paper turning green, together with the characteristic smell, indicates sulphur dioxide, and hence a *sulphite* (or *thiosulphate*), and the staining of lead acetate paper indicates hydrogen sulphide and hence a *sulphide*. A *thiosulphate* gives a precipitate of sulphur on treatment with the acid as well as releasing sulphur dioxide.

(b) ACTION WITH CONCENTRATED SULPHURIC ACID
(heat if necessary)

Gas evolved	*Indication*
Hydrogen chloride	chloride
Hydrogen nitrate	nitrate
Sulphur dioxide	sulphite (sulphate(IV)), thiosulphate or reducing agent
Hydrogen bromide, bromine and sulphur dioxide	bromide
Hydrogen iodide, iodine and hydrogen sulphide	iodide
Carbon monoxide and carbon dioxide	oxalate (ethanedioate)
Carbon monoxide only	formate (methanoate)
Hydrogen acetate	acetate (ethanoate)

If an organic acid is present, the substance should be ignited, extracted with dilute hydrochloric acid and filtered before proceeding with main group separation. N.B. This is one of the situations where someone who does not do the tests in the recommended order will encounter great difficulty.

(c) TESTS FOR ACIDIC RADICALS IN SOLUTION

It is essential, before testing a solution for acidic radicals, to remove any heavy metals, because they would interfere with the tests. To take a simple case, the test for a sulphate radical in solution is to add dilute hydrochloric acid and barium chloride solution, when a white precipitate of barium sulphate indicates the presence of a sulphate.

$$Ba^{2+} + SO_4^{2-} \rightarrow BaSO_4\downarrow$$

But if the solution already contains the silver ion, then, on the addition of hydrochloric acid, there would be observed a white precipitate of silver chloride, and any further test for a sulphate would be worthless.

$$Ag^+ + Cl^- \rightarrow AgCl\downarrow$$

Possibilities such as this make it necessary to eliminate all heavy metal radicals before testing for acidic radicals, leaving only sodium, potassium or ammonium in solution.

This is done by boiling the finely divided solid (or a solution of it) with sodium carbonate solution. Heavy metals are then precipitated as carbonates, or by hydrolysis, as the hydroxides. The precipitates are filtered or centrifuged off. Occasionally, as with copper, a soluble double carbonate may be formed, but such cases are rare. The acidic radicals, originally combined with the heavy metals, are now in solution as the sodium salts, ionic association usually having taken place, e.g. if a mixture contained barium chloride and calcium nitrate, the reactions would be

$$Ba^{2+} + CO_3^{2-} \rightarrow BaCO_3\downarrow$$

$$Ca^{2+} + CO_3^{2-} \rightarrow CaCO_3\downarrow$$

The filtrate can now be tested, without interference, for these acidic radicals. It is certain to be alkaline with the excess of sodium carbonate used, and for the tests it must be made acidic with the acid appropriate for each test as given below. Test the solution with litmus paper after acidifying. Effervescence will, of course, occur because of the excess of sodium carbonate.

$$CO_3^{2-} + 2H^+ \rightarrow H_2O + CO_2$$

The acid to use is that which contains the same acidic radical as that of the testing reagent which you are adding, e.g. with barium chloride use hydrochloric acid, with silver nitrate nitric acid, and so on. In this way you are certain not to add the radical you are testing for, a mistake often made by beginners. If the solutions were not made acidic, the sodium carbonate present would again precipitate, as heavy metal carbonate, the metal of the testing reagent.

The procedure is as follows: Take two spatula-loads of the original powder, finely ground, put it into a small flask or dish (if soluble, dissolve it in a little water), add concentrated sodium carbonate solution (or three spatula-loads of the solid and a little water) and boil for a few minutes. Filter or centrifuge. The solution should be alkaline, showing excess of sodium carbonate. Use a *portion* of the filtrate for each of the following tests:

1. *Sulphate radical.* Add excess dilute hydrochloric acid, and then barium chloride solution. A white precipitate of barium sulphate proves sulphate radical present.

$$Ba^{2+} + SO_4^{2-} \rightarrow BaSO_4\downarrow$$

2. (i) *Chloride radical.* Add excess dilute nitric acid, followed by silver nitrate solution. A white curdy precipitate of silver chloride, readily soluble in ammonia solution, proves the presence of chloride radical.

$$Ag^+ + Cl^- \rightarrow AgCl\downarrow$$
$$AgCl + 2NH_3 \rightarrow \underset{\text{soluble diamminesilver chloride}}{[Ag(NH_3)_2]^+} + Cl^-$$

To confirm chloride, prepare chromyl chloride (chromium(VI) dichloride dioxide) as in Experiment 165.

(ii) *Bromide radical.* Test as for (i). A pale yellow precipitate (often only seen to be yellow by comparison with the chloride) of silver bromide, sparingly soluble in ammonia, proves the presence of bromide radical.

$$Ag^+ + Br^- \rightarrow AgBr\downarrow$$

(iii) *Iodide radical.* Test as for (i). A yellow precipitate of silver iodide, insoluble in ammonia, proves the presence of iodide.

$$Ag^+ + I^- \rightarrow AgI\downarrow$$

To confirm bromide or iodide, heat the *solid* with a little manganese(IV) oxide and concentrated sulphuric acid. The element bromine (dark red vapour) or iodine (violet) will be seen.

(iv) Other radicals may conveniently be detected at this stage as follows: remove the precipitate obtained (if any) and to the solution add ammonia solution drop by drop. (This test depends on the presence of excess silver nitrate in the filtrate.)

Precipitate	*Indication*
yellow	phosphate
crimson red	chromate
white	oxalate (ethanedioate)

3. *Nitrate radical.* Place some of the filtrate in a test-tube and add a slight excess of dilute sulphuric acid. The solution must be quite cold. Add to it excess freshly prepared iron(II) sulphate solution. The depth of solution in the boiling-tube should now be adjusted to about one-fifth the depth of the tube. Hold the tube firmly in a slanting position and pour concentrated sulphuric acid (*Care!*) down the side of the tube to form a separate layer underneath the solution, and to be about one-tenth the depth of the tube. The acid must not mix with the solution or heat will be generated and the test spoiled. A brown ring will presently be seen at the junction of acid and solution, proving the presence of a nitrate. Nitrites give a brown coloration before the concentrated acid is added.

Explanation: The nitrate in the presence of a dilute acid is reduced by some of the iron(II) sulphate to nitrogen oxide. This reacts with more iron(II) sulphate to form the brown compound seen at the junction of the two liquids as a ring.

$$3Fe^{2+} + NO_3^- + 4H^+ \rightarrow 3Fe^{3+} + 2H_2O + NO$$
$$[Fe(H_2O)_6]^{2+} + NO \rightarrow [Fe(H_2O)_5(NO)]^{2+} + H_2O$$

Complete oxidation of the iron(II) salt yields a pale yellow solution of an iron(III) salt.

Nitrate in presence of bromide or iodide. If a bromide or iodide is also present, a ring due either to free bromine or to free iodine will be formed. (The presence of iron(II) sulphate has no part in this reaction.) If the ions of either of these halogens have been shown to be present, first add a solution of silver *sulphate* to precipitate the halogen as its silver salt and test the filtrate for the nitrate ion.

Nitrate in presence of nitrite. If a nitrite has been shown to be present by the action of dilute hydrochloric acid in the preliminary tests, a brown ring (more diffuse than that for a nitrate) will be formed on performing the nitrate test. To find whether a nitrite and a nitrate are present together, add a concentrated solution of urea, then dilute sulphuric acid, and warm until there is no further effervescence of nitrogen. (See Experiment 361, p. 444.) The nitrite ion having been eliminated, test the solution for nitrate ion.

Confirm the presence of the nitrate radical by warming a mixture of the original *solid* with a little copper and a few drops of concentrated sulphuric acid. A yellow or brown gas indicates nitrate. The nitrate with concentrated sulphuric acid forms nitric acid which, with copper, gives nitrogen dioxide.

$$Cu + 4HNO_3 \rightarrow Cu(NO_3)_2 + 2H_2O + 2NO_2\uparrow$$

4. *Phosphate radical.* Although provision is made later for the test for a phosphate, it may quite conveniently be done at this stage.

To a little of the solution of the unknown substance in a test-tube add dilute nitric acid and ammonium molybdate solution, so that the latter is in a large excess. *Warm gently* but *do not boil.* A yellow coloration, with probably a precipitate on standing, indicates a phosphate. The precipitate when washed with dilute ammonium nitrate solution becomes one of the formula

$$(NH_4)_3PMo_{12}O_{40}$$

5. *Acetate radical.* Take a portion of the filtrate from the sodium carbonate extract in a test-tube, neutralize with dilute nitric acid and ammonia solution, then add iron(III) chloride solution. A blood-red coloration, which is discharged by hydrochloric acid, indicates acetate (ethanoate).

6. *Oxalate radical.* The presence of an oxalate (ethanedioate) will have been shown in the preliminary tests by the action of concentrated sulphuric acid. To confirm, make a solution of the substance with water, add excess calcium chloride solution and boil. Wash the precipitate of calcium oxalate by decantation and warm it with dilute sulphuric acid. While still warm add a few drops of potassium permanganate solution which is decolorized.

$$2MnO_4^- + 16H^+ + 5C_2O_4^{2-} \rightarrow 2Mn^{2+} + 8H_2O + 10CO_2$$

7. *Chromate radical.* Acidify with dilute nitric acid, add ammonia solution until *just* alkaline, then boil. Divide the solution into two parts. To one part add silver nitrate solution. A crimson red precipitate, soluble in dilute nitric acid, indicates chromate.

$$2Ag^+ + CrO_4^{2-} \rightarrow Ag_2CrO_4\downarrow$$

To the other part add barium chloride solution. A yellow precipitate soluble in hydrochloric acid confirms chromate.

$$Ba^{2+} + CrO_4^{2-} \rightarrow BaCrO_4\downarrow$$

8. *Confirmatory tests for certain acid radicals:* (i) *Acetate radical.* To a little of the solid add an equal bulk of ethanol and then a few drops of concentrated sulphuric acid. Warm gently and smell the vapour. The fruity smell of ethyl acetate indicates the presence of an acetate (ethanoate) in the original substance, e.g.

$$\underset{\text{sodium acetate}}{CH_3COONa} + C_2H_5OH + H_2SO_4 \rightarrow CH_3COOC_2H_5 + NaHSO_4 + H_2O$$

(ii) *Carbonate and hydrogencarbonate.* To a solution of the original substance, add magnesium sulphate solution. A white precipitate in the cold confirms carbonate. No precipitate in the cold, but a white precipitate on boiling, confirms hydrogencarbonate. Note: if the original solid is insoluble in water, an aqueous suspension of it may be boiled. If carbon dioxide is evolved a hydrogencarbonate is indicated.

(iii) *Chromate.* A portion of the sodium carbonate extract is acidified with dilute sulphuric acid. A few drops of diethyl ether (ethoxyethane) are added, followed by hydrogen peroxide solution. On shaking and allowing to settle, a blue coloration in the ether confirms chromate. The blue colour is due to the formation of a chromium peroxide (see Experiment 87).

(iv) *Sulphide.* On addition of lead acetate solution to a solution of the original solid, a black precipitate will appear if sulphide is present.

(v) *Sulphite.* To a solution of the original solid, add barium chloride solution. A white precipitate, soluble in hydrochloric acid, confirms sulphite (sulphate(IV)).

(vi) *Nitrate.* If nitrate is suspected to be present (e.g. because nitrogen dioxide appeared to be evolved during ignition tests) and if nitrite has also been found, take two spatula-loads of the sample in a crucible, half fill with dilute hydrochloric acid and heat to dryness. Any nitrite will be eliminated and the brown ring test as described in part II, p. 125, can be done without any risk of confusion.

$$2NO_2^- + 2H^+ \rightarrow H_2O + NO_2 + NO$$

The presence of bromide or iodide does not interfere with the following test for nitrate. Transfer the solid from the crucible to the boiling-tube. If ammonium salt has been shown to be present in the sample, add some sodium hydroxide solution and heat until all ammonia has been expelled, otherwise omit this operation. Add a portion of zinc dust or Devarda's alloy (see Experiment 139), add some more sodium hydroxide and warm the mixture. Test for ammonia. Nitrates (and nitrites) are reduced to ammonia by zinc and alkali.

$$9Zn + 16OH^- + 2NO_3^- + 14H_2O \rightarrow 9[Zn(OH)_4]^{2-} + 2NH_3 + H_2$$

The observations during the three operations of ignition, action of acid and tests with the prepared solution should now be considered and correct deductions made.

A TEST FOR THE POTASSIUM ION

Some of the prepared solution for the above tests for acidic radicals may conveniently be used at this point to test for potassium. (If ammonium salts are known to be present in the sample, all traces of ammonia must be expelled during the boiling with sodium carbonate.) Take three drops of the prepared solution on the watch-glass, acidify with acetic (ethanoic) acid and add a small crystal of sodium hexanitrocobaltate(III); a yellow precipitate confirms *potassium*, the flame colour of which has probably already been observed in the preliminary tests. The flame test may be repeated here using the prepared solution.]

43
Tests for Metallic Radicals

CHEMISTRY OF GROUP SEPARATION

The system of qualitative analysis is based on specific differences between the properties of one metallic radical and those of another. The table used for elementary work contains twenty metals and a first separation is made into groups by selecting a reagent which precipitates a few of the metals but leaves the remainder to be precipitated by a different reagent or reagents. A precipitate obtained in a given group may contain one (or more, but rarely so at this stage) of the metals of that group and it is then that use is made of specific differences of properties of the metals likely to be present. Properties of metals and their compounds are given in Part II and it may be considered advisable to study them in conjunction with the system of analysis, e.g. the properties of silver, mercury(I) and lead might be studied prior to precipitating them as their chlorides in Group I.

It is certainly advantageous to precede group separations by showing how each metal in the group behaves towards the various reagents used, and to illustrate the procedure the following experiments are given for Group I; the student will be able to suggest similar experimental work for the subsequent groups.

For the preparation of a solution for the cation separation, see p. 376.

PRELIMINARY EXPERIMENTS BEFORE THE SEPARATION OF GROUP I METALS, SILVER, MERCURY(I), AND LEAD

(*a*) To test-tubes one-quarter full of solutions of silver nitrate, mercury(I) nitrate (in dilute nitric acid because of hydrolysis) and lead acetate (ethanoate) respectively, add an equal volume of dilute hydrochloric acid. The white precipitate in each case is the insoluble metal chloride.

$$Ag^+ + Cl^- \rightarrow AgCl\downarrow$$
$$Hg_2^{2+} + 2Cl^- \rightarrow Hg_2Cl_2\downarrow$$
$$Pb^{2+} + 2Cl^- \rightarrow PbCl_2\downarrow$$

(*b*) Decant the liquid from each tube, retaining as much as possible of the solids. Fill each tube half-full with distilled water. Show that lead chloride is easily soluble in boiling water, but that mercury(I) chloride and silver chloride are both apparently insoluble. Add a few drops of potassium chromate to the clear solution of lead chloride; the yellow precipitate is lead chromate.

$$Pb^{2+} + CrO_4^{2-} \rightarrow PbCrO_4\downarrow$$

(*c*) Decant the liquid from the precipitates of silver chloride and mercury(I) chloride, then to each add a dilute solution of ammonia and shake. Silver chloride dissolves because of the formation of a complex ion, a soluble silver ammine.

$$AgCl + 2NH_3 \rightarrow [Ag(NH_3)_2]^+ + Cl^-$$

Mercury(I) chloride turns black owing to the formation of mercury; mercury(II) aminochloride (white) is also formed, but its colour is masked by the mercury.

$$Hg_2Cl_2 + 2NH_3 \rightarrow Hg(NH_2)Cl + NH_4Cl + Hg\downarrow$$

GROUP I INSOLUBLE CHLORIDES

$$PbCl_2,\ AgCl,\ Hg_2Cl_2$$

Precipitate may contain lead(II), silver(I) and mercury(I) chlorides. Wash two or three times with a little **cold** water. **Discard** washings. Then wash the precipitate through into a beaker. Boil with water. Centrifuge **hot**.

<table>
<tr><td colspan="2">RESIDUE
Wash well with hot water. Pour warm ammonia solution on to the solid.</td><td rowspan="2">SOLUTION
contains lead(II) chloride (white crystals may separate on cooling). Add potassium chromate solution. Yellow precipitate of lead chromate proves presence of lead.</td></tr>
<tr><td>RESIDUE
if black, due to presence of finely divided metallic mercury. This proves mercury(I) originally present.</td><td>SOLUTION
contains the silver. Acidify with nitric acid. White precipitate of silver chloride turning violet on exposure to light proves silver present.</td></tr>
</table>

SEPARATION INTO GROUP

To a quarter of a test-tube of solution of sample (p. 376) add a little **dilute hydrochloric acid**. Centrifuge in the cold or filter. Ensure no further precipitation.

<table>
<tr><td rowspan="5">RESIDUE
Group 1
All white
$PbCl_2$
$AgCl^-$
Hg_2Cl_2</td><td colspan="5">SOLUTION
Take test portion and add one drop of hydrogen sulphide solution (or pass in gas). If no precipitate, dilute with water. If still no precipitate, discard test portion and use main portion for Group III. If precipitate is formed, treat the main portion with hydrogen sulphide. Centrifuge. Ensure no further precipitation.</td></tr>
<tr><td rowspan="4">RESIDUE
Group II
(a) Black HgS
Black CuS
Black PbS (if not eliminated in Group I)
Brown Bi_2S_3
Yellow CdS
(b) Orange Sb_2S_3
Brown SnS
Yellow SnS_2
Group II(a) insoluble in $(NH_4)_2S_x$
Group II(b) soluble in $(NH_4)_2S_x$</td><td colspan="4">SOLUTION
If solution contains hydrogen sulphide, boil to expel it or use new portion if examining single salt. Test drops of original solution for iron. If iron is present, oxidize by boiling with three drops of concentrated nitric acid.[1] Add three heaped spatula-loads of ammonium chloride. Make alkaline with ammonia solution. Stir well and boil.[2] Centrifuge. Ensure no further precipitation.</td></tr>
<tr><td rowspan="3">RESIDUE
Group III
Red-brown $Fe(OH)_3$
White $Al(OH)_3$
Green-blue $Cr(OH)_3$</td><td colspan="3">SOLUTION
Take test portion of buffered solution and add hydrogen sulphide solution. If no precipitate, discard test portion and use main portion in Group V. If precipitate is formed, treat main portion with hydrogen sulphide. Centrifuge. Ensure no further precipitation.</td></tr>
<tr><td rowspan="2">RESIDUE
Group IV
Black NiS
Black CoS
Flesh-coloured MnS
Dirty white ZnS</td><td colspan="2">SOLUTION
To buffered solution add ammonium carbonate solution. Bring to boil. Centrifuge. Ensure no further precipitation.</td></tr>
<tr><td>RESIDUE
Group V
Both white
$BaCO_3$
$CaCO_3$</td><td>SOLUTION
Group VI
Contains Na^+, k^+ or Mg^{2+}</td></tr>
<tr><td>Wash solid: examine as described on p. 392</td><td>Wash solid: examine as described on pp. 396 and 398</td><td>Wash solid: examine as described on p. 399</td><td>Wash solid: examine as described on p. 400</td><td>Wash solid: examine as described on p. 402</td><td>Examine solution as described on p. 403</td></tr>
</table>

[1] A phosphate separation is never required. The use of nitric acid is vital only if iron(II) cation or an organic anion (alternative, see Chapter 42 (*b*)) is present.
[2] If the solution is not boiled some chromium may remain in solution as $[Cr(NH_3)_6]^{3+}$.

CONFIRMATORY TESTS WITH ORIGINAL SOLUTION OR SOLID

Silver — Add potassium chromate solution to neutral solution of silver salt, brick red precipitate obtained.

Mercury — Heat solid mixed with anhydrous sodium carbonate in tube. Grey metallic mirror obtained which can be scraped together to form bead of metallic mercury.

Lead — With potassium iodide solution, solutions of lead salts give a yellow precipitate sparingly soluble in boiling water.

Also see Chapter 45 for confirmatory tests.

EQUATIONS

Precipitation by hydrochloric acid:

$$Pb^{2+} + 2Cl^- \rightarrow PbCl_2\downarrow \qquad Ag^+ + Cl^- \rightarrow AgCl\downarrow$$

$$Hg_2^{2+} + 2Cl^- \rightarrow Hg_2Cl_2\downarrow$$

Action of potassium chromate on lead chloride:

$$Pb^{2+} + CrO_4^{2-} \rightarrow PbCrO_4\downarrow$$

Action of ammonia on silver and mercury(I) chlorides:

$$AgCl + 2NH_3 \rightarrow [Ag(NH_3)_2]^+ + Cl^-$$

$$Hg_2Cl_2 + 2NH_3 \rightarrow \underset{\text{mercury(II) aminochloride}}{Hg(NH_2)Cl\downarrow} + NH_4Cl + Hg\downarrow$$

Precipitation of silver chloride by nitric acid:

$$[Ag(NH_3)_2]^+ + Cl^+ + 2H^+ \rightarrow AgCl\downarrow + 2NH_4^+$$

$$2AgCl \rightarrow 2Ag + Cl_2$$

Confirmatory test for silver:

$$2Ag^+ + CrO_4^{2-} \rightarrow \underset{\text{brick red}}{Ag_2CrO_4\downarrow}$$

Confirmatory test for mercury:

$$\underset{\text{(for example)}}{2HgCl_2} + 2Na_2CO_3 \rightarrow 2Hg + 4NaCl + 2CO_2 + O_2$$

Confirmatory test for lead:

$$Pb^{2+} + 2I^- \rightarrow PbI_2\downarrow$$

GROUP II SULPHIDES INSOLUBLE IN DILUTE HYDROCHLORIC ACID

$HgS, PbS, Bi_2S_3, CuS, CdS$ Group IIa

Sb_2S_3, SnS, SnS_2 Group IIb

Keep the precipitate covered with water to minimize oxidation of copper sulphide to copper sulphate which would be washed out.

Wash the precipitate well with hot water and then wash it through into an evaporating basin. Add **sodium hydroxide** solution and a few drops of **yellow ammonium sulphide**. Warm for a few minutes. Centrifuge.

RESIDUE Group II(a)	SOLUTION Group II(b)

EQUATIONS

Precipitation of the sulphides:

$$Pb^{2+} + S^{2-} \rightarrow PbS\downarrow \quad \text{(similarly } Hg^{2+}, Cu^{2+}, Cd^{2+}, Sn^{2+}\text{)}$$
$$2Bi^{3+} + 3S^{2-} \rightarrow Bi_2S_3\downarrow \quad \text{(similarly } Sb^{3+}\text{)}$$
$$Sn^{4+} + 2S^{2-} \rightarrow SnS_2\downarrow$$

The red precipitate produced at first from a lead salt on adding hydrogen sulphide is $PbS \cdot PbCl_2$. It is decomposed by more hydrogen sulphide to black lead sulphide. The precipitate from mercury(II) chloride may first be white ($HgCl_2 \cdot 2HgS$). It changes to yellow, brown and black by gradual decomposition of this compound into black mercury(II) sulphide by more hydrogen sulphide.

Action of sodium hydroxide and ammonium sulphide: the sulphur of the polysulphide, ammonium sulphide, oxidizes the lower sulphide and forms a thio-salt, e.g.

$$Sb_2S_3 + 3S^{2-} + \underset{\text{from polysulphide}}{2S} \rightarrow \underset{\text{thioantimonate(V) ion}}{2SbS_4^{3-}}$$

$$SnS + S^{2-} + S \rightarrow \underset{\text{thiostannate(IV)}}{SnS_3^{2-}}$$

A white suspension of sulphur may occur due to the oxidation of hydrogen sulphide by nitrate ions, etc.

$$H_2S + 2H^+ + 2NO_3^- \rightarrow S + 2H_2O + 2NO_2$$

GROUP IIa SULPHIDES INSOLUBLE IN YELLOW AMMONIUM SULPHIDE

HgS, PbS, Bi_2S_3, CuS, CdS

Wash the precipitate well with hot water, and then wash it into a beaker. **Add dilute nitric acid** and boil for a few minutes. (Sulphur will usually remain here.) Transfer the whole to a boiling-tube, add **dilute sulphuric acid** and a little **ethanol** to complete the precipitation of lead sulphate. (Omit this if lead not found present in Group I.) Allow to stand for a few minutes. Centrifuge.

<table>
<tr><th colspan="2">RESIDUE</th><th colspan="2">SOLUTION</th></tr>
<tr><td colspan="2">may contain lead(II) sulphate, mercury(II) sulphide and sulphur. Wash with hot water, and then wash precipitate into a boiling-tube and boil with ammonium acetate solution. Centrifuge.</td><td colspan="2">may contain bismuth, copper and cadmium as nitrates. Add ammonia solution in excess.</td></tr>
<tr><th>RESIDUE</th><th>SOLUTION</th><th>RESIDUE</th><th>SOLUTION</th></tr>
<tr><td>is mercury(II) sulphide and sulphur. Wash, warm in a dish in a fume chamber with concentrated hydrochloric acid and potassium chlorate until dissolved. Dilute and add tin(II) chloride solution. White silky precipitate of mercury(I) chloride proves mercury (II) originally present.</td><td>contains lead ions. Add potassium chromate solution. Yellow precipitate of lead chromate proves presence of lead.</td><td>is white bismuth hydroxide. Wash; dissolve by pouring a little warm, dilute hydrochloric acid through the filter paper. Pour solution into beaker nearly full of water. Turbidity (perhaps delayed) due to bismuth(III) oxychloride proves presence of bismuth.</td><td>If blue: acidify with dilute acetic acid, add potassium hexacyanoferrate(II) solution. Brown colour or precipitate due to copper hexacyanoferrate(II) proves presence of copper.
If colourless: pass in hydrogen sulphide. Yellow precipitate of cadmium sulphide proves presence of cadmium.</td></tr>
</table>

CONFIRMATORY TEST WITH ORIGINAL SOLUTION

Copper — Ammonia solution gives a pale blue precipitate which dissolves in excess to give a deep blue solution.

Cadmium — Ammonia solution gives white precipitate easily soluble in excess to give a colourless solution.

Also see Chapter 45 for confirmatory tests.

EQUATIONS

Action of dilute nitric acid:

$$CuS + 2H^+ \rightarrow Cu^{2+} + H_2S\uparrow \quad \text{(similarly CdS, PbS)}$$
$$Bi_2S_3 + 6H^+ \rightarrow 2Bi^{3+} + 3H_2S\uparrow$$

together with some oxidation to sulphate in each case, e.g.

$$CuS + 8HNO_3 \rightarrow CuSO_4 + 8NO_2 + 4H_2O$$

Mercury(II) sulphide is unchanged.

Action of sulphuric acid:

$$Pb^{2+} + SO_4^{2-} \rightarrow PbSO_4\downarrow$$

Action of ammonium acetate:

$$PbSO_4 + 2CH_3COONH_4 \rightarrow (CH_3COO)_2Pb + (NH_4)_2SO_4$$

Action of ammonia solution on bismuth, copper and cadmium nitrate solutions:

$$Bi^{3+} + 3OH^- \rightarrow Bi(OH)_3\downarrow$$
$$Cu^{2+} + 4NH_3 \rightarrow [Cu(NH_3)_4]^{2+} \quad \text{(similarly } Cd^{2+}\text{)}$$

Action of acetic acid:

$$[Cu(NH_3)_4]^{2+} + 4H^+ \rightarrow Cu^{2+} + 4NH_4^+$$

(similarly Cd^{2+})

$$2Cu^{2+} + [Fe(CN)_6]^{4-} \rightarrow Cu_2Fe(CN)_6\downarrow$$

Action of hydrogen sulphide:

$$Cd^{2+} + S^{2-} \rightarrow CdS\downarrow$$

Confirmatory test for mercury(II): the concentrated hydrochloric acid and potassium chlorate mixture oxidizes the mercury(II) sulphide by a complex reaction into soluble mercury(II) chloride. Then, on adding tin(II) chloride solution,

$$2Hg^{2+} + Sn^{2+} + 2Cl^- \rightarrow Hg_2Cl_2\downarrow + Sn^{4+}$$

Confirmatory test for lead:

$$Pb^{2+} + CrO_4^{2-} \rightarrow PbCrO_4\downarrow$$

Confirmatory test for bismuth:

$$Bi(OH)_3 + 3H^+ \rightarrow Bi^{3+} + 3H_2O$$
$$Bi^{3+} + H_2O + Cl^- \rightarrow BiOCl\downarrow + 2H^+$$

Confirmatory test for copper:

$$Cu^{2+} + 2OH^- \rightarrow Cu(OH)_2\downarrow$$

$$Cu(OH)_2 + 4NH_3 \rightarrow [Cu(NH_3)_4]^{2+} + 2OH^-$$

Confirmatory test for cadmium:

$$Cd^{2+} + 2OH^- \rightarrow Cd(OH)_2\downarrow$$

$$Cd(OH)_2 + 4NH_3 \rightarrow [Cd(NH_3)_4]^{2+} + 2OH^-$$

GROUP IIb SULPHIDES SOLUBLE IN YELLOW AMMONIUM SULPHIDE

Sb_2S_3, SnS, SnS_2

If the Group II precipitate was soluble in sodium hydroxide and ammonium sulphide, Group IIb was present: acidify with **dilute hydrochloric acid**. If a precipitate appears, pass hydrogen sulphide to complete the precipitation, warm and centrifuge (discard solution). Wash the precipitate with hot water and then wash the precipitate through into a beaker. Add **concentrated hydrochloric acid** and boil for a few moments. Divide the solution into two parts. Cool.

Part I	*Part II*
Dilute with its own volume of water. Add **hydrogen sulphide** solution. Orange-red precipitate of antimony(III) sulphide proves **antimony**.	Add **zinc foil** in a dish until no further effervescence. If **tin** is present, it will be as a grey precipitate round the *top* of the dish. Collect it with a rod and dissolve it in a test-tube with **concentrated hydrochloric acid.** Pour this solution into a solution **mercury(II) chloride**. White silky precipitate at once or on standing is mercury(I) chloride, proving presence of **tin**.

CONFIRMATORY TESTS WITH ORIGINAL SOLUTION OR SOLID

Study the original substance to see if tin(II) or tin(IV) is present. Also see Chapter 45 for confirmatory tests.

EQUATIONS

The addition of dilute acid precipitates the higher sulphide:

$$2SbS_4^{3-} + 6H^+ \rightarrow Sb_2S_5\downarrow + 3H_2S\uparrow$$
$$SnS_3^{2-} + 2H^+ \rightarrow SnS_2\downarrow + H_2S\uparrow$$

The sulphides are then redissolved in concentrated acid.

$$SnS_2 + 4H^+ \rightarrow Sn^{4+} + 2H_2S\uparrow$$
$$Sb_2S_5 + 6H^+ \rightarrow 2Sb^{3+} + 3H_2S\uparrow + 2S\downarrow$$

Confirmatory test for antimony:

$$2Sb^{3+} + 3S^{2-} \rightarrow Sb_2S_3\downarrow$$

Confirmatory test for tin:

$$Zn + 2H^+ \rightarrow Zn^{2+} + H_2\uparrow$$
$$2Zn + Sn^{4+} \rightarrow 2Zn^{2+} + Sn\downarrow$$
$$Sn + 2H^+ \rightarrow Sn^{2+} + H_2\uparrow$$
$$Sn^{2+} + 2Hg^{2+} + 2Cl^- \rightarrow Sn^{4+} + Hg_2Cl_2\downarrow$$

GROUP III INSOLUBLE HYDROXIDES

$Fe(OH)_3$, $Cr(OH)_3$, $Al(OH)_3$

The precipitate may contain iron(III), chromium(III) and aluminium hydroxides. Wash well with hot water, and then wash the precipitate into a boiling-tube. Add **sodium hydroxide** solution and **hydrogen peroxide** solution and boil for a few minutes (oxygen may be released). Centrifuge.

<table>
<tr><td rowspan="3">RESIDUE
is brown iron(III) hydroxide. Dissolve in **dilute hydrochloric acid**, and add **potassium hexacyanoferrate(II)** solution. Blue precipitate proves presence of **iron.** To the original solution add (i) **potassium hexacyanoferrate(III)** solution. Blue precipitate proves **iron(II)** present; and (ii) **potassium thiocyanate** (KSCN) solution. Red coloration proves **iron(III)** present. A little manganese(II) hydroxide may occur here, but may be neglected. It will appear in bulk in Group IV.</td><td colspan="2">SOLUTION
may contain chromium and aluminium as sodium chromate and sodium aluminate. If **yellow**, chromium is present. Divide into two parts.</td></tr>
<tr><td>Part I
If solution is colourless, omit this. Add **acetic (ethanoic) acid** in excess, and then **lead acetate** (ethanoate) solution. Yellow precipitate of lead chromate proves **chromium** present.</td><td>Part II
Add **litmus.** Add **dilute hydrochloric acid** in excess, then **ammonia** solution just in excess. Shake. Allow to stand. **Blue lake** of aluminium hydroxide and litmus proves **aluminium** present.</td></tr>
</table>

CONFIRMATORY TESTS WITH ORIGINAL SOLUTION OR SOLID

Iron As above.

Aluminium Heat solid on charcoal block with sodium carbonate. Add a few drops of cobalt(II) nitrate solution and heat again. Blue residue obtained.

Chromium Fuse with sodium carbonate and a little potassium nitrate in a porcelain crucible. Dissolve in water, add acetic (ethanoic) acid and lead acetate (ethanoate) solution. Yellow precipitate obtained.

Also see Chapter 45 for confirmatory tests.

EQUATIONS

Precipitation of hydroxides.

$$Fe^{3+} + 3OH^- \rightarrow Fe(OH)_3\downarrow$$
$$Cr^{3+} + 3OH^- \rightarrow Cr(OH)_3\downarrow$$
$$Al^{3+} + 3OH^- \rightarrow Al(OH)_3\downarrow$$

Action of sodium hydroxide and hydrogen peroxide. Iron(III) hydroxide unchanged.

$$2Cr(OH)_3 + 3H_2O_2 + 4OH^- \rightarrow 2CrO_4^{2-} + 8H_2O$$
$$Al(OH)_3 + OH^- \rightarrow Al(OH)_4^-$$

Confirmatory test for iron:

$$Fe^{3+} + SCN^- \rightarrow FeSCN^{2+}$$

Prussian blue can be formed in two ways:

$$\left.\begin{array}{l} 4Fe^{3+} + 3[Fe(CN)_6]^{4-} + 14H_2O \\ \text{or} \\ 4Fe^{2+} + 4[Fe(CN)_6]^{3-} + 14H_2O \end{array}\right\} \rightarrow Fe_4[Fe(CN)_6^{4-}]_3 \cdot 14H_2O$$

(and in the second case $[Fe(CN)_6]^{4-}$ formed also)

Confirmatory test for chromium:

$$Pb^{2+} + CrO_4^{2-} \rightarrow PbCrO_4\downarrow$$

Confirmatory test for aluminium:

$$Al(OH)_4^- + 4H^+ \rightarrow Al^{3+} + 4H_2O$$
$$Al^{3+} + 3OH^- \rightarrow \underset{\text{adsorbs litmus}}{Al(OH)_3\downarrow}$$

GROUP IV SULPHIDES PRECIPITATED BY HYDROGEN SULPHIDE IN ALKALINE SOLUTION

ZnS, MnS, CoS, NiS

Precipitate may contain zinc, manganese, cobalt and nickel sulphides and sulphur. If it is not black, cobalt and nickel are both absent. Wash the precipitate and then add **very dilute hydrochloric acid.**

<table>
<tr>
<td rowspan="2">RESIDUE
may contain cobalt or nickel as sulphides. Transfer it to a dish, add concentrated hydrochloric acid and a crystal of potassium chlorate. Heat until dissolved and then evaporate nearly to dryness.

Nickel. Solution greenish-yellow, crystal deposit yellow. Add alkaline solution of dimethylglyoxime (butanedione dioxime): red coloration or precipitate shows nickel.
Cobalt. Solution will be pink and deposit of crystals blue.
To confirm, apply borax bead test.
Blue bead cobalt.
Brown bead nickel.</td>
<td colspan="2">SOLUTION
may contain zinc and manganese chlorides. Boil in a dish for several minutes to remove hydrogen sulphide. (If it is still turbid, finely divided sulphur is suspended in it. Add a little potassium chlorate and boil until clear.) Cool. Add excess sodium hydroxide solution. Filter.</td>
</tr>
<tr>
<td>RESIDUE
is manganese hydroxide. Turns brown on filter paper. Wash through into a boiling-tube, allow to settle, pour off as much water as possible. Add concentrated nitric acid and lead(IV) oxide or sodium bismuthate(V) and boil. Dilute and allow to settle. Crimson coloration, due to permanganic acid proves presence of manganese.</td>
<td>RESIDUE
contains zinc as sodium zincate. Pass hydrogen sulphide.

White precipitate (often discoloured) of zinc sulphide proves zinc present.
For confirmatory test, filter and dissolve precipitate in a little concentrated nitric acid. Add a little cobalt nitrate solution, evaporate to concentrate, and soak a filter paper in the mixture. Ignite. A green ash (Rinmann's green) confirms zinc.
(The green substance is a compound of zinc and cobalt(II) oxides.)</td>
</tr>
</table>

CONFIRMATORY TESTS WITH ORIGINAL SOLUTION OR SOLID

Zinc, cobalt, nickel As above.

Manganese Fuse with sodium carbonate and a little potassium nitrate in porcelain crucible. Bluish-green residue obtained.

Also see Chapter 45 for confirmatory tests.

EQUATIONS

Precipitation of sulphides.

$$\begin{aligned} Zn^{2+} + S^{2-} &\rightarrow ZnS\downarrow \\ Mn^{2+} + S^{2-} &\rightarrow MnS\downarrow \\ Co^{2+} + S^{2-} &\rightarrow CoS\downarrow \\ Ni^{2+} + S^{2-} &\rightarrow NiS\downarrow \end{aligned}$$

If ammonium sulphide is used, some sulphur may also occur because ammonium sulphide has not really the simple formula $(NH_4)_2S$, but contains also polysulphides of the type $(NH_4)_2S_2$. A typical reaction would be

$$Zn^{2+} + S_2^{2-} \rightarrow ZnS\downarrow + S\downarrow$$

Action of very dilute hydrochloric acid.

Cobalt and nickel sulphides are unchanged.

$$\begin{aligned} ZnS + 2H^+ &\rightarrow Zn^{2+} + H_2S\uparrow \\ MnS + 2H^+ &\rightarrow Mn^{2+} + H_2S \end{aligned}$$

Action of excess sodium hydroxide solution on zinc and manganese chlorides.

$$\begin{aligned} Zn^{2+} + 2OH^- &\rightarrow Zn(OH)_2\downarrow \\ Zn(OH)_2 + 2OH^- &\rightarrow Zn(OH)_4^{2-} \\ Mn^{2+} + 2OH^- &\rightarrow Mn(OH)_2\downarrow \end{aligned}$$

On exposure to air the manganese(II) hydroxide turns into a brown substance.

$$4Mn(OH)_2 + O_2 \rightarrow 4MnO\cdot OH + 2H_2O$$

Confirmatory test for zinc:

$$2[Zn(OH)_4]^{2-} + 8H^+ + 2S^{2-} \rightarrow 2ZnS\downarrow + 8H_2O$$

Confirmatory test for manganese:

$$2Mn(OH)_2 + 5PbO_2 + 8H^+ \rightarrow 2MnO_4^- + 6H_2O + 5Pb^{2+}$$

GROUP V INSOLUBLE CARBONATES

$CaCO_3$, $BaCO_3$

Precipitate may contain calcium and barium carbonates. Wash well with hot water and then add some warm **dilute acetic (ethanoic) acid**. To a small portion of the solution add **potassium chromate** solution and boil. If there is a precipitate add potassium chromate solution to the whole and boil. If there is no precipitate discard sample and treat whole solution. Centrifuge.

RESIDUE	*SOLUTION*
is pale yellow barium chromate. This precipitate proves presence of **barium.** Confirm by flame test.	may contain calcium as its acetate. Add an excess of **ammonia** solution and **ammonium oxalate** solution. White precipitate of calcium oxalate proves presence of **calcium.**

CONFIRMATORY TESTS WITH ORIGINAL SOLID

Barium	Flame test	Light green
Calcium	Flame test	Brick-red (light green through blue glass)

Also see Chapter 45 for confirmatory tests.

EQUATIONS

Precipitation of carbonates:

$$Ba^{2+} + CO_3^{2-} \rightarrow BaCO_3\downarrow$$

$$Ca^{2+} + CO_3^{2-} \rightarrow CaCO_3\downarrow$$

Action of dilute acetic acid:

$$BaCO_3 + 2H^+ \rightarrow Ba^{2+} + H_2O + CO_2\uparrow$$
$$CaCO_3 + 2H^+ \rightarrow Ca^{2+} + H_2O + CO_2\uparrow$$

Action of potassium chromate: (Calcium chromate is not precipitated in the presence of acetic acid.)

$$Ba^{2+} + CrO_4^{2-} \rightarrow BaCrO_4\downarrow$$

Test for calcium:

$$Ca^{2+} + C_2O_4^{2-} + H_2O \rightarrow CaC_2O_4 \cdot H_2O$$

GROUP VI MAGNESIUM, SODIUM AND POTASSIUM

IF CALCIUM WAS FOUND IN GROUP V, add to the solution a little **ammonium oxalate** solution and boil. Centrifuge, and reject the precipitate of calcium oxalate. If calcium was absent from GROUP V, omit this. Divide the solution into two parts.

Part I	*Part II*
Add **ammonia** solution and **disodium hydrogenphosphate** solution. Shake well and put away to stand. A white crystalline precipitate of ammonium magnesium phosphate proves the presence of **magnesium.**	Evaporate to dryness in a dish. Heat until no more fumes from dissociating ammonium compounds are seen. **Examine residue by flame test.** Persistent golden-yellow flame proves presence of **sodium.** Lilac flame proves presence of **potassium.** If sodium present, examine the flame through blue glass for potassium.

CONFIRMATORY TEST WITH ORIGINAL SUBSTANCE

Magnesium Heat solid on charcoal block with sodium carbonate. Add a few drops of cobalt(II) nitrate solution and heat again. Pink residue obtained.

Potassium See p. 390.

Also see Chapter 45 for confirmatory tests.

Equation.

Test for magnesium:

$$Mg^{2+} + HPO_4^{2-} + 6H_2O + NH_3 \rightarrow MgNH_4PO_4 \cdot 6H_2O$$

44
The Theory of the Separation of Cations into Groups

THE SOLUBILITY PRODUCT

In a saturated solution of a sparingly soluble electrolyte (at a fixed temperature) there is an equilibrium between the excess solid (whose mass is effectively constant) and the ions formed by the dissociation of the solute.

$$\underset{\substack{\text{solid}\\\text{solute}}}{XY} \rightleftharpoons X^+ + Y^-$$

Then, similarly to the derivation of equilibrium constant (see p. 70),

$$[X^+][Y^-] = K_s$$

i.e. in a saturated solution, the product of the concentrations of the ions is a constant mathematical value. Consequently, if relatively large concentrations of the ions are brought together into the same solution, ions of X^+ and Y^- will be precipitated as a solid until the concentrations of the remaining ions have such values that the product of the concentrations equals the specific constant (the solubility product).

For the salt $(X^{n+})_m(Y^{m-})_n$, the solubility product is given by

$$[X^{n+}]^m[Y^{m-}]^n$$

and the units are $(\text{mol dm}^{-3})^{m+n}$. The values of solubility product that are quoted in chemical literature are so variable that they have been rounded off to the nearest power of 10; they are given at 25 °C.

GROUP I

In this group lead, silver and mercury(I) are precipitated as their chlorides by introducing chloride ions from the fully ionized hydrochloric acid. In terms of the ionic theory, the concentrations of, say, silver and chloride ions which can remain in solution are small, and, in consequence, when a solution of a silver salt (containing an appreciable concentration of silver ions) is mixed with hydrochloric acid, the bulk of the silver and chloride ions unite to form solid silver chloride and leave the solution until the remaining ions attain the equilibrium:

$$K_s = [Ag^+][Cl^-] = 10^{-10}\ \text{mol}^2\ \text{dm}^{-6}$$

It should also be realized that the *product* has the constant value, so that by adding a reasonable excess of chloride ions the concentration of the silver ions may be reduced to a negligible quantity.

Of all the metal chlorides only those of silver, lead and mercury(I) have low solubility products and consequently the ions of other metals remain in solution, under the given conditions, in the presence of reasonably high concentrations of chloride ions. The value of K_s for lead(II) chloride is 10^{-5} mol^3 dm^{-9} and that for mercury(I) chloride is 10^{-18} mol^3 dm^{-9}.

GROUP II

We may consider hydrogen sulphide to be ionized.

$$H_2S \rightleftharpoons 2H^+ + S^{2-}$$

$$\frac{[H^+]^2[S^{2-}]}{[H_2S]} = 10^{-22}\ \text{mol}^2\,\text{dm}^{-6}$$

In neutral solution the concentration of sulphide ion is low because hydrogen sulphide is a weak electrolyte. The concentration of hydrogen ion is also low. In acid solution, as used for Group II, the concentration of hydrogen ions is greatly increased by the presence of the strong acid. Therefore, to maintain the value of the constant, the concentration of the sulphide ion is reduced much below its already small value in neutral solution. The amount of sulphide ion is, however, still great enough to allow the solubility products of the sulphides of mercury(II), lead(II), copper(II) and bismuth(III) momentarily to be exceeded. The same is true for cadmium sulphide if the acid is not too concentrated. Therefore all the sulphides of these metals precipitate.

Compound	*Solubility product* (mol^2 dm^{-6})
lead(II) sulphide	10^{-28}
copper(II) sulphide	10^{-35}
mercury(II) sulphide	10^{-52}
cadmium(II) sulphide	10^{-26}

Table continued

Compound	*Solubility product* ($mol^2\ dm^{-6}$)
tin(II) sulphide	10^{-26}
	($mol^5\ dm^{-15}$)
bismuth(III) sulphide	10^{-72}
antimony(III)	10^{-93}

The concentration of sulphide ion in acid solution is not great enough to allow the higher (though still comparatively low) solubility products of the sulphides of manganese, zinc, cobalt or nickel to be reached with any possible concentration of metal ion, so that these sulphides do not precipitate. They come down later in Group IV, where the precipitating agent may be the highly ionized salt, ammonium sulphide, and the concentration of the sulphide ion from it is correspondingly high.

GROUP III

The precipitating agent in Group III is a solution of ammonia in the presence of ammonium chloride (a buffer solution).

$$NH_3 + H_2O \rightleftharpoons NH_4^+ + OH^-$$

$$\frac{[NH_4^+][OH^-]}{[NH_3]} = 10^{-5}\ mol\ dm^{-3}$$

Ammonia is a weak base, and therefore does not ionize to a very great extent: most of the ammonia will be dissolved but not ionized. Suppose ammonia solution is added to the solutions of soluble salts of zinc, manganese, cobalt and nickel to make the solutions 0.5 M with respect to ammonia. The degree of ionization of ammonium hydroxide at this dilution is 0.006, so that the concentration of the ion is $0.5 \times 0.006\ mol\ dm^{-3}$.

This concentration, together with the concentration of the metal ion, is great enough to exceed momentarily the low solubility products of these hydroxides, and they precipitate. Supposing that ammonia alone were the precipitating agent for Group III, then, together with the hydroxides of iron, aluminium and chromium, we should find those of zinc, manganese, cobalt and nickel. Suppose now that ammonium chloride solution and ammonia solution are added to soluble salts of these metals to make the solutions 0.5 M with respect to both ammonia and ammonium chloride. Ammonium chloride is a salt and is highly ionized, its degree of ionization being about 0.8. Then

$$[NH_4^+] \text{ from ammonium chloride} = 0.5 \times 0.8$$
$$= 0.4 \text{ mol dm}^{-3}$$

Neglecting the $[NH_4^+]$ from the ammonia solution in comparison with that from the ammonium chloride,

$$\frac{0.4[NH_4^+]}{0.5} = 10^{-5} \text{mol dm}^{-3}$$

whence $$[OH^-] = 1.25 \times 10^{-5} \text{ mol dm}^{-3}$$

Thus the concentration of hydroxide ions in 0.5 M ammonia solution is 3×10^{-3} mol dm^{-3}, and in ammonia solution, which is also 0.5 M with respect to ammonium chloride, it is reduced to 1.25×10^{-5}. The addition of ammonium chloride has reduced the concentration of the hydroxide ion to less than one-hundredth of its former value. In general, therefore, the effect of the addition of ammonium chloride is to depress greatly the ionization of ammonia and so reduce the hydroxide ion concentration in the solution.

The small hydroxide ion concentration in a solution of ammonia which is also fairly concentrated with respect to ammonium chloride is still large enough to cause precipitation of the hydroxides of iron(III), chromium and aluminium, but not great enough to precipitate those of zinc, manganese, cobalt and nickel. Manganese hydroxide may precipitate slightly if the concentration of ammonium chloride is not sufficiently great.

Compound	*Solubility product*
	(mol^4 dm^{-12})
aluminium hydroxide	10^{-32}
iron(III) hydroxide	10^{-39}
chromium(III) hydroxide	10^{-30}
	(mol^3 dm^{-9})
manganese(II) hydroxide	10^{-14}

GROUP IV

In the explanation of precipitation of metallic sulphides in Group II it was shown that the presence of hydrogen ions from the added acid reduced the concentration of sulphide ion, but that even this reduced value was still great enough to allow the solubility products of the metallic sulphides in the group to be exceeded. In Group IV, hydrogen sulphide is added to a solution which has been made alkaline with ammonia, and which therefore contains excess of hydroxide ion. Then

$$\underset{\text{from } H_2S}{H^+} + OH^- \rightarrow H_2O$$

and the removal of hydrogen ions causes an increased value for the concentration of sulphide ion. In Group IV those metal sulphides which were not precipitated in Group II because of their relatively high solubility products, are here precipitated. This is a good example of the control which may be exercised on ionic concentration (here of the sulphide ion) by variation of the concentration of the ion with which it is associated (here the hydrogen ion).

Compound	*Solubility product* ($mol^2\ dm^{-6}$)
zinc sulphide	10^{-24}
manganese(II) sulphide	10^{-15}
nickel sulphide	10^{-26}
cobalt sulphide	10^{-26}

GROUP V

The metals still remaining in solution include barium, calcium, magnesium, sodium and potassium. Of these, the first two are precipitated as their carbonates in alkaline solution because of the low values of the solubility products: 10^{-9} and 10^{-8} $mol^2\ dm^{-6}$ respectively. The solubility product of magnesium carbonate is low (10^{-5} $mol^2\ dm^{-6}$) and in neutral solution would be precipitated, but its precipitation is avoided in this group by the presence of ammonium ions, chiefly from the ammonium chloride added prior to Group III. In the presence of the large ammonium ion concentration from this source, the concentration of carbonate ions in the expression

$$\frac{[NH_4^+]^2[CO_3^{2-}]}{[(NH_4)_2CO_3]} = K$$

is reduced below the required concentration to precipitate magnesium as its carbonate.

GROUP VI

The solubility product of ammonium magnesium phosphate is 10^{-3} $mol^3\ dm^{-9}$. The solution from which it is precipitated contains ammonia and ammonium chloride.

45
Organic Reagents in Analysis

INTRODUCTION

Many organic substances have been suggested for use in spot tests for metallic ions. Their use in the accompanying tables is confined to confirmatory tests, and even for this purpose many of them are superfluous. For example, the presence of iron in the group precipitate and in the original mixture is seldom in doubt after the usual confirmatory tests. On the contrary, some confirmatory tests take a considerable time and may be masked by the presence of other materials present in a mixture. Aluminium, the Group V metals and magnesium are sometimes missed and sodium is returned when traces only are present. It is suggested that in cases of doubtful identity organic reagents may be used.

The use of organic reagents presents certain drawbacks. They may be costly, or unstable in solution, or not specific in action. Their use should, therefore, be accompanied by a certain amount of 'research' to determine the conditions of use and the extent to which other metals interfere.

The table of confirmatory tests included in this chapter contains many tests using organic reagents. In using it, the following points are relevant.

(*a*) An asterisk indicates a test which has been found to be satisfactory in elementary analysis.

(*b*) The quantities given are those which have usually been selected for accurate quantitative analysis. They serve as a very useful guide when the solution made up is to be used for detection only.

A few inorganic substances useful for spot tests are also included in this chapter.

TABLE OF CONFIRMATORY TESTS FOR METALLIC IONS IN ORDER OF ANALYTICAL GROUPS

GROUP I

LEAD *Sodium rhodizonate* 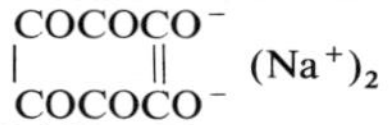 COCOCO⁻ | ‖ COCOCO⁻ $(Na^+)_2$ 0.1 % *m/V* aqueous solution.	Add one or two drops of reagent to portion of Group I precipitate, still wet with acid. *Violet colour* shows lead.	Prepare FRESH solution of reagent if it has decolorized (12 hours).
MERCURY(I) 4-*dimethylaminobenzalrhodanine* 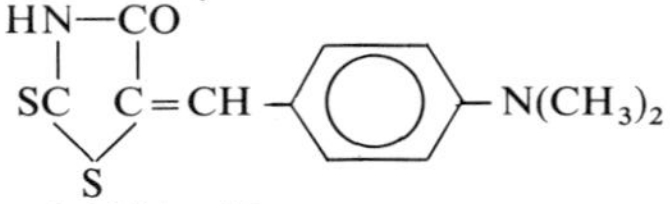 HN—CO | SC C=CH–C_6H_4–$N(CH_3)_2$, S bridging SC and C 0.03 % *m/V* acetone.	Test may be performed by adding reagent to Group I precipitate. *Brilliant purple coloration* shows mercury(I).	Silver interferes, copper(I) gives violet colour and must be removed.
SILVER* as for mercury(I).	Use Group I precipitate. *Red coloration* shows silver.	Will detect silver chloride in solution in water.

GROUP II *a*

MERCURY(II) as for mercury(I)	Warm Group IIa precipitate with dilute nitric acid. *Red coloration* shows mercury(II).	Copper interferes.
COPPER *Rubeanic acid* (dithiooxamide) $NH_2CSCSNH_2$ saturated ethanolic solution (0.5% m/V)	Test with Group II precipitate. Dissolve copper sulphide in dilute nitric acid and neutralize with sodium hydroxide solution. 2 cm^3 of neutralized copper salt solution + 0.5 cm^3 of dilute acetic acid + a few drops of reagent. *Greenish black precipitate.*	Nickel and cobalt interfere.
BISMUTH* *Thiourea.* NH_2CSNH_2 10% m/V aqueous solution.	Use Group IIa precipitate. 1 cm^3 bismuth salt solution + 1 cm^3 dilute nitric acid + 1 cm^3 reagent. *Yellow coloration.*	Antimony gives weak reaction. Mercury and silver give precipitate soluble in excess.
CADMIUM *Diphenylcarbazide.* $CO(NHNHC_6H_5)_2$ Saturated ethanolic solution.	Use solution in dilute nitric acid in Group separation. Add few drops of reagent to cadmium salt solution to give *violet coloration.*	If copper present (blue solution) first saturate reagent with potassium thiocyanate and add crystal of potassium iodide. Copper(II) is reduced and does not interfere.

GROUP II*b*

Reagent	Procedure	Remarks
ANTIMONY *Gallocyanine* COOH, N, O, O, $NH(CH_3)_2^+Cl^-$, OH. 0.05% *m*/*V* dilute hydrochloric acid.	Use Group IIb precipitate dissolved in concentrated hydrochloric acid and diluted. To one drop of antimony solution on filter paper, add one drop of reagent. Colour change—wine red to *blue* shows antimony.	
TIN* *Cacotheline.* $C_{21}H_{21}O_7N_3$ Saturated aqueous solution.	Tin must be in tin(II) condition in dilute hydrochloric acid. Add a few drops of reagent. *Violet coloration* shows tin.	Stability of reagent about fourteen days. Copper, cobalt, nickel, chromium and iron interfere.

GROUP III

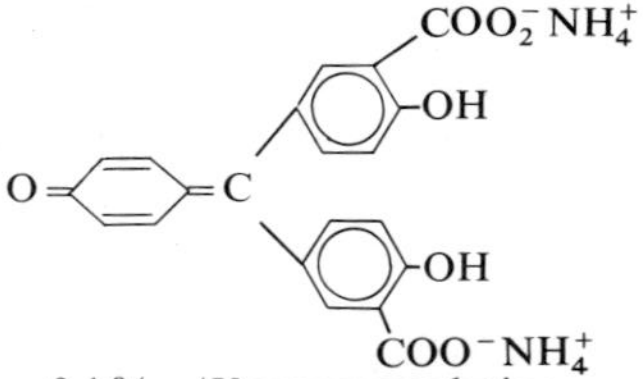 **ALUMINIUM*** *'Aluminon'*. (structure: O=C₆H₄=C with two rings, each bearing –OH and $COO^-NH_4^+$) 0.1 % m/V aqueous solution.	Use Group III precipitate dissolved in dilute hydrochloric acid. To 1 cm^3 of slightly acidic solution add ammonium acetate solution followed by reagent. *Red coloration* or precipitate shows aluminium.	Iron gives positive test and aluminium cannot be shown in its presence. Chromium gives an effect but removed by ammonia and ammonium carbonate solution.
IRON *'Cupferron'*. $C_6H_5N(NO)O^-NH_4^+$ 5 % m/V aqueous solution.	Add reagent to strongly acidic (hydrochloric) solution. *Brown red compound* shows iron.	Filter if reagent turbid. Unstable over long periods. A piece of solid ammonium carbonate added to reagent will delay decomposition.
CHROMIUM (as chromate) *Diphenylcarbazide*. $CO(NH \cdot NH \cdot C_6H_5)_2$ 0.2 % m/V solution in one part glacial acetic (pure ethanoic) acid and nine parts ethanol.	Use in Group III when chromium in form of chromate. See p. 399. Render chromate solution acidic with acetic or sulphuric acid. Add reagent. *Deep violet red coloration* shows chromate.	

GROUP IV

Reagent	Procedure	Interference
MANGANESE *Benzidine* (4,4′-biphenyldiamine) 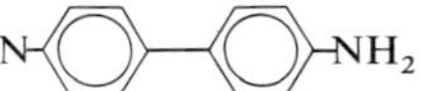 0.05% m/V solution in dilute acetic (ethanoic) acid.	Dissolve Group IV precipitate in very dilute acid and use solution (reject any solid). To one drop of solution on filter paper add one drop of very dilute sodium hydroxide solution then one drop reagent. *Blue colour* shows manganese.	
ZINC Ammonium tetrathiocyanatomercuriate(II) in presence of cobalt chloride. A. 2.7 g mercury(II) chloride and 3 g ammonium thiocyanate in 100 cm^3 water. B. 0.02% m/V cobalt chloride solution.	Dissolve Group IV precipitate in very dilute acid and use solution (reject any undissolved solid). Add 1 cm^3 cobalt chloride solution (B) to 1 cm^3 of test solution followed by 1 cm^3 of ammonium tetrathiocyanatomercuriate(II) solution (A). *Blue crystals* show zinc.	
COBALT *1-nitrosonaphth-2-ol.* NO OH 1 g in 50 cm^3 acetic (ethanoic) acid. Dilute to 100 cm^3.	Use in Group IV when in solution after treatment with potassium chlorate and acid. Alternatively, use solution after Group III. Add reagent to neutral or slightly acid solution. *Brown coloration* shows cobalt.	Copper, iron, tin, silver, chromium and bismuth all interfere.
NICKEL* *Dimethyl-glyoxime* (*butanedione dioxime*) $CH_3C{=}NOH$ | $CH_3C{=}NOH$ 1% m/V ethanol.	Use as for cobalt. Slightly acidic test solution warmed, reagent added followed by ammonia solution until alkaline. *Red precipitate* shows nickel.	Bismuth interferes.

GROUP V

BARIUM *Sodium rhodizonate.* COCOCO$^-$ | ‖ $(Na^+)_2$ COCOCO$^-$ 0.1 % *m/V* aqueous solution.	Use Group V precipitate after solution in dilute acetic acid. One drop of solution on filter paper. Add one drop reagent. *Reddish brown spot* shows barium. Add one drop of dilute hydrochloric acid: spot is *intensified*.	Prepare FRESH solution of reagent if it has decolorized (12 hours).

GROUP VI

MAGNESIUM* *Titan yellow* (complex dye) 0.1 % m/V aqueous solution.	Test Group VI solution. Add 6 drops of dilute potassium hydroxide solution to 2 cm^3 of water and then add 2 drops of solution. Boil to remove ammonium ions and add 2 drops Titan yellow. *Red coloration* or precipitate shows magnesium.	Ammonium ions interfere and must be removed.
POTASSIUM* *Sodium perchlorate (chlorate(VII))* $Na^+ClO_4^-$ 20 % m/V solution in equal parts water and ethanol.	Test Group VI solution (concentrated by evaporation and cooled). Add reagent to equal volume of test solution. *White precipitate* of potassium perchlorate shows potassium.	May be performed on glass plate above black background.
SODIUM* *Uranyl magnesium acetate.* $UO_2(CH_3COO)_2$, $Mg(CH_3COO)_2$ Saturated aqueous solution.	As for potassium. Add reagent to cold solution. *Yellow precipitate* shows sodium.	

AMMONIUM RADICAL

AMMONIUM RADICAL* *Nessler's solution.* $(K^+)_2HgI_4^{2-}$ Potassium tetraiodomercuriate(II) in sodium hydroxide solution.	Add reagent to original solution. *Brown coloration* or precipitate shows ammonium ion.	Very sensitive. Be certain that test solution contains no ammonium ions added during analysis.

PART VI

Gravimetric Analysis

46
Introduction

GENERAL REMARKS ON TECHNIQUE

In the following estimations, which depend for success on accurate weighing, it is evident that the strictest attention must be given to the accuracy of the balance and weights (masses), the cleanliness of apparatus, and the avoidance of loss during the process of collection and disposal of precipitates. When performed with care, a gravimetric estimation is capable of a high degree of accuracy.

The aim of gravimetric analysis is to obtain a final compound of a known composition from which the mass of the element (or radical) which is being estimated may be calculated. Thus, in Experiment 338 the whole of the iron in the given compound appears in the final residue of iron(III) oxide which is known to have the formula Fe_2O_3 and to contain $\frac{112}{160}$ of its mass of iron. The original compound is caused to undergo such reactions as will produce a compound which will remain unchanged under the conditions of final treatment. Often the last stage is incineration (ignition), e.g. iron(III) hydroxide is decomposed by continued heating to iron(III) oxide; in other cases, the final product is dried in an air oven at about 110 °C to constant mass in a filter paper of known mass.

WEIGHING

To test a two-pan balance for mechanical reliability, raise the lever and note the freedom of swing of the pointer. Any balance should respond readily to a change of load; with a load of 100 g in both pans the balance should be sensitive to 5 mg or less. One-pan balances can be tested similarly.

The substance which is to be analysed should be weighed from a stoppered weighing bottle. Weigh the bottle with the substance, transfer what is judged to be about the required quantity into a beaker and weigh the bottle again to find, by difference, the mass of substance taken for analysis.

FILTRATION

Because efficient filtration depends on the porous nature of the paper, the early introduction of solid matter which slows the filtration should be avoided. First decant the supernatant liquid through the filter, using a glass rod, held to the lip of the beaker and pointing to the apex of the filter, to prevent loss down the outside of the beaker.

The transfer of precipitates from beaker to filter funnel usually leaves traces in the beaker; the wash-bottle assists the transfer to be complete. The apparatus consists of a flask (about 250 cm^3) fitted with tubes as shown in Figure 83. The beaker is held almost inverted over

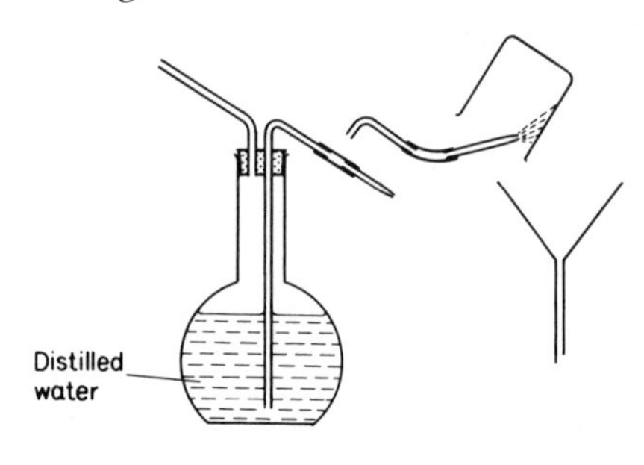

FIG. 83

the filter funnel and a jet of water is directed on to the adhering particles of solid until every trace has been sent into the filter. Keep most of the precipitate below an imaginary line 1 cm from the edge of the paper. Wash the precipitate at least twice by filling the filter to the limit indicated with distilled water, and test a few drops of the filtrate from time to time with suitable reagents until the absence of the substance in the filtrate shows that washing has been completed.

IGNITION OF PRECIPITATE

Draw the filter paper slightly away from the narrow end of the funnel, then place the funnel and contents in a steam oven to dry. Meanwhile, the crucible and lid should be heated to redness and placed while hot into a desiccator to cool. Their mass is found and assumed to remain constant during later ignitions: this is referred to as a 'tared' crucible. After ignitions, the crucible and contents should invariably be transferred, while hot, to the desiccator and left to cool. *On no account must a hot crucible be placed on the bench or anywhere where it would gather impurities on its base.*

Place the crucible on a piece of glazed paper (about 15 cm square) and transfer the dried precipitate from the filter paper as completely as possible. (A fairly stiff camel-hair brush or a feather is useful in removing the adhering particles.) With care there should be no spilling but if this does occur the particles should now be added to the crucible. Roll the filter paper into a tight roll, wrap a piece of platinum or nichrome wire round it as a holder, light the paper and allow it to burn to a white ash. Carefully shake the ash into the crucible, again adding any particles which may have fallen on the glazed paper. Cover the crucible with the lid, place on a pipe-clay triangle, and heat to redness for about ten minutes, lifting the lid from time to time by means of tongs. Allow to cool in a desiccator and weigh. Reheat to redness for a few minutes, cool and weigh as before, repeating until the mass is constant.

As ignitions of precipitates must include the ash of the paper, the mass of the ash must be subtracted from the final mass of the ignited mass. The average mass of paper ash is given in the data supplied by the manufacturer. Where this is not available, proceed as follows. Take a tared crucible and lid weighed accurately. Take six filter papers each made into a tight roll and the whole bound in a length of wire with one end long enough to hold. Burn the paper so that the ash falls into the crucible. When all the ash has been added to the crucible replace the lid and heat to redness on a pipe-clay triangle, lifting the lid from time to time. When all traces of carbon have disappeared, cool in a desiccator and weigh. Find the mass of *one* filter-paper ash.

In the following estimations the foregoing technique is implied and the various operations are therefore given only briefly.

TYPICAL GRAVIMETRIC ESTIMATIONS

Experiment 338 Estimation of iron in hydrated ammonium iron(II) sulphate

Material: ammonium iron(II) sulphate-6-water.

Weigh accurately about 2 g of ammonium iron(II) sulphate, and dissolve in 50 cm^3 of distilled water in a beaker. Heat to boiling, add about 2 cm^3 of concentrated nitric acid and continue boiling for a minute or two. Allow to cool and add ammonia solution, stirring during the addition, until precipitation is complete (test the liquid with litmus). Heat again to boiling and filter, using the stirring rod held to the lip of the beaker as a guide to the apex of the paper. Use the wash-bottle to remove traces of iron(III) hydroxide from beaker and rod, and finally wash the precipitate into a smaller bulk at the bottom of the filter. Leave the filter and contents to dry (preferably in a steam oven). Weigh a prepared crucible and transfer the dry paper and residue. Ignite, cool and weigh. Again heat to redness, cool and weigh until the mass remains constant.[1]

---

[1] This method of ensuring complete ignition should be followed in all subsequent estimations.

Relative molecular mass of iron(III) oxide $= (2 \times 56)+(3 \times 16)$
$= 160$

Specimen readings:

Mass of hydrated ammonium iron(III) sulphate $= 1.96$ g

Mass of iron(III) oxide (after subtracting mass of ash) $= 0.40$ g

Mass of iron in iron(III) oxide $= \frac{112}{160} \times 0.40$ $= 0.28$ g

Percentage mass of iron in hydrated ammonium iron(III) sulphate $= \frac{0.28}{1.96} \times 100$
$= 14.3$

Experiment 399 Estimation of aluminium in hydrated aluminium sulphate

Material: aluminium sulphate (this is probably 16- or 18-water or aluminium potassium sulphate-12-water).

Weigh accurately about 5 g of aluminium sulphate, dissolve it in about 50 cm^3 of distilled water, add a little dilute sulphuric acid and heat to boiling. Add excess of ammonia solution and bring again to the boil. Filter, dry, ignite in a tared crucible, cool and weigh.

Relative molecular mass of aluminium oxide $= (2 \times 27)+(3 \times 16)$
$= 102$

Calculation:

Mass of aluminium sulphate $= a$ g

Mass of aluminium oxide $= b$ g

Mass of aluminium in aluminium oxide $= \frac{54}{102} \times b$ g

Percentage mass of aluminium in aluminium sulphate $= \frac{1}{a} \times \frac{54 \times b}{102} \times 100$

Experiment 340 Estimation of sulphate in hydrated sodium sulphate

Material: sodium sulphate-10-water.

Weigh accurately about 3 g of sodium sulphate, dissolve it in about 50 cm^3 of distilled water, add about 5 cm^3 each of dilute hydrochloric acid and ammonium chloride solution and boil. The object of these additions is to assist the precipitation of barium sulphate. Add excess barium chloride solution and bring to the boil. Leave to settle and then carefully decant *the liquid* into the filter. Finally wash the precipitate into the filter. Dry in an oven. Add a few drops of concentrated nitric acid and heat the crucible to redness. Cool and weigh.

$$\text{Relative molecular mass of barium sulphate} = 137+32+(4\times 16) = 233$$

Calculation:

$$\text{Mass of sodium sulphate crystals} = a \text{ g}$$

$$\text{Mass of barium sulphate} = b \text{ g}$$

$$\therefore \text{Mass of sulphate in barium sulphate} = \frac{96}{233}\times b \text{ g}$$

$$\text{Percentage mass of sulphate in hydrated sodium sulphate} = \frac{1}{a}\times\frac{96\times b}{233}\times 100$$

Experiment 341 Estimation of magnesium in hydrated magnesium sulphate

Material: magnesium sulphate-7-water; 0.5 M ammonia solution.

Weigh accurately about 2 g of magnesium sulphate and dissolve in about 50 cm^3 of distilled water. Add about 10 cm^3 of ammonium chloride solution and then ammonia solution until alkaline after stirring. *Magnesium hydroxide should not be precipitated; if it is, more ammonium chloride should be added.* Heat and add disodium hydrogenphosphate solution in excess, stir, and allow to settle. Filter, wash the precipitate with ammonia solution until all chloride ions have been removed (test with silver nitrate solution acidified with nitric acid). Heat the solid very gradually, finally strongly, then cool and weigh.

$$Mg^{2+} + NH_3 + HPO_4^{2-} \rightarrow \underset{\text{ammonium magnesium phosphate}}{MgNH_4PO_4}$$

$$2MgNH_4PO_4 \rightarrow \underset{\text{magnesium pyrophosphate (heptaoxodiphosphate(V))}}{Mg_2P_2O_7} + 2NH_3 + H_2O$$

Relative molecular mass of magnesium pyrophosphate

$$= (2\times 24)+(2\times 31)+(7\times 16) = 222$$

Calculation:

Mass of magnesium sulphate $= a$ g

Mass of magnesium pyrophosphate $= b$ g

$\therefore$ Mass of magnesium in the pyrophosphate $= \frac{48}{222} \times b$ g

Percentage mass of magnesium in hydrated magnesium sulphate $= \frac{1}{a} \times \frac{48 \times b}{222} \times 100$

Experiment 342 Estimation of calcium in calcium carbonate

Material: marble.

Weigh about 1.5 g of marble accurately. Place it in a beaker with about 10 cm^3 of distilled water and add about 5 cm^3 of concentrated hydrochloric acid. Warm to dissolve and dilute to about 50 cm^3. Add ammonia solution until the solution is alkaline and then bring to boiling point. Add 3 g of crushed ammonium oxalate, stir, and again heat to boiling. Allow to settle and carefully decant the liquid into the filter, then wash the precipitate into the filter. Wash the precipitate with water until the filtrate, when tested as in Experiment 341, is shown to be clear of chloride ions. Dry the precipitate of calcium oxalate, ignite in a muffle furnace, cool and weigh. The final residue is calcium oxide.

$$CaCO_3 + 2HCl \rightarrow CaCl_2 + CO_2 + H_2O$$
$$CaCl_2 + (NH_4)_2C_2O_4 \rightarrow CaC_2O_4 + NH_4Cl$$
$$2CaC_2O_4 + \underset{\text{(from air)}}{O_2} \rightarrow 2CaO + 4CO_2$$

Relative molecular mass of calcium oxide $= 40 + 16 = 56$

Calculation:

Mass of calcium carbonate $= a$ g

Mass of calcium oxide $= b$ g

$\therefore$ Mass of calcium in calcium oxide $= \frac{40}{56} \times b$ g

Percentage mass of calcium in marble $= \frac{1}{a} \times \frac{40 \times b}{56} \times 100$

Experiment 343 Estimation of tin in solder

Material: solder.

Use a rough file to obtain fairly finely divided solder and weigh out about 1 g. Transfer to an evaporating basin and add about 10 cm^3 of concentrated nitric acid. Heat gently (preferably in a fume chamber) and when action has apparently ceased, add a drop or two of nitric acid and warm again, repeating until there is no further evolution of nitrogen dioxide. Dilute the contents to about 50 cm^3 and filter. Wash the precipitate with dilute nitric acid. Dry, ignite, cool and weigh.

The tin was first oxidized to hydrated tin(IV) oxide. The formula of this is doubtful, probably $H_2Sn_5O_{11}$ or better, $SnO_2 \cdot xH_2O$.

$$Sn + 4HNO_3 + (x-2)H_2O \rightarrow SnO_2 \cdot xH_2O + 4NO_2\uparrow$$
$$SnO_2 \cdot xH_2O \rightarrow SnO_2 + xH_2O\uparrow$$

Relative molecular mass of tin(IV) oxide $= 119 + 32 = 151$

Calculation:

Mass of solder $= a$ g

Mass of tin(IV) oxide $= b$ g

$\therefore$ Mass of tin in tin(IV) oxide $= \dfrac{119}{151} \times b$

Percentage mass of tin in solder $= \dfrac{1}{a} \times \dfrac{119 \times b}{151} \times 100$

PART VII

Biochemistry

The food requirements of the animal body are of seven kinds:

(*a*) Carbohydrates, as an immediate source of expendable energy and for storage for later energy requirements.

(*b*) Fats, as a means of storing energy and for conveying into the body fat-soluble vitamins A and D.

(*c*) Proteins, for the growth and maintenance of the body, and as a source of energy.

(*d*) Vitamins, containing radicals, some of which the body cannot manufacture, but which are essential in small quantities.

(*e*) Water, the chief constituent of living protoplasm, and the medium in which all chemical reactions of living protoplasm occur.

(*f*) Salts of a number of metals with the important radicals chloride, iodide and phosphate. Deficiency in these causes numerous disorders.

(*g*) Air, i.e. oxygen, to oxidize fats, proteins and carbohydrates, thus providing energy to maintain the body temperature and enabling work to be done and growth or maintenance to occur.

47
Carbohydrates

These compounds contain carbon, hydrogen and oxygen; the hydrogen and oxygen are present in the proportions in which they are found in water. They have the general formula $C_xH_{2y}O_y$ and, in order of complexity, may be grouped as follows:

(i) Simple sugars (monosaccharides) of formula $C_6H_{12}O_6$. The group includes the important sugars, glucose, fructose, and galactose.

(ii) More complex sugars (disaccharides) of formula $C_{12}H_{22}O_{11}$. Sucrose (cane-sugar), maltose (malt-sugar) and lactose (milk-sugar) are examples.

(iii) Polysaccharides, of formula $(C_6H_{10}O_5)_n$, where n is large. Starches, dextrins, and celluloses are included in this group.

MONOSACCHARIDES

The most important simple sugars are the six-carbon sugars (hexoses) and the best-known members are glucose and fructose. Glucose has the properties of an aldehyde; fructose has properties of a ketone.

```
CH2OH                    CH2OH
|                        |
CHOH                     CHOH
|                        |
CHOH                     CHOH
|                        |
CHOH                     CHOH
|                        |
CHOH   glucose           C=O      fructose
|                        |
CHO                      CH2OH
```

Both sugars:

(*a*) reduce Fehling's solution,
(*b*) reduce Barfoed's reagent,
(*c*) give the Molisch reaction,
(*d*) give the same osazone.

Experiment 344 Reactions of simple sugars

Material: Fehling's solution; copper acetate; 1-naphthol; phenylhydrazine; glucose; fructose.

(*a*) *Reduction of Fehling's solution.* Prepare about 5 cm^3 of Fehling's solution (see Experiments 110 and 226). Add a few small crystals of glucose and boil the solution. Copper(I) oxide is formed. Repeat using fructose in place of glucose.

If M represents the first four carbon groups of both glucose and fructose (see formulas given on p. 431) the similarity of reactions can be explained in terms of the interconvertibility of the remainder of the molecules:

$$\underset{\text{glucose}}{\begin{array}{c}\text{M}\\|\\\text{CHOH}\\|\\\text{CHO}\end{array}} \rightleftharpoons \begin{array}{c}\text{M}\\|\\\text{COH}\\\|\\\text{CHOH}\end{array} \rightleftharpoons \underset{\text{fructose}}{\begin{array}{c}\text{M}\\|\\\text{CO}\\|\\\text{CH}_2\text{OH}\end{array}}$$

The positive action here is due to the ease of oxidation:

$$\text{—COCH}_2\text{OH} \rightarrow \text{—COCHO} \rightarrow \text{—COCOOH}$$

(*b*) *Reduction of Barfoed's reagent.* Dissolve 4 g of copper(II) acetate and 1 cm^3 of acetic acid in 100 cm^3 of distilled water; this is Barfoed's reagent. Take about 5 cm^3 of the reagent, add a few crystals of glucose, and boil the solution. Note the reduction to copper(I) oxide. Repeat using fructose, and show that reduction again occurs.

(*c*) *Molisch reaction.* Take enough 1-naphthol to cover the bottom of a test-tube and dissolve it in about 10 drops of ethanol. Add 2 or 3 drops of a glucose solution and mix well. Pour a few drops of concentrated sulphuric acid down the side of the tube to form a layer at the bottom of the ethanolic solution. Note the violet coloration at the junction of the two layers. Repeat, using fructose.

(*d*) *Preparation of the osazone.* Take enough phenylhydrazine to cover the bottom of a test-tube. Add 1.5 times as much sodium acetate, and then the least amount of water to make a solution. Make a solution of glucose by covering the bottom of a test-tube with the sugar and filling the tube to the top with water. Take one-quarter of this solution and add it to the phenylhydrazine solution. Mix well, and filter. Heat the clear solution, in a test-tube standing in boiling water, for half an hour. Remove the Bunsen, and leave the solution to cool slowly in the water-bath. The yellow crystals are the osazone of glucose. Examine under a microscope and note the characteristic shape of the crystals—resembling a bunch of twigs. Repeat the experiment using fructose in place of glucose, and show that the osazone is identical with that of glucose.

The reason for the identity of the two products is that osazone-formation occurs in three stages. First, the hydrazone is formed (a different hydrazone in each case), then the adjacent group is oxidized to an aldehyde or ketone group by some of the hydrazine, and finally, a second molecule of hydrazine condenses with the aldehyde or ketone group. If M represents the first four carbon groups of both glucose and fructose (see formulas given on p. 431), the reactions may be shown as

$$\begin{array}{ccccc} M & & M & & M \\ | & & | & & | \\ CHOH & \rightarrow & CHOH & \rightarrow & C=O \\ | & & | & & | \\ CHO & & CH=NNHC_6H_5 & & CH=NNHC_6H_5 \\ \text{glucose} & & \text{hydrazone} & & \text{by oxidation} \end{array} \searrow$$

$$\begin{array}{c} M \\ | \\ C=NNHC_6H_5 \\ | \\ CH=NNHC_6H_5 \\ \text{osazone} \end{array}$$

$$\begin{array}{ccccc} M & & M & & M \\ | & & | & & | \\ C=O & & C=NNHC_6H_5 & & C=NNHC_6H_5 \\ | & & | & & | \\ CH_2OH & \rightarrow & CH_2OH & \rightarrow & CHO \\ \text{fructose} & & \text{hydrazone} & & \text{by oxidation} \end{array} \nearrow$$

Experiment 345 Distinguishing test for fructose

Material: resorcinol (1,3-benzenediol); ethanol; glucose; fructose.

Put about 2 cm^3 of concentrated hydrochloric acid in a test-tube, add an equal volume of water and a few crystals of resorcinol. Add a few drops of a dilute solution of fructose, and warm the mixture. Note the red coloration, and the red-brown precipitate which dissolves on the addition of ethanol to give a red solution. The fructose formed a furfural derivative which condensed with resorcinol to form the red-coloured compound. This is *Selivanoff's* test. Repeat using glucose in place of fructose and show that the test gives a negative result.

Experiment 346 Estimation of a reducing sugar with Benedict's solution

Material: potassium thiocyanate; sodium citrate (2-hydroxy-propane-1,2,3-tricarboxylate); glucose; fructose.

Benedict's solution has certain advantages over Fehling's solution for quantitative estimations of sugars; it is more stable, and is not affected by light or by the presence of protein as an impurity.

Benedict's solution is prepared by dissolving 125 g of potassium thiocyanate, 200 g of sodium carbonate-10-water and 200 g of sodium citrate in 800 cm^3 of hot water, and filtering. While this solution is cooling, a solution is prepared by dissolving 18 g of copper(II) sulphate in water and making up to 100 cm^3, and the solution is added gradually, and with thorough shaking, to the first solution. 0.5 g of potassium hexacyanoferrate(II) is then dissolved in about 10 cm^3 of water and added to the solution which is then made up to 1 dm^3.

25 cm^3 of the solution are reduced by (i.e. equivalent to) 0.05 g of glucose.

Measure 25 cm^3 of Benedict's solution (*Care! This is poisonous*) from a burette into a conical flask and add 5 to 6 g of anhydrous sodium carbonate and a few pieces of porous pot. Fill a burette with the sugar solution and arrange the apparatus so that the solution can be run into the conical flask while the latter is being heated.

Boil the Benedict's solution over a small flame and run the sugar solution in slowly. When the white precipitate of copper(I) cyanide appears, add the sugar solution more slowly and allow intervals of 30 seconds between additions. The end-point is the disappearance of the blue colour. Water may be added during the boiling to replace loss due to evaporation.

DISACCHARIDES (AND OXALIC ACID)

The important disaccharides are sucrose (cane-sugar), maltose, and lactose. They all have the formula $C_{12}H_{22}O_{11}$. When boiled with aqueous solutions of mineral acids they yield simple sugars by hydrolysis.

$$C_{12}H_{22}O_{11} + H_2O \rightarrow C_6H_{12}O_6 + C_6H_{12}O_6$$

Sucrose gives glucose and fructose.
Maltose gives glucose only.
Lactose gives glucose and galactose (an aldehyde).

Experiment 347 Hydrolysis of sucrose: preparation of glucose

Apparatus: flask (1 dm^3); thermometer (100 °C); water-bath; measuring cylinder.

Material: sucrose; ethanol.

Measure 350 cm^3 of ethanol and 25 cm^3 of concentrated hydrochloric acid and mix in the flask. Grind 125 g of sucrose very finely.

Heat the ethanol-acid mixture on a water-bath, kept at 50 °C, and add the sugar in portions of about 5 g, shaking well at each addition. When all the sugar has been added, leave on the water-bath for half an hour, then leave the flask, loosely corked, overnight. Add a few crystals of glucose to promote crystallization, shake well and set aside for a day. Test the crystals deposited, by the reactions described for glucose. (Note that fructose is more soluble in ethanol and remains in solution.)

Experiment 348 Reactions of sucrose

Material: ethanol; diethyl ether (ethoxyethane); Fehling's solution; sucrose.

(*a*) Take three test-tubes, each containing enough sucrose to cover the bottom of the tube. Half-fill the first tube with water, the second with ethanol, and the third with ether. Note the insolubility of sucrose in both ethanol and ether.

(*b*) Heat a small quantity of sucrose in a dry test-tube. Note the decomposition, finally giving carbon.

(*c*) Warm, very gently, a spatula-load of sucrose with enough concentrated sulphuric acid to dampen it. The sucrose is rapidly dehydrated to leave a residue of carbon.

(*d*) Prepare about 5 cm^3 of Fehling's solution, add a few crystals of sucrose, and boil. Show that Fehling's solution is not reduced.

Experiment 349 Oxidation of sucrose: preparation of oxalic acid

Apparatus: flask (1 dm^3); water-bath; thermometer (100 °C); measuring cylinder; large evaporating basin.

Material: sucrose.

Care! The experiment must be done in a fume chamber.

Measure 130 cm^3 of concentrated nitric acid into the flask and heat on a water-bath until the temperature of the acid is 50 °C. Add 25 g of sucrose in portions of about 5 g allowing the action to moderate between additions. Nitrogen dioxide is evolved. When the action has finished, transfer the liquid to an evaporating basin, and heat on a water-bath until the volume has been reduced to about 50 cm^3. Set the basin aside to cool, until crystallization is complete.

Decant the liquid from the crystals and recrystallize from about an equal volume of distilled water.

Experiment 350 Reactions of oxalic acid

Material: oxalic acid crystals (ethanedioic acid-2-water); anhydrous calcium chloride.

(*a*) *Dehydration by concentrated sulphuric acid.* Take enough oxalic acid to cover the bottom of a test-tube and add an equal bulk of concentrated sulphuric acid. Boil the mixture and light the gas at the mouth of the test-tube (carbon monoxide). Carbon dioxide is also formed, as may be shown by passing the mixture of gases through calcium hydroxide solution.

$$(COOH)_2 \rightarrow H_2O + CO + CO_2$$

(*b*) *Reducing action of oxalic acid.* Make a solution by dissolving a crystal of oxalic acid in about 5 cm^3 of water. Acidify with dilute sulphuric acid and heat almost to boiling point. Add potassium permanganate (manganate(VII)) solution, drop by drop, and note the decolorization as the permanganate is reduced.

$$2MnO_4^- + 16H^+ + 5C_2O_4^{2-} \rightarrow 2Mn^{2+} + 8H_2O + 10CO_2$$

(*c*) *Calcium chloride test.* Make a solution of calcium chloride by dissolving 5 g of the anhydrous substance in distilled water and making up to 100 cm^3. Boil a few crystals of oxalic acid with an excess of ammonia solution until ammonia ceases to be evolved. Add the calcium chloride solution to obtain a white precipitate of calcium oxalate. Allow to settle, and then decant the liquid. Take a little of the solid in each of two test-tubes and show that calcium oxalate is soluble in dilute hydrochloric acid, but insoluble in acetic (ethanoic) acid.

$$Ca^{2+} + C_2O_4^{2-} + H_2O \rightarrow CaC_2O_4 \cdot H_2O\downarrow$$

See also Experiments 73, p. 66, and 236, p. 232.

POLYSACCHARIDES

Starch, dextrin and cellulose are complex compounds, the molecules of which yield simple sugars and disaccharides when hydrolysed. They are therefore classed as polysaccharides, with the general formula $(C_6H_{10}O_5)_n$, where *n* has a very large value.

Experiment 351 Hydrolysis of starch by acid

Material: starch; Fehling's solution.

Take a test-tube one-quarter full of water, and add enough starch just to cover the surface. Boil to break down the starch granules. Add five drops of concentrated hydrochloric acid, boil for about three minutes, then cool under the tap. Neutralize the solution with sodium hydroxide solution (testing with litmus paper) and add about 3 cm^3 of Fehling's solution. Boil, and observe the reduction to copper(I) oxide, due to the presence of glucose. The hydrolysis of starch to sugar has been catalysed by the hydrogen ions of the acid.

Experiment 352 Hydrolysis of starch in stages

Apparatus: water-bath; measuring cylinder.
Material: starch; iodine solution.

Weigh 1 g of starch and add it to about 10 cm^3 of water in a boiling-tube. Boil, and cool under the tap. Dilute the solution to 100 cm^3. Measure 40 cm^3 of this solution into a flask. (Retain the remaining 60 cm^3 for use in Experiment 353.) Add 10 cm^3 of dilute hydrochloric acid and mix well. Divide the solution equally in four test-tubes. Test one portion of the solution with iodine solution and observe the characteristic blue starch-iodide colour. Place the other three test-tubes in boiling water in a water-bath.

After 4, 8 and 16 minutes remove a test-tube, add enough cold water to double the volume of the solution, and then a drop of iodine solution.

The red or yellow colour is due to the conversion of starch to dextrins by hydrolysis, and the final stage in the series shows the continuance of hydrolysis towards (but not completely to) glucose.

Experiment 353 Hydrolysis of starch by ptyalin

Apparatus: water-bath; thermometer (100 °C); white tile; measuring cylinder.
Material: boiled starch solution (from Experiment 352); iodine solution; sodium chloride.

Make a solution by dissolving 1 g of sodium chloride in 100 cm^3 of water. Use the solution as a mouth wash, and filter the saliva through a coarse filter paper. Fill a test-tube about a quarter full of the starch solution and add about 5 drops of saliva filtrate. Place at once into water maintained at 37 °C (this is your body temperature). After a minute remove a little of the solution on a glass rod and stir it into a drop of iodine solution on a white tile. Repeat at intervals of a minute and note the stages of hydrolysis. Compare the rapid action of ptyalin with that of acid (Experiment 352) as a hydrolysing agent.

Take a second test-tube one-quarter filled with starch solution and add saliva as before, but in this case boil the mixture before placing the test-tube into the water-bath. Test with iodine at minute intervals as described above. Note the absence of action, due to the destruction of ptyalin and the consequent loss of its enzymotic properties.

Experiment 354 Reactions of cellulose

Material: filter-paper; cotton wool; iodine solution; Fehling's solution; zinc chloride.

(*a*) *Test for cellulose.* Make a concentrated solution of zinc chloride and add a little iodine solution. Add a drop of the reagent to a piece of filter-paper or cotton wool. The blue coloration indicates cellulose. Show that the blue colour is not obtained when cellulose is treated with iodine solution alone.

(*b*) *Action of concentrated sulphuric acid.* Grind a piece of filter-paper or a little cotton wool with a few drops of concentrated sulphuric acid in a mortar. Add about a test-tube full of water, decant the solution into a beaker, and boil. Among the products of hydrolysis is glucose. First neutralize the solution with sodium hydroxide solution, then add Fehling's solution and boil. Reduction to copper(I) oxide indicates glucose. (See Experiment 344, p. 432.)

48
Fats

Fats are esters of glycerol with various fatty acids (chiefly acids of high relative molecular mass). An example is tristearin, present in natural fat.

$$\begin{array}{l} CH_2OOCC_{17}H_{35} \\ | \\ CHOOCC_{17}H_{35} \\ | \\ CH_2OOCC_{17}H_{35} \end{array} \quad \begin{array}{c}\text{1,2,3-propanetriyl} \\ \text{trioctadecanoate}\end{array}$$

When boiled with a solution of sodium hydroxide or potassium hydroxide, the ester is hydrolysed (saponified) into the free glycerol (1,2,3-propanetriol) and the sodium (or potassium) salt of the acid (stearic, i.e. octadecanoic). These salts are soaps.

$$\begin{array}{l} CH_2OOCC_{17}H_{35} \\ | \\ CHOOCC_{17}H_{35} \\ | \\ CH_2OOCC_{17}H_{35} \end{array} + 3NaOH \rightarrow \begin{array}{l} CH_2OH \\ | \\ CHOH \\ | \\ CH_2OH \\ \text{glycerol} \end{array} + \underset{\text{sodium stearate}}{3C_{17}H_{35}COONa}$$

Oils differ from fats in containing an unsaturated acid, in place of the saturated acid of a solid fat. Thus triolein (the main constituent of olive oil) is the glycerol ester of the unsaturated acid oleic (*cis*-9-octadecenoic) acid, $C_{17}H_{33}COOH$, which contains two hydrogen atoms per molecule less than stearic acid and an olefinic bond in its structural formula. Because of this unsaturation, oleic acid readily absorbs bromine and iodine.

Fats are assessed according to their 'acid value', 'iodine value' and 'saponification value'.

Experiment 355 To find the acid value of a fat

Material: fat; carbon tetrachloride (tetrachloromethane); 0.05 M potassium hydroxide.

A fat is an ester of glycerol and it should not (ideally) contain any free acid, but some free acid is often present and it is the purpose of this estimation to find its value.

Weigh accurately about 1 g of the given fat. Dissolve it in carbon tetrachloride and titrate the solution with 0.05 M alkali using phenolphthalein as indicator. Because 1 cm^3 of 0.05 M potassium hydroxide contains 2.8 mg of potassium hydroxide, the number of mg of potassium hydroxide needed to neutralize 1 g of fat may be calculated; this is the acid value of the fat.

Experiment 356 Comparison of unsaturation of fats

Material: solution of bromine in carbon tetrachloride (tetrachloromethane); butter; lard; olive oil; linseed oil.

Iodine value. This is defined as the number of grams of iodine absorbed by 100 g of fat. Fats which are esters of the saturated stearic or palmitic (hexadecanoic) acids will not absorb iodine, but a fat which is an ester of the unsaturated oleic acid will absorb iodine; the mass of iodine absorbed is a measure of the degree of unsaturation. Thus a molecule of triolein will absorb 3 molecules of iodine:

$$\begin{array}{lcl} CH_2OOCC_{17}H_{33} & & CH_2OOCC_{17}H_{33}I_2 \\ | & & | \\ CHOOCC_{17}H_{33} & +3I_2 \rightarrow & CHOOCC_{17}H_{33}I_2 \\ | & & | \\ CH_2OOCC_{17}H_{33} & & CH_2OOCC_{17}H_{33}I_2 \end{array}$$

The relative molecular mass of pure triolein is 884; 6 moles of iodine weigh 762 g. Therefore, 100 g of triolein absorbs $(762/884) \times 100$, i.e. 86 g of iodine.

Note: Unsaturation may more conveniently be shown by the decoloration of bromine.

Take four test-tubes each about one-quarter full of carbon tetrachloride and add three drops of melted butter to the first, a similar amount of melted lard to the second, and of olive oil and linseed oil to the third and fourth. Add, from a burette, enough bromine solution to give a permanent brown colour. Note the volume needed for each sample, and give the order of unsaturation of the fats. Linseed oil contains triolein; olive oil is 80% triolein; lard contains about 60% triolein and 30% tripalmitin; butter contains 30% triolein and 30% tripalmitin.

Experiment 357 Saponification of a fat: determination of the saponification value

Apparatus: flask (250 cm^3); condenser; pieces of porous pot.

Material: fat or oil; solution of potassium hydroxide in ethanol (1 M); hydrochloric acid.

Saponification value. This is defined as the number of milligrams of potassium hydroxide needed to change 1 g of fat completely into glycerol and the potassium soap.

Weigh accurately about 1 g of fat and place it in the flask. Add exactly 50 cm^3 of the potassium hydroxide solution and a few pieces of porous pot. Fit the flask with a reflux condenser and boil the solution for about forty-five minutes, by which time the solution will be clear. Titrate the solution against the hydrochloric acid to find the volume of potassium hydroxide solution which was unused. Hence find the volume of alkali required by the fat during saponification.

Because 1 cm^3 of the potassium hydroxide solution contains 56 mg of potassium hydroxide, the mass in milligrams of potassium hydroxide needed to saponify 1 g of fat may be calculated; this is the saponification value.

Experiment 358 Reactions of soap

Apparatus: water-bath; measuring cylinder.

Material: stearic (octadecanoic) acid; 1 M potassium hydroxide; soap.

(*a*) *Preparation of a soap.* Take 4 g of stearic acid in a beaker and heat on a water-bath until the acid has melted. Add 50 cm^3 of hot potassium hydroxide solution. Stir well and note the lather of potassium stearate.

$$C_{17}H_{35}COOH + KOH \rightarrow C_{17}H_{35}COOK + H_2O$$

Use the soap solution from (*a*), or a solution prepared from a good quality soap, for reactions (*b*)–(*d*).

(*b*) *Salting out of soap.* To a test-tube about half-full of soap solution add a test-tube one-quarter full of a saturated solution of sodium chloride. Soap is insoluble in brine and is thrown out of solution as a scum.

(*c*) *Action of calcium ions.* To some soap solution add a solution of calcium chloride. Ionic association occurs and calcium stearate separates as a scum. This is the familiar action of soap in hard water.

$$2C_{17}H_{35}COO^- + Ca^{2+} \rightarrow (C_{17}H_{35}COO)_2Ca\downarrow$$

(*d*) *Action of acids.* To some soap solution add a dilute acid. The solid thrown out of solution is free stearic acid.

$$C_{17}H_{35}COO^- + H^+ \rightarrow C_{17}H_{35}COOH\downarrow$$

49
Proteins

Proteins are nitrogen compounds of complex structure, and frequently exist in the colloidal state, from which, under certain conditions, they may be precipitated. A single protein molecule consists of a very large number of molecules of amino-acids linked together by elimination of water. Thus, taking the simplest amino-acid (glycine), the linkage may be shown as:

$$\underset{\text{glycine or aminoacetic acid}}{H_2NCH_2COOH + H_2NCH_2COOH} \rightarrow H_2NCH_2\underset{\text{'peptide' linkage}}{\boxed{CONH}}CH_2COOH + H_2O$$

The formulas of many proteins are not known, but the constituent amino-acids of a given protein can be found from the products of its decomposition. A number of these amino-acids are given below:

Alanine $CH_3—CH(NH_2)—COOH$

Phenylalanine $C_6H_5—CH_2—CH(NH_2)—COOH$ (benzene ring)

Tyrosine $HO—C_6H_4—CH_2—CH(NH_2)—COOH$ (benzene ring)

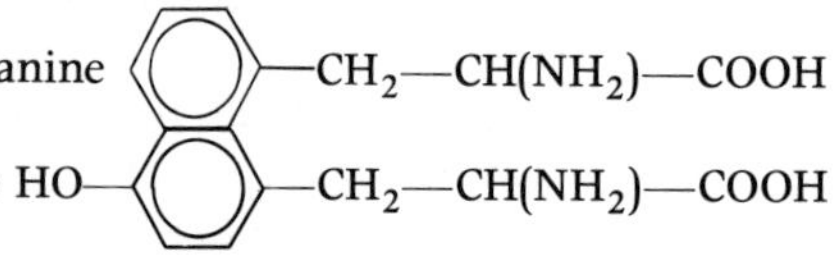

Cystine
$$\begin{array}{l} SCH_2CH(NH_2)COOH \\ | \\ SCH_2CH(NH_2)COOH \end{array}$$

Arginine $HN{=}C(NH_2)—NH—(CH_2)_3—CH(NH_2)—COOH$

Tryptophan (indole ring, N—H)—$CH_2—CH(NH_2)—COOH$

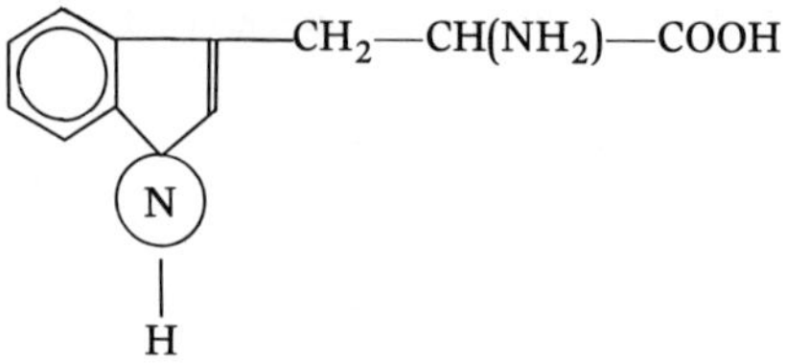

Experiment 359 Tests for proteins

Material: diluted blood serum; egg albumen; dried egg; dried milk; gelatine; reagents as given.

Blood serum may be prepared by mixing 100 cm^3 of fresh blood with a solution of 0.25 g of sodium oxalate (ethanedioate) in 20 cm^3 of water to remove the calcium ions and prevent coagulation. The blood is then centrifuged, or, failing this, left to stand for two days. The serum may be diluted to 1 dm^3 for use in these tests.

(*a*) *Biuret test.* Take a test-tube about an eighth full of a dilute solution of a protein. Add an equal volume of sodium hydroxide solution. Make a dilute solution of copper(II) sulphate (very light blue colour) and add one drop to the protein in alkaline solution. A violet colour is obtained with all proteins. Also see Experiment 265, p. 262.

Note. The following are not general tests. If a negative result is obtained the particular amino-acid is absent; it is valuable to find which amino-acids are present in, or absent from, the given material.

(*b*) *Millon's test.* Take a test-tube containing about 2 cm^3 of protein solution and add an equal volume of Millon's reagent[1] (mercury(I) and mercury(II) nitrates in nitric acid). A white precipitate, which becomes brick-red when the test-tube is heated in boiling water, indicates the amino-acid, tyrosine.

(*c*) *Sakaguchi's* (*arginine*) *test.* Take a test-tube containing about 2 cm^3 of protein solution and make it alkaline with sodium hydroxide solution. Add about 5 drops of a 2% m/V solution of 1-naphthol in ethanol, followed by a drop of sodium hypochlorite (chlorate(I)), or bleaching powder, solution. A carmine colour indicates the presence of the amino-acid, arginine.

(*d*) *Test for sulphur in proteins.* (Egg albumen is particularly suitable.) Take a test-tube containing about 3 cm^3 of protein solution. Add a few drops of lead(II) acetate solution and then sodium hydroxide solution until the precipitate of lead hydroxide, first formed, is redissolved, and boil for about a minute. A brown (or black) precipitate (of lead sulphide) indicates the presence of the amino-acid, cystine.

(*e*) *Xanthoproteic test.* Take a test-tube containing about 3 cm^3 of protein solution and add about 1 cm^3 of concentrated nitric acid. Heat to boiling and note the change of colour of the precipitate from white to yellow (*xanthos* = yellow). Cool under the tap, then make alkaline with ammonia solution. Note the change in colour

[1] Millon's reagent is prepared by warming a globule of mercury with concentrated nitric acid. Dilute the solution with twice its volume of water.

to orange. Proteins which contain an aromatic group, e.g. phenylalanine, tyrosine or tryptophan, give positive results.

Experiment 360 Amphoteric nature of a protein

Material: white feathers or white rabbit wool; eosin; methylene blue.

Isoelectric point. Proteins are amphoteric; they dissolve in alkalis and in concentrated solutions of acids. In alkaline solutions they are negatively charged, and in strongly acid solutions, positively charged. They are uncharged at the iso-electric point and are precipitated. Thus when the pH is a higher value than that of the iso-electric point, a protein acts as an acid; when the pH has a low value, the protein acts as a base. By adjusting the pH so that a protein is acting as an acid, it forms a fast colour with a basic dye such as methylene blue; if conditions are such that it is acting as a base, the protein reacts with an acid dye such as eosin.

Take four test-tubes, two containing eosin, and two containing methylene blue, each containing about 3 cm^3 of solution. Add a little rabbit wool or a feather to each. Add three drops of acetic (ethanoic) acid to one of the eosin (acid dye) solutions and three drops of concentrated ammonia solution to the other, and leave for about five minutes. Wash the wool or feather and note the fast dyeing in the acidic solution. (The protein acted as a base in the presence of acid, and reacted with the acid dye.) Add three drops of acetic acid to one of the methylene blue solutions and three drops of concentrated ammonia solution to the other. Wash the wool or feather after about five minutes. Note that it is the alkaline solution which gives the fast dyeing. (In alkaline solution the protein acted as an acid, and reacted with the basic dye.)

Experiment 361 Reactions of urea

Apparatus: thermometer (100 °C).
Material: urea; urease tablets or soya flour; sodium nitrite (nitrate(III)); sodium hypochlorite (chlorate(I)) solution; soda lime.
Formula: $CO(NH_2)_2$.
Physical properties: colourless crystals, very soluble in water. Melting point 132 °C. Urea is the chief break-down product of protein metabolism, and is found naturally in urine. Since it contains the amide group ($—CONH_2$) its properties resemble those of acetamide.

(*a*) *Biuret reaction.* Take enough urea to cover the bottom of a (dry) test-tube. Heat very gently until the liquid which forms resolidifies. The white solid is biuret.

$$2NH_2CONH_2 \rightarrow NH_2CONHCONH_2 + NH_3$$

Dissolve the biuret in about 2 cm^3 of water and use the solution as described in Experiment 359(*a*).

(*b*) *Hypochlorite reaction.* Take a few crystals of urea and dissolve in the minimum amount of water. Add a few drops of a solution of sodium hypochlorite. The gas evolved is nitrogen; carbon dioxide, which also forms, dissolves in the alkaline solution.

$$NH_2CONH_2 + 3NaOCl \rightarrow 3NaCl + 2H_2O + CO_2 + N_2$$
$$2NaOH + CO_2 \rightarrow Na_2CO_3 + H_2O$$

(*c*) *Reaction with nitrous acid.* Make a saturated solution of sodium nitrite. Add an equal volume of dilute hydrochloric acid and cool under the tap. When the effervescence has moderated, add a few drops of a solution of urea. The gas evolved is nitrogen. (Compare with a similar action with acetamide (ethanamide) and with methylamine, Experiments 255, p. 249, and 263, p. 259, respectively.)

$$NH_2CONH_2 + 2HNO_2 \rightarrow 3H_2O + CO_2 + 2N_2$$

(*d*) Heat a small quantity of a mixture of urea and soda lime. Test the issuing gas for ammonia.

$$NH_2CONH_2 + 2NaOH \rightarrow Na_2CO_3 + 2NH_3$$

(*e*) *Hydrolysis of urea with urease.* Take enough urea to cover the bottom of a test-tube and dissolve in a test-tube half-full of water. Add a tablet of urease or about a spatula-load of soya flour (which contains urease), and stand the test-tube in water maintained at 40 °C for a minute. Test for ammonia. The enzyme, urease, has hydrolysed the urea. This action is quantitative and may be used to estimate urea.

$$NH_2CONH_2 + H_2O \rightarrow 2NH_3 + CO_2$$

(*f*) *Urea as a base.* Put 2 cm^3 of a saturated solution of urea in a test-tube. Add an equal volume of concentrated nitric acid. The white precipitate is urea nitrate,

$$[NH_2{-}CO{-}NH_3]^+NO_3^-$$

50
Vitamins

These are compounds which the animal organism is unable to make for itself; they are required by the body in small quantities only, but without them health deteriorates and certain well-defined diseases result. They are, however, compounds of established chemical formulas and in no sense 'mystery' substances.

Vitamin A is a complex unsaturated alcohol, found in a high concentration in fish liver oil; the animal organism can also convert carotin (present in vegetables) into the same vitamin.

Vitamin B is a group-complex of compounds. There are no simple, satisfactory chemical tests.

Vitamin C is ascorbic acid in the form of a lactone,

$$\overbrace{CO—C(OH)=C(OH)—CH}^{O}—CH(OH)—CH_2OH,$$

and is found in citrus and other fruits. Chemically it is a powerful reducing agent, and the only compound in food capable of reducing 'indophenol'; this is the basis of its detection and estimation.

Vitamin D is found, with vitamin A, in fish oils. It has no simple chemical test.

Experiment 362 Test for vitamin A

Material: chloroform (trichloromethane); cod liver oil; halibut oil; butter; antimony(III) chloride.

Take three dry test-tubes, each containing about 2 cm^3 of chloroform. To the first add a drop of cod liver oil, to the second a drop of halibut oil, and to the third a drop of melted butter. Make a saturated solution of antimony(III) chloride in chloroform and add to each tube. Observe the blue colour due to the presence of vitamin A, and compare the intensities of the colours.

Experiment 363 Test for vitamin C

Material: orange juice; turnip juice; rose hip syrup; indophenol.

Make a 0.01% solution (0.1 g dm^{-3}) of indophenol. Take three test-tubes, each containing about 3 cm^3 of this solution. To the first add a drop of rose hip syrup, to the second a drop of orange juice, and to the third a drop of turnip juice. The observed change in each case from blue to pink (or colourless) indicates vitamin C.

Experiment 364 Estimation of vitamin C

Material: ascorbic acid; indophenol; rose hip syrup or turnip juice.

Fill a burette with 0.01% indophenol solution. Prepare a solution containing 1 g dm^{-3} of ascorbic acid (so that 1 $cm^3 \equiv 1$ mg of vitamin C). Take 10 cm^3 of the ascorbic acid solution and acidify with two or three drops of dilute hydrochloric acid. Run in indophenol solution until the solution is permanently pink.

If x cm^3 are required, 1 cm^3 of indophenol solution is equivalent to $10/x$ mg of vitamin C. Having standardized the indophenol solution, take 10 cm^3 of the given syrup or juice and treat in a similar way. Calculate (i) the mass of vitamin C in 10 cm^3 of the sample, and (ii) the volume of syrup or juice required to give the daily protective dose of 75 mg of vitamin C.

Appendix I
Molal Depression of the Freezing Point (Cryoscopic Constants)

k_f is the number of degrees Celsius by which the freezing point of the solvent is lowered by 1 mole of a non-ionizable solute in 1000 g of the given solvent (1 molal solutions).

Solvent	k_f
Acetic (ethanoic) acid	3.9
Benzene	4.9
Camphor	40.0
Naphthalene	6.86
Phenol	7.3
Water	1.86

Appendix II
Molal Elevation of the Boiling Point (Ebullioscopic Constants)

k_b is the number of degrees Celsius by which the boiling point of the solvent is raised by 1 mole of a non-ionizable solute in 1000 g of the given solvent (1 molal solutions).

Solvent	k_b
Acetic (ethanoic) acid	3.07
Acetone (propanone)	1.67
Benzene	2.67
Ethanol	1.16
Methanol	0.87
Water	0.52

Appendix III
Dissociation (Equilibrium) Constants of Acids (at 25 °C)

	$mol\ dm^{-3}$
Acetic (ethanoic) acid	1.7×10^{-5}
Formic (methanoic) acid	1.6×10^{-4}
Hydrogen sulphide	
first ionization	8.9×10^{-8}
second ionization	1.2×10^{-13}
Lactic (2-hydroxypropanoic) acid	1.4×10^{-4}
Oxalic (ethanedioic) acid	
first ionization	3.8×10^{-2}
second ionization	4.9×10^{-5}
Phenol	1.3×10^{-10}
Phosphoric acid	
first ionization	7.9×10^{-3}
second ionization	6.2×10^{-8}
third ionization	4.4×10^{-13}

Appendix IV
Dissociation (Equilibrium) Constants of Bases (at 25 °C)

	$mol\ dm^{-3}$
Ammonia	1.8×10^{-5}
Aniline (phenylamine)	4.6×10^{-10}
Diethylamine	1.3×10^{-3}
Ethylamine	5.6×10^{-4}
Urea (carbamide)	1.5×10^{-14}

Appendix V
Physical Constants of Inorganic Compounds

Where accuracy is not seriously affected, numerical values have been given to the nearest integer.

Substance	*Formula*	*Relative molecular mass*	*Solubility g in 1 kg water* 0 °C	50 °C	100 °C	*Also soluble in*
Aluminium chloride	$AlCl_3$	133.5	hydrolyses and hydrates			ether, chloroform
Aluminium sulphate	$Al_2(SO_4)_3 \cdot 16H_2O$ (approx.)	—	870	2010	113	—
Aluminium ammonium sulphate	$NH_4Al(SO_4)_2 \cdot 12H_2O$	453	40	440	3570	—
Aluminium potassium sulphate	$KAl(SO_4)_2 \cdot 12H_2O$	474	50	—	4220	—
Ammonium chloride	NH_4Cl	53.5	300	500	770	sl. soluble ethanol
Ammonium nitrate	NH_4NO_3	80	1180	—	8710	ethanol
Ammonium oxalate	$(NH_4)_2C_2O_4 \cdot H_2O$	142	40 (15 °C)	—	—	—
Ammonium sulphate	$(NH_4)_2SO_4$	132	710	840	1030	—
Ammonium iron(III) sulphate	$FeNH_4(SO_4)_2 \cdot 12H_2O$	482	400 (15 °C)	—	4000	—
Ammonium iron(II) sulphate	$Fe(NH_4)_2(SO_4)_2 \cdot 6H_2O$	392	180	400	780	—
Antimony(III) chloride	$SbCl_3$	228	6020	45310 (60 °C)	—	ethanol
Barium chloride	$BaCl_2 \cdot 2H_2O$	244	360	490	740	—
Barium hydroxide	$Ba(OH)_2 \cdot 8H_2O$	315	60	1820 (80 °C)	—	sl. soluble ethanol
Boric acid	H_3BO_3	62	50 (20 °C)	—	290	ethanol, glycerol
Bromine	Br_2	160	42	35	—	carbon disulphide
Calcium chloride	$CaCl_2$	111	600	1160	1590	ethanol
Calcium hydroxide	$Ca(OH)_2$	74	1.85	1.30	0.80	ammonium chloride
Calcium sulphate	$CaSO_4$	136	1.75	2.08	1.62	—
Chlorine	Cl_2	71	1.5 dm^3	1.36 dm^3 (40 °C)	—	alkali
Chromium potassium sulphate	$KCr(SO_4)_2 \cdot 12H_2O$	499	200	500	—	—
Copper(II) nitrate	$Cu(NO_3)_2 \cdot 3H_2O$	242	1380	6410	12700	ethanol
Copper(II) sulphate	$CuSO_4 \cdot 5H_2O$	250	145	335	755	—
Disodium tetraborate	$Na_2B_4O_7 \cdot 10H_2O$	381	30	270	2010	glycerol
Iron(III) chloride	$FeCl_3 \cdot 6H_2O$	270	2400	—	∞	—
Iron(II) sulphate	$FeSO_4 \cdot 7H_2O$	278	330	1960 (70 °C)	—	—
Lead(II) acetate	$Pb(CH_3COO)_2 \cdot 3H_2O$	379	460 (15 °C)	—	2000	—
Lead(II) chloride	$PbCl_2$	278	6.7	21	33	—
Lead(II) nitrate	$Pb(NO_3)_2$	331	390	810	1390	sl. sol. ethanol

Magnesium sulphate	$MgSO_4 \cdot 7H_2O$	246	770	—	6710	ethanol
Manganese(II) sulphate	$MnSO_4 \cdot 7H_2O$	277	1020	1380	970	—
Mercury(II) chloride	$HgCl_2$	271	60	110	540	ethanol, ether
Mercury(II) nitrate	$Hg(NO_3)_2$	324	(soluble, hydrolyses)			nitric acid
Mercury(I) nitrate	$Hg_2(NO_3)_2 \cdot 2H_2O$	561	(soluble, hydrolyses)			nitric acid
Nickel(II) sulphate	$NiSO_4 \cdot 7H_2O$	281	760 (15 °C)	—	4760	ethanol
Potassium bromide	KBr	119	530	810	1020	sl. sol. ethanol
Potassium carbonate	K_2CO_3	138	1050	1210	1560	—
Potassium chloride	KCl	74.5	280	430	565	ethanol
Potassium chlorate	$KClO_3$	122.5	30	190	540	—
Potassium dichromate	$K_2Cr_2O_7$	294	50	370	1020	—
Potassium hydroxide	KOH	56	970	1400	1780	ethanol, ether
Potassium iodide	KI	166	1280	1680	2090	ethanol, ether
Potassium nitrate	KNO_3	101	135	845	2450	—
Potassium perchlorate	$KClO_4$	138.5	10	60	200	—
Potassium permanganate	$KMnO_4$	158	30	320 (70 °C)	—	—
Potassium sulphate	K_2SO_4	174	80	160	260	—
Silver nitrate	$AgNO_3$	170	1220	4760	9400	ethanol, ether
Sodium carbonate-10-water	$Na_2CO_3 \cdot 10H_2O$	286	220	11420 (38 °C)	4210	—
Sodium carbonate (anhydrous)	Na_2CO_3	106	70	475	455	—
Sodium chloride	NaCl	58.5	357	370	392	sl. sol. ethanol
Sodium dichromate	$Na_2Cr_2O_7 \cdot 2H_2O$	298	2390	5780	12260	—
Sodium hydrogencarbonate	$NaHCO_3$	84	70	145	decom-poses	—
Sodium hydroxide	NaOH	40	420	1450	3470	ethanol, ether
Sodium iodide	NaI	150	1590	2310	3120	ethanol
Sodium nitrate	$NaNO_3$	85	730	1140	1780	—
Sodium sulphate	$Na_2SO_4 \cdot 10H_2O$	322	45	465	425	—
Sodium thiosulphate	$Na_2S_2O_3 \cdot 5H_2O$	248	525	1700	2660	—
Sulphur dioxide	SO_2	64	79.8 dm^3	15.6 dm^3 (60 °C)	—	—
Tin(II) chloride	$SnCl_2 \cdot 2H_2O$	226	1190	∞	—	alkali, ethanol
Zinc chloride	$ZnCl_2$	136	2090	—	6160	ethanol, ether
Zinc sulphate	$ZnSO_4 \cdot 7H_2O$	287	1150	2640	6340	sl. sol. ethanol

Appendix VI
Physical Constants of Organic Compounds

Substance	Formula	Relative molecular mass	Form and density ($g\ cm^{-3}$)	M.p. (°C)	B.p. (°C)	Solubility in Water	Ethanol	Diethyl ether (Ethoxyethane)
Acetaldehyde (ethanal)	CH_3CHO	44	liq. (0.8)		21	∞	∞	∞
Acetamide (ethanamide)	CH_3CONH_2	59	colourless cryst.	82	222	high	high	low
Acetanilide (N-phenylethanamide)	$C_6H_5NHCOCH_3$	135	white cryst.	112	304	low	medium	medium
Acetic (ethanoic) acid	CH_3COOH	60	liq. (1.05)	17	118	∞	∞	∞
Acetone (propanone)	CH_3COCH_3	58	liq. (0.8)		56	∞	∞	∞
Aniline (phenylamine)	$C_6H_5NH_2$	93	liq. (1.0)		184	low	∞	∞
Benzaldehyde	C_6H_5CHO	106	liq. (1.05)		180	low	∞	∞
Benzamide	$C_6H_5CONH_2$	121	colourless cryst.	128	290	low	medium	high
Benzene	C_6H_6	78	liq. (0.88)	5	80	0	∞	∞
Benzoic acid	C_6H_5COOH	122	colourless cryst.	121	249	medium in hot	medium	medium
Benzyl chloride ((chloromethyl) benzene)	$C_6H_5CH_2Cl$	126.5	liq. (1.1)		176	0	∞	∞
Bromobenzene	C_6H_5Br	157	liq. (1.5)		157	0	medium	medium
Bromoethane (ethyl bromide)	C_2H_5Br	109	liq. (1.5)		39	low	∞	∞
Carbon tetrachloride (tetrachloromethane)	CCl_4	154	liq. (1.58)		77	0	medium	medium
Chlorobenzene	C_6H_5Cl	112.5	liq. (1.1)		132	0	∞	∞
Chloroform (trichloro-methane)	$CHCl_3$	119.5	liq. (1.5)		61	low	∞	∞
1,2-Dibromoethane	CH_2BrCH_2Br	188	liq. (2.2)		131	v. low	medium	∞
Diethyl ether (ethoxyethane)	$C_2H_5OC_2H_5$	74	liq. (0.7)		35	low	∞	—
Dimethylamine	$(CH_3)_2NH$	45	gas		7	low	—	∞
Dimethylaniline	$C_6H_5N(CH_3)_2$	121	yel. liq. (0.96)	2	194	low	medium	medium
1,3-Dinitrobenzene	$C_6H_4(NO_2)_2$	168	pale yel. cryst.	90	297	0	medium	high

Ethane	C_2H_6	30	gas		−86	low	low	—
Ethanol (ethyl alcohol)	C_2H_5OH	46	liq. (0.79)		78	∞	—	∞
Ethene (ethylene)	C_2H_4	28	gas		−103	low	medium	medium
Ethyl acetate (ethanoate)	$CH_3COOC_2H_5$	88	liq. (0.9)		77	low	—	∞
Ethylamine	$C_2H_5NH_2$	45	liq. (0.69)		19	∞	∞	∞
Formaldehyde (methanal)	HCHO	30	gas		−21	medium	medium	medium
Formic (methanoic) acid	HCOOH	46	liq. (1.2)		101	∞	∞	—
Fructose	$C_6H_{12}O_6$	180	white cryst.	94		high	low	medium
Glucose	$C_6H_{12}O_6$	180	white cryst.	146		medium	low	0
Glycerol (1,2,3-propanetriol)	$CH_2(OH)CH(OH)CH_2(OH)$	92	liq. (1.26)		290	∞	∞	0
Iodobenzene	C_6H_5I	204	liq. (1.8)		190	0	medium	∞
Iodoform (triiodomethane)	CHI_3	394	yel. cryst.	119*		0	low	medium
Lactic (2-hydroxy-propanoic) acid	$CH_3CH(OH)COOH$	90	syrup		119†	∞	∞	∞
Maltose	$C_{12}H_{22}O_{11}$	342	colourless cryst.			high	low	0
Methane	CH_4	16	gas		−165	v. low	low	medium
Methylamine	CH_3NH_2	31	gas		−7	high	medium	—
Nitrobenzene	$C_6H_5NO_2$	123	yel. liq. (1.2)	5	206	v. low	high	high
Oxalic (ethanedioic) acid	$H_2C_2O_4 \cdot 2H_2O$	126	colourless cryst.	99		medium	high	low
Phenol	C_6H_5OH	94	colourless cryst.	43	183	low cold	∞	high
Salicylic (2-hydroxy-benzenecarboxylic) acid	$C_6H_4(OH)COOH$	138	colourless cryst.	158		low	medium	medium
Sucrose	$C_{12}H_{22}O_{11}$	342	colourless cryst.	160		high	low	0
Tartaric (2,3-dihydroxy-butanedioic) acid	$HOOCCH(OH)CH(OH)COOH$	150	colourless cryst.		170	medium	high	low
Toluene (methylbenzene)	$C_6H_5CH_3$	92	liq. (0.87)		111	0	∞	∞
Urea	$CO(NH_2)_2$	60	colourless cryst.	133		high	low cold	low

* With decomposition. † Sublimes at 12 mmHg pressure.

Appendix VII
Relative Atomic Masses 1971

Based on the Assigned Relative Atomic Mass of $^{12}C = 12$

The values given here apply to elements as they exist in materials of terrestrial origin and to certain artificial elements. When used with due regard to the footnotes they are considered reliable to ±1 in the last digit, or ±3 if that digit is in small type.

Name	*Symbol*	*Atomic number*	*Relative atomic mass*
Actinium	Ac	89	
Aluminium	Al	13	26.98154[a]
Americium	Am	95	
Antimony	Sb	51	121.7_5
Argon	Ar	18	39.94_8[b,c,d,g]
Arsenic	As	33	74.9216[a]
Astatine	At	85	
Barium	Ba	56	137.3_4
Berkelium	Bk	97	
Beryllium	Be	4	9.01218[a]
Bismuth	Bi	83	208.9804[a]
Boron	B	5	10.81[c,d,e]
Bromine	Br	35	79.904[c]
Cadmium	Cd	48	112.40
Caesium	Cs	55	132.9044
Calcium	Ca	20	40.08
Californium	Cf	98	
Carbon	C	6	12.011[b,d]
Cerium	Ce	58	140.12
Chlorine	Cl	17	35.453[c]
Chromium	Cr	24	51.996[c]
Cobalt	Co	27	58.9332[a]
Copper	Cu	29	63.54_6[c,d]
Curium	Cm	96	
Dysprosium	Dy	66	162.5_0
Einsteinium	Es	99	
Erbium	Er	68	167.2_6
Europium	Eu	63	151.96
Fermium	Fm	100	
Fluorine	F	9	18.99840[a]
Francium	Fr	87	
Gadolinium	Gd	64	157.2_5
Gallium	Ga	31	69.72
Germanium	Ge	32	72.5_9
Gold	Au	79	196.9665[a]
Hafnium	Hf	72	178.4_9
Helium	He	2	4.00260[b,c]
Holmium	Ho	67	164.9304
Hydrogen	H	1	1.0079[b,d]
Indium	In	49	114.82
Iodine	I	53	126.9045[a]
Iridium	Ir	77	192.2_2
Iron	Fe	26	55.84_7
Krypton	Kr	36	83.80
Lanthanum	La	57	138.905_5[b]
Lawrencium	Lr	103	
Lead	Pb	82	207.2[d,g]
Lithium	Li	3	6.94_1[c,d,e]
Lutetium	Lu	71	174.97
Magnesium	Mg	12	24.305[c]
Manganese	Mn	25	54.9380[a]
Mendelevium	Md	101	
Mercury	Hg	80	200.5_9
Molybdenum	Mo	42	95.9_4
Neodymium	Nd	60	144.2_4
Neon	Ne	10	20.17_9[c]
Neptunium	Np	93	237.0482[b,f]
Nickel	Ni	28	58.7_1
Niobium	Nb	41	92.9064[a]
Nitrogen	N	7	14.0067[b,c]
Nobelium	No	102	
Osmium	Os	76	190.2
Oxygen	O	8	15.999_4[b,c,d]
Palladium	Pd	46	106.4
Phosphorus	P	15	30.97376[a]
Platinum	Pt	78	195.0_9
Plutonium	Pu	94	
Polonium	Po	84	
Potassium	K	19	39.10_8
Praseodymium	Pr	59	140.9077[a]
Promethium	Pm	61	
Protactinium	Pa	91	231.0359[a,f]
Radium	Ra	88	226.0254[a,f,g]
Radon	Rn	86	
Rhenium	Re	75	186.2
Rhodium	Rh	45	102.9055[a]
Rubidium	Rb	37	85.467_8[c]
Ruthenium	Ru	44	101.0_7
Samarium	Sm	62	150.4
Scandium	Sc	21	44.9559[a]
Selenium	Se	34	78.9_6
Silicon	Si	14	28.08_6[d]
Silver	Ag	47	107.868[c]
Sodium	Na	11	22.98977[a]
Strontium	Sr	38	87.62[g]
Sulphur	S	16	32.06[d]
Tantalum	Ta	73	180.947_9[b]
Technetium	Tc	43	98.9062[f]
Tellurium	Te	52	127.6_0
Terbium	Tb	65	158.9254[a]
Thallium	Tl	81	204.3_7
Thorium	Th	90	232.0381[a,f]
Thulium	Tm	69	168.9342[a]
Tin	Sn	50	118.6_9
Titanium	Ti	22	47.9_0
Tungsten	W	74	183.8_5
Uranium	U	92	238.029[b,c,e]
Vanadium	V	23	50.941_4[b,c]
Xenon	Xe	54	131.30
Ytterbium	Yb	70	173.0_4
Yttrium	Y	39	88.9059[a]
Zinc	Zn	30	65.38
Zirconium	Zr	40	91.22

[a] Mononuclidic element.
[b] Element with one predominant isotope (about 99–100% abundance).
[c] Element for which the relative atomic mass is based on calibrated measurements.
[d] Element for which variation in isotopic abundance in terrestrial samples limits the precision of the relative atomic mass given.
[e] Element for which users are cautioned against the possibility of large variations in relative atomic mass due to inadvertent or undisclosed artificial isotopic separation in commercially available materials.
[f] Most commonly available long-lived isotope.
[g] In some geological specimens this element has a highly anomalous isotopic composition, corresponding to a relative atomic mass significantly different from that given.

Appendix VIII
Water Vapour Pressure

Temperature (°C)	*Pressure* (mmHg)	*Temperature* (°C)	*Pressure* (mmHg)
0	4.6	30	31.8
1	4.9	40	55.3
2	5.3	50	92.5
3	5.7	60	149
4	6.1	70	233
5	6.5	80	355
6	7.0	82	385
7	7.5	84	417
8	8.0	86	451
9	8.6	88	487
10	9.2	90	526
11	9.8	92	567
12	10.5	94	611
13	11.2	96	658
14	12.0	98	707
15	12.8	99	733
16	13.6	100	760
17	14.5	101	788
18	15.5	102	816
19	16.5	103	845
20	17.5	104	875
21	18.6	106	938
22	19.8		
23	21.0		
24	22.4		
25	23.7		

It is useful to draw graphs of water vapour pressure from 10 to 25 °C and from 86 to 106 °C for many physical chemistry experiments.

Appendix IX
Logarithms of Numbers

LOGARITHMS

	0	1	2	3	4	5	6	7	8	9	1	2	3	4	5	6	7	8	9
10	0000	0043	0086	0128	0170	0212	0253	0294	0334	0374	4	8	12	17	21	25	29	33	37
11	0414	0453	0492	0531	0569	0607	0645	0682	0719	0755	4	8	11	15	19	23	26	30	34
12	0792	0828	0864	0899	0934	0969	1004	1038	1072	1106	3	7	10	14	17	21	24	28	31
13	1139	1173	1206	1239	1271	1303	1335	1367	1399	1430	3	6	10	13	16	19	23	26	29
14	1461	1492	1523	1553	1584	1614	1644	1673	1703	1732	3	6	9	12	15	18	21	24	27
15	1761	1790	1818	1847	1875	1903	1931	1959	1987	2014	3	6	8	11	14	17	20	22	25
16	2041	2068	2095	2122	2148	2175	2201	2227	2253	2279	3	5	8	11	13	16	18	21	24
17	2304	2330	2355	2380	2405	2430	2455	2480	2504	2529	2	5	7	10	12	15	17	20	22
18	2553	2577	2601	2625	2648	2672	2695	2718	2742	2765	2	5	7	9	12	14	16	19	21
19	2788	2810	2833	2856	2878	2900	2923	2945	2967	2989	2	4	7	9	11	13	16	18	20
20	3010	3032	3054	3075	3096	3118	3139	3160	3181	3201	2	4	6	8	11	13	15	17	19
21	3222	3243	3263	3284	3304	3324	3345	3365	3385	3404	2	4	6	8	10	12	14	16	18
22	3424	3444	3464	3483	3502	3522	3541	3560	3579	3598	2	4	6	8	10	12	14	15	17
23	3617	3636	3655	3674	3692	3711	3729	3747	3766	3784	2	4	6	7	9	11	13	15	17
24	3802	3820	3838	3856	3874	3892	3909	3927	3945	3962	2	4	5	7	9	11	12	14	16
25	3979	3997	4014	4031	4048	4065	4082	4099	4116	4133	2	3	5	7	9	10	12	14	15
26	4150	4166	4183	4200	4216	4232	4249	4265	4281	4298	2	3	5	7	8	10	11	13	15
27	4314	4330	4346	4362	4378	4393	4409	4425	4440	4456	2	3	5	6	8	9	11	13	14
28	4472	4487	4502	4518	4533	4548	4564	4579	4594	4609	2	3	5	6	8	9	11	12	14
29	4624	4639	4654	4669	4683	4698	4713	4728	4742	4757	1	3	4	6	7	9	10	12	13
30	4771	4786	4800	4814	4829	4843	4857	4871	4886	4900	1	3	4	6	7	9	10	11	13
31	4914	4928	4942	4955	4969	4983	4997	5011	5024	5038	1	3	4	6	7	8	10	11	12
32	5051	5065	5079	5092	5105	5119	5132	5145	5159	5172	1	3	4	5	7	8	9	11	12
33	5185	5198	5211	5224	5237	5250	5263	5276	5289	5302	1	3	4	5	6	8	9	10	12
34	5315	5328	5340	5353	5366	5378	5391	5403	5416	5428	1	3	4	5	6	8	9	10	11
35	5441	5453	5465	5478	5490	5502	5514	5527	5539	5551	1	2	4	5	6	7	9	10	11
36	5563	5575	5587	5599	5611	5623	5635	5647	5658	5670	1	2	4	5	6	7	8	10	11
37	5682	5694	5705	5717	5729	5740	5752	5763	5775	5786	1	2	3	5	6	7	8	9	10
38	5798	5809	5821	5832	5843	5855	5866	5877	5888	5899	1	2	3	5	6	7	8	9	10
39	5911	5922	5933	5944	5955	5966	5977	5988	5999	6010	1	2	3	4	5	7	8	9	10
40	6021	6031	6042	6053	6064	6075	6085	6096	6107	6117	1	2	3	4	5	6	8	9	10
41	6128	6138	6149	6160	6170	6180	6191	6201	6212	6222	1	2	3	4	5	6	7	8	9
42	6232	6243	6253	6263	6274	6284	6294	6304	6314	6325	1	2	3	4	5	6	7	8	9
43	6335	6345	6355	6365	6375	6385	6395	6405	6415	6425	1	2	3	4	5	6	7	8	9
44	6435	6444	6454	6464	6474	6484	6493	6503	6513	6522	1	2	3	4	5	6	7	8	9
45	6532	6542	6551	6561	6571	6580	6590	6599	6609	6618	1	2	3	4	5	6	7	8	9
46	6628	6637	6646	6656	6665	6675	6684	6693	6702	6712	1	2	3	4	5	6	7	7	8
47	6721	6730	6739	6749	6758	6767	6776	6785	6794	6803	1	2	3	4	5	5	6	7	8
48	6812	6821	6830	6839	6848	6857	6866	6875	6884	6893	1	2	3	4	4	5	6	7	8
49	6902	6911	6920	6928	6937	6946	6955	6964	6972	6981	1	2	3	4	4	5	6	7	8
50	6990	6998	7007	7016	7024	7033	7042	7050	7059	7067	1	2	3	3	4	5	6	7	8
51	7076	7084	7093	7101	7110	7118	7126	7135	7143	7152	1	2	3	3	4	5	6	7	8
52	7160	7168	7177	7185	7193	7202	7210	7218	7226	7235	1	2	2	3	4	5	6	7	7
53	7243	7251	7259	7267	7275	7284	7292	7300	7308	7316	1	2	2	3	4	5	6	6	7
54	7324	7332	7340	7348	7356	7364	7372	7380	7388	7396	1	2	2	3	4	5	6	6	7
	0	**1**	**2**	**3**	**4**	**5**	**6**	**7**	**8**	**9**	**1**	**2**	**3**	**4**	**5**	**6**	**7**	**8**	**9**

LOGARITHMS

	0	1	2	3	4	5	6	7	8	9	1	2	3	4	5	6	7	8	9
55	7404	7412	7419	7427	7435	7443	7451	7459	7466	7474	1	2	2	3	4	5	5	6	7
56	7482	7490	7497	7505	7513	7520	7528	7536	7543	7551	1	2	2	3	4	5	5	6	7
57	7559	7566	7574	7582	7589	7597	7604	7612	7619	7627	1	2	2	3	4	5	5	6	7
58	7634	7642	7649	7657	7664	7672	7679	7686	7694	7701	1	1	2	3	4	4	5	6	7
59	7709	7716	7723	7731	7738	7745	7752	7760	7767	7774	1	1	2	3	4	4	5	6	7
60	7782	7789	7796	7803	7810	7818	7825	7832	7839	7846	1	1	2	3	4	4	5	6	6
61	7853	7860	7868	7875	7882	7889	7896	7903	7910	7917	1	1	2	3	4	4	5	6	6
62	7924	7931	7938	7945	7952	7959	7966	7973	7980	7987	1	1	2	3	3	4	5	6	6
63	7993	8000	8007	8014	8021	8028	8035	8041	8048	8055	1	1	2	3	3	4	5	5	6
64	8062	8069	8075	8082	8089	8096	8102	8109	8116	8122	1	1	2	3	3	4	5	5	6
65	8129	8136	8142	8149	8156	8162	8169	8176	8182	8189	1	1	2	3	3	4	5	5	6
66	8195	8202	8209	8215	8222	8228	8235	8241	8248	8254	1	1	2	3	3	4	5	5	6
67	8261	8267	8274	8280	8287	8293	8299	8306	8312	8319	1	1	2	3	3	4	5	5	6
68	8325	8331	8338	8344	8351	8357	8363	8370	8376	8382	1	1	2	3	3	4	4	5	6
69	8388	8395	8401	8407	8414	8420	8426	8432	8439	8445	1	1	2	2	3	4	4	5	6
70	8451	8457	8463	8470	8476	8482	8488	8494	8500	8506	1	1	2	2	3	4	4	5	6
71	8513	8519	8525	8531	8537	8543	8549	8555	8561	8567	1	1	2	2	3	4	4	5	5
72	8573	8579	8585	8591	8597	8603	8609	8615	8621	8627	1	1	2	2	3	4	4	5	5
73	8633	8639	8645	8651	8657	8663	8669	8675	8681	8686	1	1	2	2	3	4	4	5	5
74	8692	8698	8704	8710	8716	8722	8727	8733	8739	8745	1	1	2	2	3	4	4	5	5
75	8751	8756	8762	8768	8774	8779	8785	8791	8797	8802	1	1	2	2	3	3	4	5	5
76	8808	8814	8820	8825	8831	8837	8842	8848	8854	8859	1	1	2	2	3	3	4	5	5
77	8865	8871	8876	8882	8887	8893	8899	8904	8910	8915	1	1	2	2	3	3	4	4	5
78	8921	8927	8932	8938	8943	8949	8954	8960	8965	8971	1	1	2	2	3	3	4	4	5
79	8976	8982	8987	8993	8998	9004	9009	9015	9020	9025	1	1	2	2	3	3	4	4	5
80	9031	9036	9042	9047	9053	9058	9063	9069	9074	9079	1	1	2	2	3	3	4	4	5
81	9085	9090	9096	9101	9106	9112	9117	9122	9128	9133	1	1	2	2	3	3	4	4	5
82	9138	9143	9149	9154	9159	9165	9170	9175	9180	9186	1	1	2	2	3	3	4	4	5
83	9191	9196	9201	9206	9212	9217	9222	9227	9232	9238	1	1	2	2	3	3	4	4	5
84	9243	9248	9253	9258	9263	9269	9274	9279	9284	9289	1	1	2	2	3	3	4	4	5
85	9294	9299	9304	9309	9315	9320	9325	9330	9335	9340	1	1	2	2	3	3	4	4	5
86	9345	9350	9355	9360	9365	9370	9375	9380	9385	9390	1	1	2	2	3	3	4	4	5
87	9395	9400	9405	9410	9415	9420	9425	9430	9435	9440	0	1	1	2	2	3	3	4	4
88	9445	9450	9455	9460	9465	9469	9474	9479	9484	9489	0	1	1	2	2	3	3	4	4
89	9494	9499	9504	9509	9513	9518	9523	9528	9533	9538	0	1	1	2	2	3	3	4	4
90	9542	9547	9552	9557	9562	9566	9571	9576	9581	9586	0	1	1	2	2	3	3	4	4
91	9590	9595	9600	9605	9609	9614	9619	9624	9628	9633	0	1	1	2	2	3	3	4	4
92	9638	9643	9647	9652	9657	9661	9666	9671	9675	9680	0	1	1	2	2	3	3	4	4
93	9685	9689	9694	9699	9703	9708	9713	9717	9722	9727	0	1	1	2	2	3	3	4	4
94	9731	9736	9741	9745	9750	9754	9759	9763	9768	9773	0	1	1	2	2	3	3	4	4
95	9777	9782	9786	9791	9795	9800	9805	9809	9814	9818	0	1	1	2	2	3	3	4	4
96	9823	9827	9832	9836	9841	9845	9850	9854	9859	9863	0	1	1	2	2	3	3	4	4
97	9868	9872	9877	9881	9886	9890	9894	9899	9903	9908	0	1	1	2	2	3	3	4	4
98	9912	9917	9921	9926	9930	9934	9939	9943	9948	9952	0	1	1	2	2	3	3	4	4
99	9956	9961	9965	9969	9974	9978	9983	9987	9991	9996	0	1	1	2	2	3	3	3	4
	0	**1**	**2**	**3**	**4**	**5**	**6**	**7**	**8**	**9**	**1**	**2**	**3**	**4**	**5**	**6**	**7**	**8**	**9**

Index